REENGINEERING

SYSTEMS INTEGRATION SUCCESS

Michael A. Mische, *Editor*

CRC Press
Taylor & Francis Group
Boca Raton London New York

CRC Press is an imprint of the
Taylor & Francis Group, an **informa** business

First published 1996 by Auerbach Publications
Taylor & Francis Group
6000 Broken Sound Parkway NW, Suite 300
Boca Raton, FL 33487-2742

Reissued 2018 by CRC Press

© 1996 by RIA Group, Inc.
CRC Press is an imprint of Taylor & Francis Group, an Informa business

No claim to original U.S. Government works

Chapter 2-3, "The Business of Hub-and-Spoke Systems," by Johan Vinckier. All rights owned by McKinsey & Co.

Chapter 3-1, "Defining Business Process Reengineering," by Michael A. Mische and Warren Bennis. All rights owned by Jossey-Bass.

Chapter 4-5, "Lessons Learned: Ethics and Failure in SI," by Dr. Effy Oz. All rights owned by the ACM.

Chapter 6-3, "Integrating the C/S Environment," by R. Jack Munyan and Gordon Couturier. All rights owned by the Decision Sciences Institute.

Chapter 6-5, "Grasping the Promise of Client/Server Computing," by Damon Beyer, Marvin Newell, and Ian Hurst. All rights owned by McKinsey & Co.

A Library of Congress record exists under LC control number: 98640121

Publisher's Note
The publisher has gone to great lengths to ensure the quality of this reprint but points out that some imperfections in the original copies may be apparent.

Disclaimer
The publisher has made every effort to trace copyright holders and welcomes correspondence from those they have been unable to contact

ISBN 13: 978-1-138-55050-6 (hbk)
ISBN 13: 978-1-138-56168-7 (pbk)
ISBN 13: 978-0-203-71051-7 (ebk)

Visit the Taylor & Francis Web site at http://www.taylorandfrancis.com and the CRC Press Web site at http://www.crcpress.com

Contributors

Warren Bennis
Distinguished Professor of Business Administration, University of Southern California, and the Founding Chairman of the Leadership Institute, Los Angeles CA

Damon M. Beyer
Engagement Manager, McKinsey & Company, Inc., Houston TX

James Bradley
Chief Information Officer, United HealthCare Corp., Minneapolis MN

Gordon Couturier
Associate Professor, Computer Information Systems Department, The University of Tampa, FL

Jagdish R. Dalal
Vice-President of Information Management, Xerox Corp., Rochester NY

Peter G. Davison
President and Founder of APEX Information Management Consultants, Inc., Rochester NY

Rui J. Figueiredo
Director, the Corporate Information Management and Technology Group, Bausch & Lomb Inc., Rochester NY

Leora Frocht
Vice-President, the Capital Markets Technology Group, Smith Barney, New York NY

Steve Harkola
Director of Applications Development, MOOG Inc., New York NY

Ian Hurst
Senior Engagement Manager, McKinsey & Co., Inc., New York NY

Roger Kahn
Private Consultant and former Senior Manager and Partner with KPMG Peat Marwick and Arthur Young, Miami FL

Michael A. Mische
President, Synergy Consulting Group, Inc., Boalsburg PA

R. Jack Munyan

Chair and Professor of Computer Information Systems, the College of Business, The University of Tampa, FL

Marvin J. Newell

Principal, McKinsey & Co., Inc., Dallas TX

Effy Oz

Coordinator of the MIS Program, School of Business Administration, Wayne State University, Detroit MI

Eugene J. Romeo

Director of Information Management and Technology, Personal Products Division of Bausch & Lomb, Inc., Rochester NY

John A. Sanders

Senior Vice-President, Management Information Services for Spelling Entertainment Group Inc., Los Angeles CA

Sundar Subramaniam

President, Cambridge Technology Enterprises, Cambridge MA

Mark T. Turnbull

Director of Information Services, Murphy Family Farms, NC, and Past Manager, Enterprise Systems Development, Hershey Chocolate North America, Hershey PA

Johan Vinckier

Manager, McKinsey & Co., Inc., Brussels

Michael Wybo

Assistant Professor of Information Systems, McGill University, Canada

Contents

Contents

Introduction

In the words of Bill Gates, "Software is the scarce element in the equation. And its value is only in its uniqueness. If two people have the same software, it's worthless."

The stakes of the systems integration process are high. Integration has come to mean much more than the melding of different technologies and application systems. Today, the definition has evolved to include the integration of business processes, organizational structures, and the human element of the organization. The system is the sum of all the parts that make organizations work. It is simply no longer feasible to consider integration the sole focus of technology. Organizations are reengineering themselves and are finding that systemic changes at the operational, organizational, and cultural levels are necessary for optimizing their investments and efforts. They are using technology as both a catalyst and enabler of the process.

As systems integration projects take on mega-dollar proportions—approaching budgets of $20 million, $50 million, and even more than $100 million—and span years of effort, there is little room for error or incompetent management. The complexities of the systems integration process, when combined with rapidly changing technology, dynamically evolving business practices, and the need for reengineering the enterprise, make the risks and returns enormously challenging to manage. Since the advent of mega-integration projects, organizations have amassed thousands of person-years of systems development and business integration experience.

Unfortunately, it is all too common to read media accounts or hear stories about systems integration and reengineering projects that have, in fact, gone awry. In fact, the results are starting to come in. *ComputerWorld* reported in 1994 that 85% of executives are dissatisfied with the results of their reengineering efforts, and many pointed to IT as a problem. Empirical evidence is suggesting that many reengineering efforts end in failure, at least in terms of satisfying their original objectives or sustaining any momentum. All too often, systems integration and business transformation efforts are reported that are past due, over budget, underperforming, understaffed, and poorly managed, or they involved too many (or not enough) consultants and contractors, or the vendor exaggerated the capabilities of the product. In the contemporary world of real-time business, mobile work forces, ever-demanding customers, and global business alliances, however, successfully melding the systems integration and reengineering processes are not just technical challenges that are restricted to the confines of the MIS group or technical personnel; they demand an enterprisewide perspective and cross-functional commitment. They are activities that are essential to achieving long-term competitive advantage, viability, and sustained value creation.

There are, in fact, many phenomenally successful integration and reengineering efforts, such as those of Chrysler Corp., First Boston, United Health Corp., John Hancock, Taco Bell, and Saturn, which serve as examples of well-executed projects.

As documented in many studies, however, technology alone does not necessarily provide significant strategic or competitive advantage to the organization, regardless of whether the organization has invested in leading-edge, state-of-the-art, mainstream systems or is one or more generations behind current computing platforms and system software. Conversely, how an organization chooses to deploy and use its technology assets, resources, and data to enable business processes does create competitive advantage. This convergence of purpose creates extraordinary results.

Therein lies the essence of systems integration and reengineering for competitive advantage. Successful systems integration and reengineering require the forging of strong business partnerships between technology providers, service suppliers, functional management, and users, as well as the creation of new business processes. To improve competitive position and shareholder value, systems integration must go beyond the historically defined parameters of retrofitting applications and replacing aging computing platforms and system software; systems integration must progress further into the realm of organizational design, employee empowerment, process enabling solutions, and process reengineering. In essence, they must architect the business for integration. Today, a major delay or problem in the system integration process can be potentially devastating, not only financially, but also to the credibility and public image of the organization. An inadequately designed and poorly performed business integration effort can not only cost millions of dollars but can be potentially devastating to the organization in its implications. Distribution channels can be disrupted, vendor relationships affected, sales opportunities missed, employee overtime increased, and customers lost. More significantly, in an era of higher expectations and performance standards, most organizations can ill afford the public humiliation, shareholder repurcussions, and Wall Street analyst wrath that invariably result from a poorly executed systems integration project. For example, a spectacular integration debacle occurred recently in Florida. In an effort managed by the State of Florida and outside contractors, this Chernobyl of systems integration projects saw the taxpayers' costs escalate needlessly from an initial estimate of $20 million to over $120 million—all in the name of improved service delivery.

The aspect that causes the most concern in today's systems integration and development work is that problems, delays, budget overruns, miscalculations, misrepresentations, and significant operational disruptions continue. It is not uncommon to read in the news media or professional literature of million-dollar projects quickly escalating to the tens of millions. Likewise, it is not unusual to learn of a $10 million project ballooning to a $30 million or $50 million fiasco. One century-old Fortune 500 company has at least two such debacles. The first involved the integration of new operating and manufacturing systems for one division, which saw a clearly underbudgeted and undermanaged MRP project grow more than 300% and its 18-month effort expand to some three years to complete. The company's second fiasco involved an international subsidiary that experienced a five-fold increase in costs from an initial estimate—all spent on traditional mainframe computers and proprietary applications for a company of about $100 million in annual sales and all without changing the basic business processes of the enterprise.

The list is long: another Fortune 500 company was sold on client/server integration for its worldwide operations. This company is watching its initial $10 million client/server integration effort escalate past $40 million, with estimates to complete approaching $70 million.

Why, after so many years of systems development and integration experience, do organizations continue to experience solution-delivery problems? The answers are neither obvious nor convenient, as they lay deep within the practices of information technology, organizational cultures, consulting doctrines, management practices, and project management processes. It is not the technology that is at fault. Rather, many of the problems are related to the organization's processes, culture, and management practices. Still others are the direct result of the quality and caliber of the people involved. Some are victims of monolithic technology, management practices, and slick marketing campaigns. Many projects simply succumb to poor project management, internal politics, low discipline, and misguided expectations.

When I first set out to create this book, I was amazed to discover how much had been said about systems and business integration and how little was formalized. The subject is well documented but under-represented in the context of meaningful work and hands-on reference material. Surprisingly, some of the best-known names in the integration business shied away from contributing to this book. I attributed this to several factors, including the lack of any formal standards surrounding systems and business integration and a need, especially among vendors, to avoid any formal definition for the subject. In a sense, systems integration and reengineering have come to mean just about anything, and there is tremendous marketing value in the loose application of the term.

This book is about systems and business integration and the effective management of the integration process. It has been designed as a practical blueprint and road map for successfully architecting, managing, and executing the systems integration process and the reengineering of business processes that many organizations require. The book is a collection of contributions from many authors and is a compilation of experiences, proven tactics and techniques, and project-management practices for systems integration. It is intended to be an effective tool and resource for the business executive and the MIS professional, student, and user. This book's objective is to recognize the complexities of the systems integration and business process reengineering efforts and present a coherent and results-driven framework for the process of integrating systems to exploit the use of technology more fully.

This book has been named, with careful consideration, *Reengineering the Systems Integration Process.* Reengineering has become a euphemism for many things. Most recently, reengineering has come to mean the radical change of business processes. Unfortunately, it also has the unjustified connotation of downsizing the organization or migrating off the mainframe or to client/server machines. From a systems perspective, that may be partially the case. Applications are being downsized through the de-massification of mainframe systems; legacy systems are being replaced or renovated. Organization structures are becoming flatter and more networked through process reengineering and the use of enabling technology and greater employee empowerment. The continuous introduction of enabling technology and integrated systems undoubtedly facilitates efforts to become more competitive and viable through the use of technology. Therefore, it is only appropriate that a book of this type be published.

The integration of business processes and enterprisewide initiatives with enabling applications, technologies, and data is a tremendously large and diverse field. No one author or organization could possibly address the subject completely. For this work, I researched the industry and identified key individuals who have not only direct experience with integration projects but a unique perspective on reengineering and systems integration. What I found is that systems integration and reengineering are highly touted and trendy terms, but very few professionals or vendors are willing or fully prepared to really define them in any formal or meaningful context. That experience has led to some interesting and exciting conclusions about the subject and industry. The most significant conclusion is that, despite the exceedingly rich history of this subject, its hype, and proclaimed benefits, there is no universal definition for integration, nor are there any standards for performing or successfully melding integration and reengineering. Furthermore, there are no de facto ways of proving and measuring the benefits of the integration and reengineering processes. Integration and any benefits that may result seem to be purely situational and relative to the original technology base and the final technology base—both of which must be assessed within the context of the organization, its strategies, and business processes. What one organization and vendor or consultant may call integration may have—and usually does have—an entirely different meaning in another organization. Hence, there is little agreement on what really constitutes a $20 billion industry.

The collection of authors who came together to contribute to this book are all proven MIS and business integration professionals. Each one has direct knowledge of systems integration and reengineering and has experienced the trials of the process. Most importantly, each of them brings a unique personal perspective and individual professional experience to this work. Collectively, they share their perspectives with respect to the tools, techniques, and methods and management practices necessary to complete an integration project successfully. I hope that you will agree that the time is appropriate for a standard for systems integration and its relationship to reengineering. I trust that you will find this book beneficial and will use this collection of knowledge, experiences, and techniques in furthering your interests and efforts.

Michael A. Mische
Boalsburg, Pennsylvania

Section 1
Introduction to Reengineering: Systems Integration Success

Historically the exclusive domain of the information technology professional, systems integration has been at the center of the information technology movement for a number of years. Integration meant making diverse applications, proprietary operating systems, and computing platforms work together; however, as many organizations strive to become more competitive and profitable, systems integration has taken on a far more extensive meaning than just technology, applications, and data. Today, the term means an integrated system of technology, human beings, processes, and organizational structures that form a unified operating architecture for the management and operation of the enterprise. Thus, as technology has become increasingly more important and integral to the operations of the organization, systems integration has evolved to become far more encompassing and pervasive.

This section explores the market for systems integration and offers several working definitions for systems integration. Dominated by a number of mega-firms, such as Andersen Consulting, CSC, EDS, and IBM, and software providers, such as SAP and AG, the systems integration market is extensive, highly dynamic, and growing at double-digit rates. In addition, literally hundreds, if not thousands, of smaller specialty and contract firms are providing systems integration services. The definitions offered in this section are by no means the final word on the subject. Rather, they present more of a perspective that provides a working construct for what systems integration is, the size of the systems integration market, and the need for systems in a contemporary and dynamic context.

1

1-1

The Systems Integration Marketplace

MICHAEL A. MISCHE

Systems integration and business process reengineering have become mainstays of the executive's vocabulary during the 1990s. Major systems integration efforts are being performed in virtually every organization, as the private and public sectors attempt to become more competitive through new processes, greater empowerment, and the use of enabling technology and applications. Technological advances, together with the migration to client/server and midrange computing options are only a few of the many factors driving the need to integrate business systems.

Historically, organizations' ideas concerning systems integration normally were confined to the technical aspects of hardware and the interconnectivity of components. Integration had a very mechanical connotation, to mean making different pieces of equipment work together in a piecemeal fashion. However, as technology and users have evolved, systems integration has come to mean much more. Integration now involves far more than technology, it involves people, strategies, business processes and the organizational structure of the enterprise. It is systemic to the organization.

SYSTEMS INTEGRATION ISSUES

The purpose of this book is to develop a broad view, description, and context for systems integration. Key issues relating to system integration and reengineering business processes are addressed by authors who have direct experience with and knowledge of the integration process. The intent is to place the systems integration industry, process, and project in a workable perspective. In doing so, this book addresses a number of important questions:

- What is the definition of systems integration?

- What are the key issues in systems integration and the management of integration efforts?
- What typifies failed projects, and how can failure be avoided?
- What is the role of the systems integrators, and how does an organization best select and manage them?
- How are systems integration projects and resources best managed?
- What are the main project management issues and processes surrounding the integration of hardware, software, data, and the migration from traditional technologies?
- What is the impact of systems integration on the organization, and what are the consequences of success and failure?
- How can failure be avoided, and what are the early warning signs and tendencies of failed projects?

The most logical place to start is with the marketplace for systems integration and the objectives and forces driving the need to integrate.

THE SYSTEMS INTEGRATION MARKETPLACE

Systems integration is a vast industry within an ever-changing landscape. The need for integrating technologies and enabling business processes is constantly growing, and the pressures on information technology management to quickly deliver cost-effective services and solutions are forever mounting. The Gartner Group estimates the market for systems integration to be over $10 billion, exclusive of hardware and software, and predicts a 14% compounded growth rate, which will bring the totals in excess of $19 billion, through 1999. Factoring in the sales of hardware, software, recurring licensing and maintenance fees, and communications costs, those market figures quickly escalate to an excess of $100 billion dollars! Systems and business integration is perhaps the most dominant and growth-oriented segment of the professional business services marketplace.

The growth in the systems integration marketplace is being driven by six major factors:

- The need to reengineer, reinvent, or invent new business processes and organizational models.
- The need to make applications work in a harmonious and synchronized manner to support new or reengineered business processes and less hierarchical organizations.
- The elevated urgency to accelerate the delivery of technology-enabling solutions . . . today, not in five years.

- The need to fully exploit new and emerging technologies and optimize the return on the investment in information technology.

- Increased acceptance by senior management to off-load risks and help ensure adequate and timely results by outsourcing the integration process to third-party systems integrators and consultants.

- The recognition that many organizations lack sufficient internal resources to effectively manage. reduce the risk of, and perform large-scale systems and business integration projects.

Undoubtedly there are many others, but there is little doubt that these six are pervasive and dominant influences in the integration process.

In response to these forces, the marketplace for systems integrators from which perspective clients can choose is larger and more diverse than ever, as there is a multitude of hardware and software vendors and consultants that profess to be systems integrators, providers of integration solutions, and solvers of enterprisewide computing needs. Based on Gartner Group estimates, the largest of these, with approximately 15% market share, is Andersen Consulting. EDS is the second largest, with 13% of the market. According to the Gartner Group, three of the Big Six accounting and consulting firms, Price Waterhouse, Ernst & Young, and Deloitte & Touche, collectively account for another 5% of the market. The remaining two-thirds of the market is composed of hardware vendors and a number of other integration firms. such as IBM. DEC, Hewlett-Packard, TSC, and SHL Systemhouse.

THE OBJECTIVES AND RATIONALE BEHIND SYSTEMS INTEGRATION

Just as there are many motivations for the integration process, there are a range of far-reaching motivations and objectives on the part of users and information management and technology (IM&T) organizations. Some of these objectives clearly revolve around technological issues and the need to reduce maintenance costs. Others are related to the need to migrate to newer technologies and the de-massing of legacy systems. Still others are driven by the need to provide reengineered process-enabling solutions that allow for more agile organizations and operations. Some of the more prevalent objectives and rationale for systems integration include:

- The need to align technology and systems solutions more effectively with new business processes and to provide more functionally robust systems.

- The decoupling of systems and data from the traditional mainframe and legacy environments through the migration to client/server technology and midrange processors.

- The need to develop and deliver systems solutions rapidly to an increasingly demanding and more fluid user base. Business needs are changing more rapidly than traditional methods of systems delivery can satisfy.
- The need to create applications and data that are more seamless and harmonious, with common access and under uniform IM&T architectural standards.
- The need to supply simpler and more user-friendly solutions to a more knowledgeable and empowered work force.

These objectives, when linked to the driving factors discussed previously, combine to make a compelling argument for integration, especially when organizations are addressing strategic goals and competitive issues in the context of process reengineering and organizational transformation issues.

DEFINING SYSTEMS INTEGRATION

The industry and process for integrating systems are extremely difficult both to define and measure with any level of precision. Much rhetoric is produced about systems integration, but comparably less material exists on exactly what integration is and how to perform such a process. Despite its heritage and popularity, systems integration remains a relatively ambiguous area. For example, there is no uniform or de facto definition for what constitutes systems integration or a systems project or effort.

The term enjoys a high level of popularity among vendors and consultants, both of which have an undeniable interest in its use to mean just about anything, including outsourcing. As Michael Wybo discusses in Chapter 1-2, "Defining the Reengineering Process", many large integration firms define systems integration in a contractual context that stresses the role of the integrator, not necessarily the process of systems integration.

In contrast, users tend to define the integration process as functional, whereas many IM&T professionals still tend to view integration as more of a traditional technical issues. There is, quite simply, a tremendous disparity of opinion and dissimilar knowledge surrounding the subject of systems and business integration.

MYTH VERSUS REALITY

The foregoing notwithstanding, systems integration has been a mainstay process since the commercialization of the computer. Virtually every information technology project involves some element of systems integration. Consequently, there are a number of myths and misconceptions regarding systems integration and systems integrators. These provide an appropriate starting point for developing a working definition of systems integration:

- *Myth.* Systems integration is purely a technical issue.
- *Reality.* Systems integration is situational to the organization attempting to integrate. Systems integration may be predominantly an issue of technology, but this is not necessarily so in all cases. Systems integration involves many other factors, including people, applications, data, communications, business processes, and how the organization effectively deploys, manages, and uses information technology to gain competitive advantage.
- *Myth.* Systems integration means business process reengineering.
- *Reality.* As discussed in section 3, systems integration is not, by itself, business process reengineering. Systems integration can be the catalyst and enabling agent of business process reengineering, but in and of itself, it is not process reengineering.
- *Myth.* Systems integration projects are driven by application development or the acquisition of third-party software, or both.
- *Reality.* Systems integration projects usually involve the development and enhancement of applications and the melding of several applications together, but software development and the acquisition of third-party software do not exclusively drive the systems integration process. Sections 6 and 7 provide insight into the role of applications and technology in systems integration.
- *Myth.* Systems integration projects can be performed without any special skills, such as project management, change management, journey navigation, or technical skills.
- *Reality.* Systems integration projects are large, complex, and risky and demand discipline and a methodology. As Effy Oz discusses in Chapter 4-2, "Systems Integration and Corporate Strategy: A Tale of Success, a Tale of Failure," many factors influence the outcome of the integration process. Jack Munyan discusses some of the tendencies and warning signs of potential failure in integration in Chapter 4-3, "Case Studies of Terminally Ill Integration Projects." Systems integration projects are inherently risky and demand special skills and a dedicated staff.
- *Myth.* All system integrators and integration firms are common.
- *Reality.* As discussed in section 7, there are a number of differences among systems integrators. Organizations considering the use of an integrator have a large and varied selection from which to choose.

7

THE FIVE LEVELS OF INTEGRATION

Systems integration involves five dimensions of technical and functional integration:

1. Technology integration.
2. Applications and software integration.
3. Data and data repository integration.
4. Communications network integration.
5. The integration of new business processes, organizational structures, and measurements with new technology capabilities.

In the light of these attributes systems integration is clearly far more encompassing and systemic to the organization than a contractual relationship with a vendor or the outsourcing of systems and personnel. Systems integration requires an organizationwide perspective that integrates business processes and people. Typically, the minimum threshold investment for a systems integration effort is $1 million.

This book defines systems integration in the following manner.

> Systems integration is the melding of divergent and often incompatible technologies, applications, data business processes, and communications into an enabling and uniform architecture and functional structure.

Integrated systems have many distinguishing characteristics but share five essential attributes.

1. Functional and technical compatibility is provided to support essential business processes.
2. The technology used to support business processes and process applications and data are transparent to their users. Integration can be achieved at any level and using any technology. The issue is selecting the best technology that optimizes several key criteria: process integration, user utility, technology longevity, adaptability, scalability and speed of solution delivery.
3. The ergonomical and functional aspects of application systems, data, access paths to data, and graphical user interfaces (GUIs) are harmonized and standardized for the user. That is, they look the same, function the same way, and are intuitive to a new user.
4. All enterprisewide data is rationalized with respect to structure, accessibility, definitions, and timing. Data means the same thing throughout the organization, from system to system, business process to business process, and across applications.
5. All enterprisewide applications and computing environments are

scalable and portable to a variety of needs. That is, technologies and applications can be rapidly deployed and tailored for use in the organization. Essential application code and data structures are replicable and reproducible, not remade.

These five characteristics define the integration process and the integrated system. In this context, systems integration is achieved when the processing environment, technologies, and business processes all align with the strategies of the organization and function in a harmonious and congruent manner.

SUMMARY

Systems integration is very significant business. The integration of systems has broad ramifications to the organization, as it involves much more than technology. To integrate and optimize both their uses of technology and their competitive performance, organizations attempting integration must transcend the normal boundaries of technology and address their fundamental uses and deployment of technology. They must integrate not only systems but business processes and organizational structures. In doing so, they must accelerate the delivery of systems solutions to the user, while providing a uniform portfolio of technology, technology solutions, and technology services. Thus, integration in its truest sense must be achieved in a multidimensional framework that includes technology, operational processes, and the organizational structure of the enterprise.

This chapter has provided an opening glance into the integration marketplace and covered some of the key issues regarding systems integration. This chapter also defines systems integration and explores some of its major attributes and qualities as well as some of the more prevalent driving factors for systems integration. The remaining chapters provide additional insights into the systems integration process and a framework to follow to help ensure a successful integration effort.

1-2
Defining the Reengineering Process

MICHAEL WYBO

A t a recent conference, the representative of a large systems consulting firm was asked whether his company thought of itself as a systems integrator. His response—"That depends on what you mean by systems integration"—provided an important insight into what has become a multibillion-dollar industry. The term *systems integration* has been adopted by hardware vendors hoping to open additional marketing channels for commodity-like computing hardware, by large corporate IS departments seeking to generate external hard dollars by selling the expertise they've gained from years of purchasing and installing a variety of technologies, and by systems consulting firms that now use *systems integration* to describe the activities that have been their mainstay for years.

So what is systems integration? Is it a marketing ploy, or is there some significant characteristic of systems in the 1990s that makes the notion of integration appealing and even necessary? Companies are spending big money on systems integration. Are these crucial investments to launch them into the next century or desperate attempts to "do something" about systems? And just what is this money buying? How does systems integration fit into organizations' larger information and systems management needs? How can a company determine whether its integrator's definition, approach, and skill set match or complement its own?

DEVELOPING A WORKING DEFINITION FOR SYSTEMS INTEGRATION

As the systems integration industry grows, and as more providers offer a variety of skills and services under the systems integration moniker, it will become increasingly important for systems integration consumers to specify precisely what they mean by systems integration. The fact that the responsibility for defining the concept lies with users is driven home by a recent working definition proposed by a group of large systems integrators. This group describes it as "a service whereby a prime contractor acts as a single point of accountability for the delivery of a totally customized system solution that satisfies specific functional and performance requirements of a customer for a predetermined price." This definition is interesting for what it both does and does not say. It specifies the who and the how—an outside prime contractor—but not the what. This focuses attention away from integration onto the contracting arrangement. From a business perspective, it is much more important to understand first what needs to be done and later to use this information to decide who will do it. Reference to a predetermined price implies that consumers have a predetermined specification or understanding of the work to be done and are able themselves to determine what is a fair and competitive price for this work. Finally, the emphasis on a totally customized systems solution raises an interesting irony. The common use of the term *integration* to mean bringing diverse objects together is often associated with shared standards, open systems, and noncustom solutions. The term *totally customized* invokes images of proprietary hardware and software, unique solutions, and difficulties in integrating new technologies and applications.

Because systems integrators have defined systems integration as a contracting arrangement devoid of specific content, it is imperative that systems integration consumers be able to describe and understand the services they require to properly choose a qualified supplier and to write a meaningful and realistic contract for these services. This chapter presents one way of thinking about the what of systems integration. The definition developed here starts with the common understanding of integration as making diverse components work together. Who does this work or how the arrangement is managed is not especially relevant for this definition. Understanding what is to be done will go a long way toward answering these later questions.

Integration in general refers to making a set of independent components function or work together as a whole. Systems integration (i.e., making a collection of different technologies, applications, and data bases work together) is becoming an extremely important issue for business. The concepts of rightsizing and client/server computing imply a mix of platforms and applications chosen for their relative strengths in a particular role in the overall information system. Making these diverse components work together is an integration problem.

The increasing use of electronic funds transfer as well as other forms of electronic data interchange (EDI) in a variety of industries implies the integration of systems across organizational boundaries. At the other end of the spectrum, the increased power of microcomputers, portable computers, and personal digital assistants combined with the increasing sophistication of users and user-friendly development tools create a pressure for integrating a diverse set of applications and platforms developed within the organization.

Furthermore, the notion of systems integration is not a single, monolithic idea. Alternative perspectives on what it means to make a collection of different technologies, applications, and data bases work together imply different integration functionality, costs, and benefits for the company and different skills on the part of the systems integrator. Thus it is important for the organization considering such a project to understand at the outset what it hopes to achieve and how to communicate its needs to an outside contractor.

The objective of this chapter is to present and clarify alternative interpretations of the term *systems integration* and provide a vocabulary for a more precise definition and communication of systems integration needs.

At the most general level, systems integration is defined as the set of activities required to make a collection of technologies, applications, and data bases work together. This definition does not specify the locus of expertise in performing these activities, thus they may originate in the company or be acquired externally. The assumption is that these differences are not trivial. The consultant that added three DOS machines from different vendors to a PC LAN is not a systems integrator; the one that is adding 12 new Macintoshes and a RISC machine is. Finally, there is no assumption that systems integration is inherently valuable or better than nonintegration. Integration is simply a characteristic of systems whose value needs to be assessed in light of their contribution to business needs.

At a more detailed level, a definition of systems integration must specify what working together means. There are three alternatives here. At the level of specifying functionality, a systems integration project may be defined as one that interconnects these components, makes them interoperable, or results in semantically consistent data. In larger projects, these alternative definitions might apply to subprojects so that in the global sense, a systems integration project includes activities aimed at providing interconnectivity, interoperability, and semantic consistency. Understanding the essential differences among these alternatives assists the organization to assign a scope to its integration projects, develop realistic expectations about costs and benefits, and evaluate the capabilities and competencies of alternative systems integrators to perform each task. Thus a consumer-based definition of systems integration can be stated as follows:

> Systems integration is the process of making diverse and potentially incompatible technologies, applications, and data work together,

when working together can mean being interconnected, interoperable, or providing semantic consistency.

AN ILLUSTRATIVE EXAMPLE

The case of TelCor highlights the differences between interconnectivity, interoperability, and semantic consistency as well as the implications that each of these definitions has for the company. TelCor is a medium-size manufacturer of electrical equipment. The firm has two divisions that manufacture complementary products. TelCorSwitch (TCS) manufactures electrical switches. TelCorBox (TCB) manufactures junction boxes. Each division has traditionally been run as a profit center, with decisions concerning information technology and systems investments being made by each division vice-president according to his or her particular needs. TCS runs all systems on a small mainframe with a proprietary operating system (OS). TCB uses a cluster of minicomputers running a different proprietary OS. TelCor Corporate has decided that to cut costs and to serve customers better, it must replace each division's sales departments with a single, corporate sales group. The new sales group will manage customer records and order entry on a newly installed OS/2 network. Corporate management has instructed the IS manager to propose a project that would permit the electronic transfer of orders from the sales group's network to each division's production planning systems. The IS manager recognizes that her already overextended staff requires outside assistance in implementing such a project and receives authorization to contract the work to a systems integrator. In the process of drafting the request for proposal (RFP), she is faced with precisely defining the terms of reference for the contract. In a more general sense, her task is to define what systems integration means for TelCor. This chapter first considers these as alternative definitions, then views them as different components of an overall systems integration concept. The argument is that all three might be present to some greater or lesser degree within the same project.

The alternative definitions relate to different types of integration functionality that the systems at TelCor will eventually exhibit. Thus in specifying the terms of reference for the systems integration RFP, the systems manager must determine whether her systems integration project is defined as interconnectivity, interoperability, or semantic consistency. Exhibit 1-2-1 illustrates these definitions.

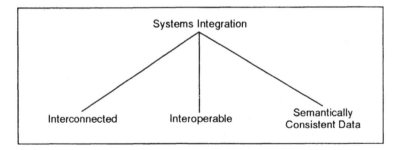

Exhibit 1-2-1. Three Definitions of Systems Integration

SYSTEMS INTEGRATION AS INTERCONNECTIVITY: NETWORKING VARIED TECHNOLOGIES

Systems integration as interconnectivity is the definition that most often reflects the services provided by commercial systems integrators. Interconnectivity means, in practical terms, that a set of processors and peripherals are connected through some sort of network that permits sharing peripherals, such as printers and storage devices, and transferring files. In most cases the processors and peripherals come from a variety of vendors, and the hardware and software network technology over which they are connected provides the common element that permits their interconnection. Data is shared either by using standard file and encoding formats (e.g., ASCII) that each platform's applications can incorporate or by using server software with a client version for each interconnected technology. This client/server approach is most evident in E-mail systems that provide server software and client packages for a variety of platforms. The essence of interconnectivity is providing the pathways over which files can be passed among different machines, such as processors and peripherals.

How Interconnectivity Would Work at TelCor

In the case of the sample manufacturing firm, TelCor, a systems integration project defined as interconnectivity would involve connecting the TCS mainframe, TCB minicomputers, and TelCor Corporate OS/2 server on a single network. Interconnection would permit sharing peripherals, such as a high-speed printing facility, among the different divisions or instituting a companywide E-mail capability across the three sites.

Interconnecting the various hardware and software platforms would also address the business objective of electronically submitting orders from the central sales group to each division's production systems. The order-entry application might split the order according to the products made by each division and send a file to each division listing the parts and quantities needed. Alternatively, the sales personnel might log into each division's system and enter the orders directly. The important point here is that defining systems integration as interconnectivity sets the project's objectives as connecting the technologies or systems, that is, providing the electronic pathways between them as well as some rules of the road (e.g., network protocols and legal filenames) for using these pathways.

Interconnectivity is a very basic definition of systems integration with a correspondingly limited integration functionality. However, every organization should ask itself if such a limited definition is sufficient to meet its business needs. For TelCor, such a definition would be adequate if the number of orders entered per period were small, the product list relatively short, the production planning requirements simple, and the order-entry data relatively consistent across divisions. Systems integration as interconnectivity is in general an appropriate definition when the interactions between systems are not numerous or not very complex (e.g., rudimentary E-mail). However, if the number of orders is large, many products are produced, the production planning requirements are complex, or order-entry data is very different across the two divisions, then an interconnectivity definition of systems integration would likely not be appropriate for TelCor. It may be necessary to reconsider systems integration as including interoperability or semantic consistency.

The Pros and Cons of Systems Integration as Interconnectivity

Defining the objective of the systems integration project at TelCor as interconnectivity has both strengths and weaknesses. Although interconnecting the three platforms is not a trivial task, the effort can be well defined and the scope of the project well understood before beginning. Interconnectivity provides a certain minimum level of integration functionality, namely the ability to pass information between the three systems electronically. Exploiting or leveraging this functionality may require the development of interface programs to translate data from each system into and out of some common file structure and format (e.g., comma-separated, value text files). Although these translation applications represent a development and maintenance cost, this cost is likely to be small. Hence interconnectivity is a relatively low-cost, low-technology solution to the problem of passing orders from a central sales group to the divisional systems.

The principal disadvantage of the interconnectivity approach is just this limited functionality. It does not provide the ability for systems to invoke one another and to operate, at least to some extent, as one. Interconnectivity provides the essential foundation for interoperability. In addition, interconnectivity may require supplementary investments in measures to ensure that the data being created at one node makes sense when incorporated into systems residing at another. Thus interconnectivity may require investments in semantic consistency to ensure that definitions and codes used by the central sales group are valid for each of the divisional systems. Neglecting measures to ensure that it makes sense to pass data among systems can mean that interconnectivity provides just enough integration functionality to be dangerous.

SYSTEMS INTEGRATION AS INTEROPERABILITY: MAKING MULTIPLE APLLICATIONS PERFORM AS ONE

Interconnectivity focuses on creating the pathways between various systems and technologies. Interoperability, on the other hand, is focused on making different systems capable of automatically exploiting one another's functionality. The notion of interoperability incorporates the idea of independent software applications working together as if they were one, able to exploit one another's functionality by passing commands and data in formats that each understands.

Interoperability is evident in most recent releases of PC-based, office-automation software. In many of these applications, one specialized program, such as word processing, can make use of the functions of another specialized program, such as a spreadsheet, to perform some operations that it cannot. For example, a table in a word processing document may contain a figure that results from calculations conducted in a complex set of spreadsheets. The spreadsheets may be updated independently of the word processing document. To ensure that the most current figure appears in the word processing table, the word processing program is able to invoke the spreadsheet, asking it to carry out a series of calculations and to pass back a figure in the format used by the word processing program. Thus the user need not be aware of the complexity of the underlying calculations, or even that the underlying spreadsheets exist. The word processor and the spreadsheet are interoperable. In the case of the PC, the programs exist on the same physical processor, but they could also reside on interconnected processors.

How Interoperability Would Work at TelCor

In the TelCor example, defining the systems integration project as interoperability means that the order-entry system used by the sales group

17

would itself invoke each division's production planning system, enter the data in the appropriate format for each system, and incorporate any response that the production systems might supply, for example, a projected shipping date. Automating these intersystem information dependencies implies that the order-entry system itself could perform additional and iterative complex processing in conjunction with the individual production systems. For example, it might compare expected shipping dates for different components of an order and attempt to change each division's production schedule so that all parts for the same order are completed at approximately the same time. This is indeed how a single, corporatewide production planning system would operate. The difference here is that TelCor is not authorizing the replacement of its existing production planning systems but instead desires to integrate and make interoperable different, existing systems.

Obviously, the skills, effort, and expense involved in implementing the systems integration project defined as interoperability differs considerably from those required to implement the project defined as interconnectivity. The functionality that such an interoperable system would deliver would also be very different from the functionality that would be realized by only interconnecting the systems. Interoperability allows a set of independent systems to function as one, automating the translation of procedure calls from one system into another. The user operates on the set of interoperable systems as if they were one and need not necessarily be aware of the location, command language, or even the existence of the various component systems.

The Pros and Cons of Systems Integration as Interoperability

The principal benefit of defining the systems integration project as interoperability is the enhanced functionality exhibited by the resulting integrated system. Because the independent systems are able to work together as one, a greater number of more complex tasks can be assigned to the system. This additional functionality could provide additional information to sales staff that might enhance their performance and the services they provide customers. The effects of such functionality might also be felt in smoother production operations, lower inventories, and overall lower costs to the divisions and the company.

The principal downside of taking an interoperability approach is that the systems integration project itself, as well as the resulting information system, becomes much more complex in terms of its design, estimation of costs, development, and maintenance. There may also be a significant training component required for users, to enhance the ability to exploit this increased functionality to the advantage of the company. To the extent that no mechanisms for ensuring semantic consistency are in place, employees need to be very aware of the semantics of the underlying data to ensure that their analyses make sense. Additional training may also be required for systems staff as they attempt to maintain and enhance the system over time. The greater the extent to which the skills necessary to carry out the project come from outside the company, the more important the long-term relationship with the systems integrator becomes and the greater the company's dependence is on the integrator. All of this translates into a higher-risk, higher-cost project.

Finally, defining systems integration as interoperability will likely imply additional effort in ensuring that data across the interoperable system is consistent or that some means of ensuring consistency is in place. As the interfaces between systems are automated, there will be an increasing need for data consistency (i.e., consistent data element names, record layouts, domains, constraints, and indexing criteria) across systems.

SYSTEMS INTEGRATION AS SEMANTIC CONSISTENCY: ENSURING CORRECT AND VALID RESPONSES

Interconnectivity and interoperability concern the integration of technologies and applications and as such operate on the hardware and software and applications elements of an information system. Systems integration projects defined as semantic consistency of data operate on the data and people elements. People tend to interpret and classify the events and objects that they experience in different ways. Human users of an information system implement these varied interpretations and classification schemes in those systems, often as what data base programmers would recognize as inconsistent data element names, definitions, domains, and integrity constraints. The value of interconnected or interoperable systems is often compromised by the inability to combine this inconsistent data in a meaningful way.

Semantically consistent data is data for which these inconsistencies have either been eliminated or resolved and therefore can be meaningfully compared, contrasted, and combined. A systems integration project defined as semantic consistency focuses on developing a shared understanding of meaning and use across the various human factors in the organization. One approach is to standardize data definitions, structures, and codes across some portion of the organization's users and systems. Another approach is to build translation applications that map meaning, structure, codes, and integrity constraints from one system or application into equivalent meaning, structure, codes, and integrity constraints in another. A third approach is to leave these inconsistencies visible to users and to rely on their understanding of the nature and implications of these

19

differences to arrive at a meaningful synthesis of the alternative perspectives.

Standardization is most appropriate for large-volume transaction systems or when there is little need for alternative data design across systems and users. Mapping is useful when it is desirable to maintain differences in local systems but to provide an automated, consistent view of the organization as a whole. This also assumes that valid transformations are possible, an assumption that does not always hold. Assigning responsibility for resolving inconsistencies to users may be most appropriate when transaction volumes are low, where users can extract important insights from being exposed to different perspectives, or where standards or mapping functions are too difficult or expensive to develop.

How Semantic Consistency Would Work at TelCor

Because at TelCor sales had traditionally been an activity conducted at the division level, no attempt had ever been made to standardize product codes over the entire organization. The result is that some product codes are used by each division to represent very different products. For example, product number 974 is a "light-duty grounded switch" for TCS and a "six-entry exterior junction box" for TCB.

Defining systems integration at TelCor as semantic consistency implies an effort to ensure correct use of codes for products across both divisions. This may take the form of data standards applied to both divisions and the sales group's application as well as potentially significant modifications to other divisional applications that use these codes. Of course, these modifications will be difficult because data semantics are managed by application systems rather than by the data management system itself. Alternatively, it may be possible to develop a global coding scheme for use by the sales group, whose elements would be translated into the codes used by each division's applications. A third alternative would be to assign responsibility for ensuring that correct codes are used to the individual salespeople as they enter orders.

The Pros and Cons of Systems Integration as Semantic Consistency

Defining systems integration as semantic consistency is qualitatively different than interconnectivity or interoperability. Rather than focusing on the concrete elements of hardware and software, semantic consistency deals with the less tangible meanings and shared understandings of events in the real world. The benefits of standardized data include the ability to pass data among applications and systems with the confidence that the results of manipulations, comparisons, and aggregations are meaningful. Standardization also helps to develop among the human resources of the organization a common language for describing objects and events in the real world.

The negative aspects of data standards include the effort needed to maintain this shared understanding (i.e., ensuring adherence to the standard as well as ensuring that a standardized means of describing and recording events does not eliminate alternative or additional definitions or codings). Thus the more complex or dynamic the business environment, the less likely that a data standards approach will be adequate.

Mapping attempts to relieve the restrictiveness of data standards by allowing individual systems to adopt their own definitions and codings. Inconsistencies are resolved when data is aggregated or compared across systems. Although the benefits of allowing local autonomy in defining and coding data may be important, one must recognize that they are paid for by increasing the complexity of the overall information system. Each mapping or transformation module is an application that needs to be developed and maintained. Although some transformations may be trivial (e.g., converting Fahrenheit to Celsius), others are not (e.g., seemingly obvious currency translations need to account for the time of the transaction and the exchange rate in effect at that time). Still others may not be possible at all, as in the case of merging household- and individual-level data for marketing purposes. A major cost of the mapping approach is the potential loss of detailed data, as objects may be defined at a very abstract level.

Finally, some responsibility for resolving inconsistencies may be assigned to users, adding high levels of flexibility in environments in which business rules and definitions of objects change rapidly. This technique is also useful for dealing with exceptions, thus avoiding the need to build additional complexity into the organizational data model or the mapping functions. However, assigning responsibility for semantic consistency to users is probably not appropriate for high-volume transaction-processing applications.

ALTERNATIVE DEFINITIONS OF SYSTEMS INTEGRATION

To highlight their differences, interconnectivity, interoperability, and semantic consistency have been referred to as three alternative definitions of systems integration. Although it is useful to think of them this way, it is also true that any one systems integration project may include all three definitions. The integration project design may likely require efforts to provide some interconnectivity, some interoperability, and some semantic consistency. Determining how much of each is the challenging part of designing the project. For example, which of the organization's systems needs to be interconnected? Which functions of which systems should be made interoperable? To what extent should which data be standardized or mapped, or inconsistencies left visible?

Exhibit 1-2-2 depicts this situation. Each of the circles represents the three alternative definitions. The project may clearly fall within areas 1, 2, 4, 5, or 7. Areas 3 and 6 are not feasible, as they represent interoperability without interconnection. The area outside the circles represents standalone, semantically inconsistent information systems. In area 1, the project would focus on ensuring the semantic consistency of data across standalone or manual systems. Data transfer across systems may be achieved by passing diskettes or paper-based forms. Integration projects in Area 2 focus on interconnecting systems to provide electronic file transfer. Area 4 projects ensure that data can be passed among systems electronically and that there is a shared understanding of this data across all nodes. Projects in Area 5 focus on interoperability. This is a danger zone. Interoperability provides the facilities by which apples and oranges can be easily and unwittingly compared or combined, or both. Staying in this area for any significant period of time poses risks. Extreme care must be taken to ensure that results are meaningful. Finally, Area 7 represents the most ambitious type of systems integration project, one that comprises all three definitions and requires expertise in all three domains.

Techniques for assessing the costs and benefits of alternative configurations of interconnectivity, interoperability, and semantic consistency are not well developed. The nature of the costs and benefits of various configurations is a topic of debate. Ongoing research is attempting to understand the implications of various levels of each as well as their interaction. In some cases the need is clear: certain systems must be connected or certain data must be standardized. In others, the answer is not so apparent. At TelCor, for example, simply transmitting orders to each division by E-mail message may suffice. Interoperable order-entry and production-planning systems may be overkill for the level of sophistication and production of the company. Efforts to ensure semantic consistency may not be necessary as long as the product mix stays relatively simple. The costs of providing more sophisticated integration functionality may exceed any benefits the firm can derive from these investments. In addition, the definition of systems integration may evolve as the organization gains experience with particular components or as its needs change. The definition provided here is useful in identifying the current state of the company's systems and what the organization might like to achieve in terms of additional or expanded integration functionality.

SUMMARY

Three alternative definitions of systems integration were presented. Interconnectivity means connecting different hardware and applications so

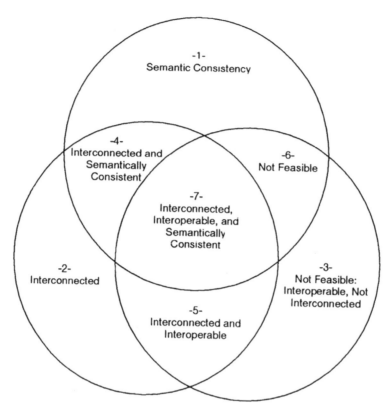

Islands of Automation
(Stand-alone, semantically inconsistent systems)

Exhibit 1-2-2. Combining Interconnectivity, Interoperability, and Semantic Consistency

that some degree of communication and resource sharing can take place. Interoperability means that applications operate as one, invoking each other's functionality as needed and passing commands and data in the format each expects to receive. Semantic consistency of data means that potential inconsistencies in the naming, definition, structure, and constraints imposed on similardata elements will be identified and resolved and that users will be provided with a consistent and correct view of the organization's data. A single systems integration project may have components that reflect each of these perspectives. One of the principal challenges for systems and business managers will be to determine how much of each type of integration is required for their particular organization.

Section 2

Systems Integration in the Global Enterprise

There is little doubt that we are living in a global economy. The trend in searching for global markets, sourcing services, and parts on a global basis and in creating worldwide trading and business relationships will continue. The rate at which this trend moves, in addition, will only be accelerated as technology enables the smallest of organizations to access markets and customers that were once the exclusive domain of the largest organizations. Technology breaks down the geographical boundaries and the cultural, economic, and natural physical barriers to globalization. The use of electronic funds transfer systems and electronic mail make hard currencies, check clearing, and postal services obsolete. Today, any one company or person can be doing business simultaneously almost anywhere in the world. This makes the need for systems integration that much more imperative to the competitive posture of many organizations. In short, we are moving away from a global community of independent markets and companies to a global village, and technology is the key enabler of the process. The physical borders between nations and economies will continue to exist. but technology will transcend them to make access to markets, customers, and business relationships borderless and transparent.

This section provides some unique insights into the technical, organizational, procedural, and managerial issues surrounding systems integration in the global enterprise. These issues are addressed not only from a technology perspective, but also from the human and chief information officer viewpoint. Cultural issues, the need for integration, and the implications of systems integration on the enterprise are explored. In doing so, the authors take us on a journey that is issue based, provocative, and insightful. The insights provided are useful regardless of the size of the organization or its geographical location, and the observations are sure to spark creative thought and dialogue.

Section 2
Systems Integration in the Global Enterprise

2-1

The Role of Systems Integration in the Global Enterprise: A CIO's Perspective

JAMES P. BRADLEY

M uch has been written about the information age and the way that technology is transforming modern industry. As a result, technology and the people that form the information technology (IT) infrastructure are becoming increasingly important in terms of the overall strategy of the organization. Yet, I must admit as I begin to write this chapter on the role of systems strategy and the enterprise, I had a bit of "writer's block." This seemed strange in that I am the chief information officer (CIO) for a major health care company and have written many articles regarding the impact of technology on the health care industry. Yet, when pressed to apply the same lessons to systems integration and business in general, I had a struggle.

The reason for the struggle became clear to me as I began to consider the fact that the lines are blurred between the role of an industry itself and the technology that supports it. As a technology strategist in the health care industry, my role has become much more that of a health care executive who uses technology and less that of a technologist applying my skills to any particular industry that happens to need them at the moment. My challenge and the purpose of this chapter is to discuss the role of IT and systems integration in the "real world" context of business and to cross-apply the lessons to any enterprise that relies on technology for its current and future survival.

The Gartner Group, a major consulting enterprise that studies the IT industry, categorizes users of technology by how quickly they adopt new technology and how intensively they apply it to their business enterprises. For example, some businesses operate in an industry that traditionally has made little use of technology or have little capital to invest to make any significant change. Other businesses operate in industries that may have traditionally made little use of IT but are now using that technology to transform their competitive stance within the industry. A third type of industry is one for which competition is information based. This type of organization usually applies technology the most aggressively.

As part of the management team for an organization, the senior system executive's job is to assess the environment within which his or her business competes, determine the role of technology in developing competitive advantage, and then arrive at the specifics of how aggressively the organization embraces technology and organizes these efforts. Part of that strategy must specifically address the degree of integration that is necessary, either within or across business-unit boundaries. Crossing those boundaries also implies the potential to integrate systems with those of suppliers and customers.

A HISTORICAL OVERVIEW OF IT

As the roles of systems strategy and integration are considered in the global enterprise, it is perhaps worthwhile to review a bit of history about IT deployment. Traditional data processing was often applied first to the back office and traditional overhead functions. Early applications in data processing were applied to payroll or other accounting functions. Early batch systems made use of punch cards to automate paper processes. As these systems evolved to online derivatives, the business process remained much the same, with the user of such a system more directly involved through a CRT display device.

These traditional systems, whether batch or online, generally used large mainframe technology and centralized processing. The challenge of integration was far simpler than today, as the entire system operated on the same physical computer system and used the same proprietary software. There was limited or no use of networking or data communications to distribute processing to other, off-premise computer systems. Any integration of systems running on this single, centralized system was accomplished by passing intermediate files or sharing access to the same physical data base.

Minicomputer technology introduced by such companies as Digital Equipment Corp. and Data General began to change the landscape of business computing. These systems provided departmentalized solutions by enabling individual business departments to buy their own minicomputers, often over the objection of data processing departments. In many cases, a different brand of minicomputer could be found from department to department within the same organization. The challenge of integration became more burdensome because business opportunity usually took precedence over technical uniformity. The political certainty of this shift was reinforced by the business unit executives making their arguments in the chief executive officer's office, whereas the data processing managers tried to react from their midlevel management positions.

The PC Enters the Marketplace

All of this began to change in the early 1980s through organizational and technological shifts. IBM introduced its personal computer (PC) and legitimized the small desktop computer as a tool of business. The PC not only gave the end user power at the desktop but also empowered that same end user to learn more about systems and to start to prescribe his or her own personal use of technology in getting the job done. The organizational shift occurred as the role of the data processing manager began to change, becoming today's CIO. As business units began to exploit their newfound technological empowerment to meet customer needs, data processing managers found themselves in a conflict between user control versus empowerment and coordination of strategy. Many had difficulty with such a transition, and even today the often-cited lack of stability within the CIO's position often can be traced to the failure to fill the role of strategist. The number-one strategy challenge within the modern organization is to develop a balance between the control required to implement an appropriate integration strategy, and empowerment of business units to satisfy customers.

Combined with this organizational and cultural dilemma, the PC became the ultimate complicating factor in systems integration. Each end user could store data and perform business processes—all totally independent of the company's centralized mainframe complex. This has been exacerbated as PC technology and power has improved at exponential rates, placing much of today's processing power on the desktop. This trend promises to continue, and PCs interconnected on a peer-to-peer basis through local area network (LAN) technology promise to change the way business is done, even for traditional transactional applications.

Communications Advances

The continued advancement of global communications and "smart" networks enables an approach to work flow that would have been impossible just a few years ago. Now a piece of paper can be captured as a digital image and processed in multiple parallel paths, as opposed to serially, in such a way that the business transaction can be accomplished in far less real time. These communications networks also enable the concept of a virtual corporation, in which people and work units come together for a finite project, spanning traditional department or external business boundaries. For example, reengineered processes can span departments or even corporation boundaries to include an organization's suppliers or customers.

The Technology Components That Remain Constant

Still, some things remain the same when it comes to technology. Those mainframes that appeared to become obsolete a few years ago are still alive and well. In fact, those who loyally watched their IBM stock decrease in value have had their patience rewarded as IBM demonstrated a major profit resurgence, primarily as a result of strong mainframe sales. This is true because of several reasons. The first is that mainframe software technology exists in massive software portfolios in most major corporations and is far more difficult to decentralize than many imagine. Second, mainframes are still the best large data servers, as they can handle massive amounts of data with the needed economies of scale. Third, in the absence of mature technology to manage systems made up of decentralized cooperating processors, the mainframe still provides organization to this sometimes disorganized approach.

Also, COBOL programmers continue to earn a living maintaining this traditional code. In fact, as decentralized systems are developed, the absence of trained personnel limits their deployment. As a result, client/server technology is still in its infancy. There are still few system management tools that allow the same kind of robust management available on mainframe-based systems. The lack of an industrialized software development process ensures that learning curves for software development do not improve the overall effectiveness of the process. Whereas the engineer has been trained to use standardized parts in new design development, today's programmers generally fail to exploit reuse potential. As a result, many new systems today are still written from the ground up, making little use of anything learned from code written during earlier projects or by other programmers. This inefficiency results in a far higher cost for development, and qualified people remain the scarcest resource.

Therefore, even though there have been massive improvements in the power of technology, the component that drives the use of that technology—software—has not improved at anywhere near the same rate. This lack of industrialization promises to hamper the systems integration process. In today's world, the role of the systems strategist becomes very important. CIOs must first and foremost be business strategists, by learning the industry, its critical success factors, and likely future direction so that the systems function facilitates competitive advantage. The lack of industrialization of software development mentioned above and the resulting lead times to develop new major systems do not allow for any directional mistakes. If the systems strategist is incorrect in predicting the future success drivers for the business, the systems that are developed will be misaligned with competitive strengths and can lead not only to inefficiency but potentially competitive failure. Thus, the executive responsible for systems must continue to strive for shorter development cycles as well as business alignment.

As a result, the days of the technology journeyman who really understands technology but not the core driving factors for business success are limited. Whereas a sound technology strategy can shorten lead time in systems integration and development, the business environment in most industries promises to change so fast that the technology strategy, once misaligned, likely can never catch up. As long as the major architecture is established in such a way that the computing model and business organization models are aligned, the systems that result should be able to evolve to the competitive needs of the business.

During a recent speech by the CEO of a major computer hardware and software manufacturer, the speaker said that his customers had told him technology is moving at too rapid a pace. This is because many of these systems professionals are finding themselves spending more time tracking the movement of their own industry with far less time available to keep up with changes in the technological landscape. The combination of major moves in global industries with the incredible evolution in technology is almost too much for a human being to assimilate. If trade-offs must be made, strategy priorities should focus on the business at the expense of being at the leading edge of technology, if necessary, until it has proven its potential to be of value.

THE VALUE OF A TECHNOLOGY ARCHITECTURE

To organize technology, most would agree that there is a need for a technology architecture to provide a road map for integrating the various components. The architecture must be compatible with the business model, as any misalignment results in an evolution away from the competitive needs of the enterprise. Such an architecture eliminates the tendency to deploy "technology du jour."

The absence of an architecture and associated standards often causes each new project to begin with the latest technology for such items as data base management or end-user interfaces. This results in inconsistency in the company's evolving software portfolio and makes systems integration far more difficult. The age of these systems can often be determined by the appearance of the user interface, just as the rings on a tree determine its age.

The result of the absence of architecture is that incompatible systems are developed, each looks different, and each has different rules for use by end users. It is unlikely that the systems will facilitate data exchange or shared business processes. An information architecture should allow for each of the foundation components to be easily replaced when a truly improved technology comes along.

The architecture should define certain common business processes, rules for end-user interfaces, and shared data components. Such a view generally results in the development of three levels or tiers within the systems integration model. Whereas such a three-tiered model is one of numerous alternatives, it is consistent with the current technology direction for distributed and network-centric systems.

This data, process, and user-interface tiering allows openness at each level and the freedom to distribute applications at will. Such a model can provide a foundation for reengineering across business boundaries. The model can also be used with a centralized system in such a way that separate functions manage the data, provide common business processes, and display the information to the end-user. In fact, such restructuring is vital as the system is reengineered to make use of a distributed architecture. The systems architecture, therefore, is a way to organize functionality to provide for maximum flexibility in future deployment rather than a proxy for layers of types of hardware.

Such information architecture cannot be viewed as a panacea to compensate for archaic portfolios of ill-designed systems. Many of the current mainframe software systems are difficult to deploy in distributed architectures; replacement is not a matter of simply wheeling out the mainframe. This fact results from the additional need to rethink current ways of doing business when such an integrated and distributed architecture is technically and organizationally feasible. The simple downsizing and code conversion of an aged mainframe system will likely do little to enhance a company's competitive position.

Deploying Integrated Systems

Several major issues must be faced before such integrated systems can be deployed. The current systems development work force needs to be trained on the new technologies. Numerous COBOL programmers write their first distributed system using the same kind of COBOL design approach that became ingrained during the earlier parts of their career. A C++ program that is patterned after its COBOL predecessor does not represent the ideal design when distributed in a three-tier model. Such code construction can also hamper reuse in other systems.

The second major issue is that cross-tier boundaries can be very rocky with respect to integrity, performance, and security. Many tools available for mainframe-centralized systems are not available to the distributed enterprise. Not only is the design of distributed systems far more complex, but problem diagnosis can be far more difficult when the system is not performing adequately. Finally, development cost, without a learning curve, is considerably higher. Because multitier systems suggest the need for a high degree of software engineering of cross-tier interfaces, incremental work must be done that may not have been required if the system were deployed on a single computer. There is hope for a learning curve, but only when developers begin to reuse components that have been properly designed to work within the distributed systems architecture.

After a well-thought-out business strategy is in place, the most important focus must be in developing a learning curve for software development in the modern global organization. That learning curve is based on a number of factors, including interpersonal communications, the development of an appropriate architecture, and resisting the technology-du-jour approach. Further, the rigorous development of standards for data architecture process engineering and end-user interface, reuse of appropriate components, and an overall approach that industrializes software development itself will be organizational imperatives.

Perhaps a technical decomposition of the development task associated with each component of the architecture would provide value. The end-user interface has become more standardized, and a number of tools for interface development have been commercially introduced. These tools facilitate end-user prototyping in such a way that the developer and user can actually see the system evolve before their eyes. One must not make the mistake, however, of believing that once the interface has been developed, the system is complete and ready for deployment. The end-user interface must be backed up by a rigorous process layer made up of reusable modules that perform standard business functions. These processes are often industry specific, although object-oriented building blocks may be increasingly commercially available.

The data tier also follows industry guidelines in most cases. For example, in the health care industry, many industry competitors are collaborating to define how automated clinical information should be stored and interpreted. The development of such guidelines by a single business enterprise must be viewed as proprietary and would hamper the exchange of data across enterprise boundaries. All industries must come together in agreement on guidelines as they hope for more widespread use of such capabilities as electronic data interchange (EDI), which facilitates interbusiness communication—resulting in increased efficiency for all involved.

One can now summarize the technology foundation upon which most systems integration efforts are based. The hardware and data communications technology of the industry have improved at a rapid pace. The result has been increased power to end users, who are interconnected through high-speed, peer-to-peer networks that exist throughout the world.

The Pace of Software Evolution

Software evolution has not kept up with the development of hardware and communications. Most software developers today can still be considered craftspeople without the advantages of interchangeable parts and sophisticated software engineering. It is indeed amazing that employees with the most sophisticated of hardware and communications technology still use very archaic approaches in their day-to-day work efforts.

However, how will businesses use technology as part of industrialized systems? IT can be used to literally transform the business enterprise and the industry in which it exists. It can drive the business strategy itself and can enable product lines or whole new businesses. It can foster innovation and improve productivity of the enterprise itself, as well as its suppliers and customers. Obtaining new customers, or retaining existing ones, is far easier when it is more efficient to do business within the organization rather than with the competitor.

Perhaps the most promising of futures comes with the integration of the global community. Advances in communications and client/server architectures have allowed for the development of what IBM calls network-centric computing, or as AT&T puts it, "the network becomes the system." Such "smart" networks enable a level of electronic commerce and work group cooperation that would never have been possible before. Such strategic evolution provides the foundation for a reengineering effort that can truly be global in scale.

AREAS TO TARGET FOR IMPROVEMENT

Technology can do more than drive business strategy; in fact, many information companies have developed when their systems and products have become one. I will draw on an example from my industry, managed health care.

Managed health plans or health maintenance organizations (HMOs) are perfect examples of information-based companies. In many cases, the doctors provide services under contract rather than work for the company. The HMO also contracts with employers to provide health care to their employees and families. The value that is provided by the managed health care company comes in administration and the quality and cost controls that they impose. Most of these controls are based on information that they broker by providing informational feedback to the doctor, hospital, and patient. Delivery and quality of care are continually monitored through the use of sophisticated data bases and network technology. At the core of the successful HMO's business strategy is the use of information and data base technology to industrialize the practice of medicine.

Another phenomenon in information-based companies is that new product lines or entire new businesses can be enabled based on the IT that drives them. Thus, technology can have an effect on speed to market and can create shareholder value through the development of new innovative products deployed faster than the competition.

My company started a business that developed systems to pay drug claims using the concept of a list of cost-effective drugs that doctors were required, when appropriate, to prescribe. That business grew dramatically and was sold six years later for over $2 billion. In this case, none of the pharmacies was a part of the company, and none of the doctors prescribing the medications was an employee. We integrated our systems with real-time networks, processed information, and added value by reducing drug costs for our customers. The value that was created for our shareholders through the leveraging of technology was tremendous.

Enhancing Employee Productivity

Employee productivity is another area that can be significantly improved through systems development and systems integration. Examples of widely distributed yet nonintegrated systems can result in much redundancy in work effort, as well as data integrity issues from one part of the enterprise to the next. Horror stories abound of how the source data is downloaded to separate systems and manipulated to reach far different conclusions.

Thus, employee productivity should be a major goal as systems are integrated within the enterprise. Such productivity can come from the total elimination of a business process, as data is transferred from one part of the system to another, or can be achieved by streamlining a current process in such a way that the employee is able to accomplish more work in a given unit of time. Whereas step elimination is perhaps the ideal situation, another approach is to off-load the capturing of information at its source. In health care, use of EDI in doctors' offices allows for electronic billing to replace completely the paper entry step that used to be associated with the claim-entry process.

Another example to help understand productivity impact can be shown through an industry's use of a throughput statistic. Throughput measures the number or percentage of transactions that can process entirely through a system without ever being touched by a human being. Of course, the higher the number, the more value the system adds. Many industries have invested in process automation improvements in just such a way, and global connectivity can bring unprecedented levels of overall productivity to organizations with such highly automated systems.

Technology can also be used to improve supplier or vendor productivity. The automobile industry has been described in much detail, and many of the manufacturers have achieved economies by helping their suppliers to be more efficient or to provide higher quality parts or components. In the health care industry, managed care organizations are now working more closely with doctors and hospitals to streamline the ways to do business together. Regardless of industry, reengineering processes and integrating them will improve productivity across the entire system and will result in lower costs for the consumer.

Another application of systems integration is in optimizing customer satisfaction and bonding that customer to the organization. Eliminating an expense for a customer makes it easier for them to do business with the organization. The company that does this throws up a barrier for competitors that, at best, must go through the same integration process to make their systems function as effectively when viewed in a larger context. In addition, an information company can use that information to score their own performance and agree with their customers on common goals of the business relationship. The definition of common goals is one of the primary ways to achieve customer satisfaction; otherwise, satisfaction is not guaranteed, and long-term customer retention is in jeopardy. Finally, better information delivers the products that customers want.

SUMMARY

There are numerous conclusions that should be drawn from the discussion in this chapter. The role of the systems strategist has evolved to be one of business person first, technologist second. This generally means that the systems strategist resides at a higher place in the organization and is privy to strategic direction of the business at the time that it is created, not as an afterthought.

Another conclusion would be that the software development process must be industrialized to shorten lead times necessary to develop new systems—and products. Even if systems strategists slightly misjudge industry direction, such shortening of cycle times allows them to recover from their mistakes before the error is fatal either to their own careers or to their businesses in general.

Each business should agree on strategic success factors that make that business an effective competitor within its industry and then should find ways to apply systems integration to each of these success factors.

Whether they are product quality, total low cost, administrative efficiency, cycle times for product development, or quality in service and communicating with customers, such efforts should be the benefactors of the best and most consistent application of technology possible. Only with an appropriate architectural road map for such development will systems strategists see their strategic dreams operationalized in competitive, integrated systems.

2-2

Systems Integration Solutions for the Global Enterprise

STEVE HARKOLA

PETER DAVISON

The management of a global organization today is becoming more complex as pressure mounts to improve profits, reduce lead times, and improve customer service while streamlining operations. Essential to better management is an understanding of factors that differentiate worldwide strategic business units (SBUs). Solutions with a global perspective minimize customization and maximize reusability; attacking individual unit integration piecemeal without a thorough understanding of the entire range of business units comprising the total organization is futile.

Exploring the functional and technical integration challenges facing a global enterprise, this chapter presents an assessment methodology for document integration issues that in turn may lead to identifying opportunities. Following this analysis, this chapter surveys the information technology (IT) itself to determine its strategic direction. Finally, the discussion covers existing interfaces between SBUs and suggests others as a basis for further integration.

INTEGRATION CHALLENGES FOR THE GLOBAL ORGANIZATION

Purposely documenting key factors within each SBU highlights the challenges and pitfalls inherent in the integration process, as illustrated in Exhibit 2-2-1. Quantifying functional data in the form of a current environment model identifies the scope of integration potential. Global companies share classic, inherent differentiators, including:

- Organizational culture.
- Multiple languages.

Key Factor	Strategic Business Units (Data Collection) 1	2	n	Current Interfaces, Problems, and Issues Across All SBUs
Organizational Culture	Narrative			Centralized Human Resources?
Language Barriers	Primary Secondary			Screen Language Differences?
Currency Translation	Primary/Auxiliary Denominations			Multicurrency?
Local Government Requirements	Statutory Laws			Value-Added Tax/Interstat?
Autonomy/ Decentralization	Corporate Links			Local/Wide Area Networks?
Measurement Systems	Business Rules			Activity-Based Costing?
Core Products	Model Numbers/ Descriptions			Part Numbers Consistent?
Suppliers/ Customers	Customer/ Vendor Files			Supplier/Customer Numbers Consistent?

Exhibit 2-2-1. Key Factor Documentation

- Multiple currencies.
- Local government requirements.
- Degrees of autonomy and decentralization.
- Financial viability.
- Core products.
- Customers and suppliers.

Organizational Culture

By far, organizational culture differences dictated by country and regional heritage, political structures, business etiquette, and local management dictate the physical implementation as well as the pace of business integration. Although difficult to collect and quantify, organizational cultural differences are instrumental in defining integration scope, support level, and ultimate probability of success. Data-collection areas include:

- Human resource policies.
- Customer and supplier interaction and customs.
- Decision-making hierarchies or authorities.
- Reward-compensation systems.
- Capital-procurement policies.

Multiple Languages

Language barriers are particularly bothersome, not only in verbal communication but in systems and procedures as well. English is the predominant business language in Europe; however, Asian and Pacific Rim countries are not as standardized. Of concern are various end- user presentation interfaces, communications, and data content itself.

Multiple Currencies

Currency translation related to financial transactions and reporting further differentiate global business units. Similar to multilingual capabilities, currency translation places an additional overhead burden on integration efforts and should be documented.

Whereas software companies competing worldwide have addressed this issue, particularly in their financial systems, companies interested in leveraging existing domestic software, for instance by exporting program code overseas, may soon run into the translation problem wherever dollar values are involved.

Local Government Requirements

Local reporting requirements take various forms: specific financial performance, tax and employment statistics, and payroll, for example, may be tightly coupled to local core business systems. Many companies avoid home-grown payroll systems because of the high table- and rate-maintenance overhead involved. Also, operating even a packaged human-resource system on a global basis may be prohibitive. Implementing generic applications globally is difficult without a high degree of parameters set up for the system and local autonomy. In addition, SBUs may maintain multiple sets of financial ledgers, depending on the reporting target constituency, many of which are offline and undocumented.

The ability to maintain multiple views of data may be a requirement, as well.

Autonomy and Decentralization

The degree of autonomy and decentralization may be measured in such areas as capital decision-making authority, duplication of overhead support functions, including data-processing systems, and customer and supplier contact interfaces. Integrating highly autonomous and decentralized organizations presents a higher degree of difficulty than a homogeneous enterprise does, and managers should review corporate allocation schemes, transfer-pricing algorithms, and cost-versus-profit-center designations. Today's global manufacturing organizations, for example, may have distributed centers for design, low-cost manufacturing, and sales offices organized in a less-than-efficient manner, because of rapid growth, regional requirements for local content, or any number of other reasons. IT organizations often recognize inefficiencies and must take the brunt of attempting to link diverse facilities while being unsuccessful in fostering organizational realignment.

Financial Viability

Internal measurement systems and overall SBU financial health may affect implementation scenario development. Management and employees better support projects and tasks for which they are accountable. Growth environments tend to support revenue-generating ventures rather than the cost-cutting measures typical of a downsizing environment. Therefore, the IT manager must always investigate allocation schemes related to IT spending in particular.

Core Products

Core product and service identification, including competitive strengths and weaknesses by business unit, may drive major business automation variables, including:

- Make-versus-buy systems solution.
- Software availability, type, and selection.
- Hardware and networking infrastructure architecture.
- Implementation priority and schedule.
- Integration and consolidation opportunities.

Because manufacturing systems in particular are extremely sensitive to product type, material resource planning (MRP) or enterprise resource planning (ERP) systems used to be tailored and marketed to specific industries based on product classification, ranging from discrete- to continuous-flow manufacturing processes. Today, more hybrid applications allow mixed-mode processing and serve a wider range of products, increasing the system's overall leverage.

Customers and Suppliers

An analysis of suppliers and competitors by SBU may highlight significant integration opportunities. Individual suppliers may be doing business with multiple SBUs without recognition for combined quantity buys. Total quality management (TQM) initiatives may dictate supplier minimization and electronic data interchange (EDI) supporting just-in-time (JIT) operations. Global sales organizations, on the other hand, are usually faced with customer- service-interface issues relative to organization by customer, product line, or geographic region. Analyzing customers for each SBU may lead to improvement identification. The Internet, for example, is used by companies to provide a common front end to the outside world in the form of a "home page." Using hypertext links, a customer viewing the home page can select a product or service and be automatically linked to the proper front-end application. Benchmarking key indicators, such as sales, cost of sales, market share, vertical and horizontal integration, and product-line profit margin, serves to position each SBU relative to the market it serves. Integration-effort payback and return on investment increase substantially when spread over a wider area of coverage.

STRATEGIC BUSINESS UNIT TECHNICAL DIFFERENTIATION

Exhibit 2-2-2 illustrates a comprehensive IT inventory matrix used to survey the hardware, software, and human resources used by each SBU. Coupled with budgetary information, a global view of the data-processing assets of the total company serves as a baseline upon which to launch strategic planning and improvement initiatives. These inventories support project planning, budgeting, asset redeployment, downsizing, mergers and acquisitions, future upgrades, and volume discounting.

Application infrastructure categories represent systems dedicated to key areas of the business in Exhibit 2-2-2. Combined, they represent the total global investment in computing software. This investment includes not only the original purchase and maintenance cost but the sum total of intellectual capital invested in project implementation, training, and most important, data.

Product systems represent applications dedicated to the design,

Application Infrastructure Category	Hardware					Software	Staffing
	Model/Version Number A	Number of Units B	Asset Type C	Amortization/Depreciation/Rent D	Maintenance E	A - E	A - E
Product Systems							
Design/Drafting Engineering—							
CAD/CAM/CAE							
Manufacturing/Methods							
Engineering							
Product Engineering							
Product Data Management							
Business Systems							
Manufacturing							
Quality							
Sales/Marketing							
Cost							
Finance							
Human Resources							
Office Automation							
Technical Publication							
Word Processing							
Spreadsheet							
Data Base							
Presentation/Graphics							
Project Management/Scheduling							
Electronic Mail							
Workflow Management							
Computing Infrastructure							
Operating Systems							
Enterprise Data Base Management							
System							
Networking							
Communications							
Software Development							
Client Platforms							
Server Platforms							
Security							
Operations							

Exhibit 2-2-2. Information Technology Inventory

manufacturing, and assembly of the products or services that generate revenue. Manufacturing industry examples include computer-aided design (CAD) and computer-aided manufacturing (CAM). These applications tend to be product-line or service-portfolio specific. Systems and data are usually distributed close to the customer and controlled for the most part by functional management. Whereas the same basic process may be conducted in each SBU, different product lines, customers, distribution channels, and local management may drive wide fluctuations relative to application solutions. Professional productivity, particularly related to the creative talent applied to product content or packaging, warrants highly automated tools and user interfaces to maintain a competitive edge. Product data management (PDM) is a growing software solution market encompassing all areas of product development, including workflow management, object vaulting, desktop editing, change control, and configuration management. PDM systems by definition integrate CAD, engineering analysis, electronic communications systems, such as E-mail and EDI, and bill of materials and may extend to suppliers and customers alike.

Business systems incorporate the required support functions, including planning, monitoring, and controlling all the various activities that constitute the business process, typically in the form of transactions. Whereas the pendulum over time swings between centralized and decentralized implementations, common applications have been proven to be more cost-effective. Planning systems, such as material requirements planning (MRP), provide economies of scale when combining part demand across multiple product lines with common subcomponents, and further economies are gained through centralized or integrated inventory supply. Business systems convert all activities into dollars for reporting, analysis, and decision making. Much effort has gone to integrating and streamlining business applications, resulting in enterprise-level systems that minimize human care and feeding, support initiatives, such as lead-time reduction and total quality control, and provide management with one overall version of the truth, relative to financial performance. Highly integrated systems usually include a single point of data entry, real-time updates of common data bases, and retrieval capabilities supporting ad hoc requirements.

Office automation accelerated in the 1980s and 1990s, fueling the explosion of personal computer (PC) hardware and software. Personal productivity products have proliferated in most organizations, raising such issues as software distribution control and piracy, standardization, maintenance, training, technical support, and network capacity. Office automation, having evolved from standalone spreadsheets to groupware technology, now enables empowered teams of workers to collaborate on projects effectively. Inherently workstation based, the control of office automation applications has remained typically outside the control of the information management services function in many companies; therefore, such applications are therefore difficult to count and manage. However, standards improve buying power, reduce support and training costs, and allow a higher degree of integration. In effect, then, "the best offense is a good defense."

Microsoft Office is an example of the new suite products that are shipped as a highly integrated set of solutions bundled in one package. Object Linking and Embedding (OLE), Dynamic Data Exchange (DDE), and Open Data Base Connectivity (ODBC) standards embedded in the software allow users or developers to customize and combine application solutions seamlessly.

Computing infrastructure represents the underlying layers of software supporting user applications. Transparent to end users in many cases, standardization of these layers is central to supporting integration efforts in the global enterprise. Because user sophistication levels have evolved beyond standalone desktop use. requirements for software connectivity across the entire range of application infrastructure categories is rising. Most would prefer a clean-sheet approach, supporting one operating system, one network protocol, and one data base engine. Most corporate inventories, however, indicate just the opposite: a proliferation of multiple platforms and proprietary systems. An open-systems approach encourages the use of standard operating environments, source languages, graphical user interfaces (GUIs), data bases, networks, and communications protocols. Open systems eliminate islands of automation and support connectivity and interoperability between various computers in a multivendor environment. Business systems that do not limit the choice of hardware or software for execution are open, thereby allowing the user to take advantage of the best-price-for-performance computing platform. Most of the hidden systems-integration costs are embedded in the requirement for infrastructure upgrades to support new initiatives. This is further compounded by globalization of the enterprise: its sheer size, geography, and functional diversity all add to the bottom line. Whereas software functionality remains largely the same, the skill sets required to manage and administer distributed computing structures are usually underestimated.

Enabling technology represents the hardware platforms, specific software packages, source language code, and staffing required to develop or support these applications. To build a pro forma global baseline budget for data processing, inventories should be taken with such collection variables as:

- *Vendor, Model, or Version Numbers.* These provide easy hardware and software classification by recording vendor name, product name, and current release levels. In most cases, data from an older software version is upwardly compatible with a new release—but not vice versa.

- *Count.* This is perhaps the most difficult variable to quantify, because of the lack of overall control of the desktop environment. Once a count is established, some controls should be put in place to maintain inventory accuracy to allow periodic inventory reevaluation and liability assessment.

- *Asset Type.* Referring to ownership. asset type describes whether the expense represents depreciation, lease amortization, or rental. Assets that are purchased outright and capitalized incur an annual expense based on a depreciation scheduled over the useful life of the hardware or software. Amortized lease expense is similar, but the buyer does not own the asset, which may be purchased at the end of the lease for a residual value. Rental is a pure month-to-month expense, with no liability commitment.

- *Annual Depreciation.* This type of amortization or rental expense represents the dollar-value budgeted cost of ownership. Cost may be rolled up by SBU, by application category, or by vendor, as required.

IT ASSESSMENT BY SBU

When evaluating IT's effectiveness, direct input from end users and managers is essential. Pursuing global integration without direct user input leaves the organization open to gross misapplication of technology. Whereas this assessment is. in many ways, qualitative, surveys of this nature serve to link IT to the SBU strategy. The risks associated with the acquisition of direct input are the bias associated with any survey instrument and the very likely possibility of negative feedback. Corrective action must be planned and budgeted to prevent aggravating problem areas further.

The first step is to identify key department and function activities and the role of IT in baseline critical success factors, as perceived by the SBU. Once defined, various departments, including sales, engineering, manufacturing, assembly, and quality, can identify initiatives and areas for improvement as they relate to supporting unit goals. Most important is the departments' assessment of the role IT plays in carrying out their mission.

The next step is to assess the support the IT function provides to the organization. This information lends itself to a high/low statistical analysis relative to the following sample items:

- Overall support level.
- Support focused in the appropriate areas.
- Accuracy of information supplied.
- Currency of information supplied.
- System availability.
- System performance.
- Office automation adequacy.
- Operational processing reliability.
- Problem-response capabilities.

Next, core business application effectiveness should be evaluated using a data base collection vehicle. Depending on the number of applications, this can represent a formidable collection effort. Common applications require only one response related to technical criteria, whereas functional assessment criteria responses may vary widely by department across each SBU:

- Application identification.
- Application type.
- Strategic impact.
- Functional quality and effectiveness.
- Flexibility and ability to support business changes.
- Ease of use.
- Number of users and shifts.
- Level of IT support or outside control of IT function.
- Return on investment.
- Technical attributes, including:

  — Number of modules and programs.
  — Percent of batch processing versus online.
  — Languages.
  — Data base access methods.
  — Processing platforms.
  — Resource consumption.
  — Age.
  — Level of maintenance.

Assessment of current applications in place is an ideal time to collect information related to new applications under development or planned for the future.

Users' perception of expenditures on IT should be documented. This may be influenced by the presence of a chargeback mechanism or budgetary allocations. IT cost may be evaluated again, according to a high/low rating or a more detailed basis, including assessment of return on investment or other cost-justification methods.

In addition. users should be allowed to rate IT personnel as a group, as candid evaluations can provide revealing insights into a user's perception of service. Depending on staffing levels in the particular organization, there are considerable criteria from which to choose. such as:

- Technical skills.
- Functional expertise.
- Project-management skills.
- Process-reengineering capabilities.
- Integration skills.
- Non-host-based system knowledge.
- Help-desk support.

Questions should be asked when describing the degree to which departments are involved in IT planning and priority setting. Are there steering committees composed of departmental managers driving the IT focus? Is there a written IT plan extending into multiple periods? What process exists to request IT services and measure progress?

Each SBU's end users should be identified as to their computer literacy. as the PC revolution has gone a long way in increasing the sophistication level of many users. To what degree are which users involved in the process of requirements definition. system prototyping and testing. ad hoc query, and overall office automation activities?

SBU INTERFACES

Unless each SBU operates in a standalone manner. there are considerable transactions between units. Minimally. financial reporting requirements dictate some level of communication. In a vertically or horizontally integrated corporation, extensive communications are required.

The lowest common denominator in interfaces are paper-based transactions, which come into play when applications systems used by each SBU are incompatible. Rather than solve fundamental process issues, IT has been applied to speed up the paper flow: faxes, E-mail, EDI, and groupware are examples. Arm's-length transactions are represented by documents, such as sales orders, purchase orders, and shipping schedules. These transactions are received and processed by clerical means by standard front ends. EDI technology, when implemented fully, attempts to link transactions directly to the target units' internal core business systems and to bypass document reentry. Interaction communications, including voice, video conferencing, and interactive multimedia systems, are becoming increasingly important, as corporations attempt to form teams across geographic boundaries. As inter-SBU activity is analyzed, business interface patterns emerge and opportunities for process improvement through integration are highlighted.

SYSTEM INTEGRATION DRIVERS

System integration efforts should closely support the corporation's goals and objectives. In their most basic form, these goals emphasize revenue enhancement, cost reduction, or overall improved competitive posture. A review of business drivers serves to establish the underlying justification to pursue systems integration initiatives further. Drivers discussed here include:

- IT benchmarking.
- Global initiatives.
- Low-cost manufacturing.
- Time to market.
- Customer service.
- Employee empowerment.
- Mergers and acquisitions.
- Maintenance, repair, and overhaul.

IT Benchmarking

A method for comparing IT expenditures and services with IT organizations in various industries, the classic benchmark is represented by the ratio of IT spending to total company revenue. Unfortunately, end-user spending on IT usually is not included in most organizations' information systems budgets. According to a Gartner Group survey of 290 companies, hidden end-user spending represents at least 28% of known IS spending. In addition, those enterprises with multiple IS groups that do not roll up cost accounting for IT experience lower ratios than expected. A lower-than-average IT spending score may indicate the organization is falling behind in the industry relative to competitive IT investment.

Global Initiatives

Serving to drive system integration in an effort to become more competitive, these programs are often sponsored by top management with serious commitment to employee education and training, to implement significant cultural change successfully. Each program requires integration by nature because participation transcends multiple departmental boundaries. Whereas departments in the past could operate rather autonomously, using separate systems with well-defined interdepartmental interfaces, newer techniques require far more teamwork, interactive communications, and shared knowledge.

- *Supply Chain Management (SCM)*. SCM includes integration focused primarily on supplier scheduling, inventory control, quality, and business-transaction documents to reduce lead time. The key focus is the effort to reduce transaction costs—the time and effort necessary to translate raw materials into a final product or service. SCM allows quicker response to customer orders, replaces inventory with information, and reduces overall inventory obsolescence.

- *Total Quality Management (TQM)*. W.E. Deming's 14 points on quality are considered the basis for all quality programs and initiatives. Employee empowerment, continuous improvement, and the removal of waste are key concepts translating into the need for integration between various functions and systems.

- *Computer-Integrated Manufacturing (CIM)*. As the name implies, CIM represents the connection of various stages of the design process leading to products that can be manufactured quicker, better, and for less cost. An example is automatic CAD translation directly into numerically controlled machines producing parts.

- *Just-in-Time (JIT)*. Whereas manufacturers for years overproduced just-in-case products to demand forecasts, JIT methodologies reduce inventories by scheduling the arrival of components just at the time they are required during the building process. This requires extensive integration, with both suppliers and customers ensuring on-time supply shipments and accurate production schedules.

- *Kanban*. Similar to JIT, Kanban is a form of shop-floor control that pulls supplies and parts through the system as needed, no sooner. Kanban has been implemented extensively in Japan, using card-based systems. Modern material requirements planning (MRP) II applications may use pull methodologies, again requiring supplier and customer integration.

- *MRP II*. The II in MRP II represents the continuing evolution of MRP; it now includes management and control of the total enterprise from the time an order is taken to the time the shipment is made. More recently referred to as enterprise resource planning (ERP) systems, these enterprisewide applications represent a series of program modules that automate and integrate each department of the company, including engineering, sales, purchasing, manufacturing, quality, and finance. The MRP module itself provides time-phased, component make or buy schedules, designed to meet customer delivery-date commitments.

- *Concurrent Engineering.* Concurrent engineering is designed to reduce front-end product cycle lead time by supporting the simultaneous start-up of the various engineering and manufacturing disciplines. Integration capabilities in the form of PDM applications allow rapid workflow management across multidepartment teams during the product's total life cycle.

- *International Organization for Standardization (ISO 9000).* ISO 9000 certification is becoming the admission price for doing business, especially in Europe. Used to ensure quality in design, development, production, installation, and service, ISO 9000 certification means that procedures have been formally documented and conformance is audited periodically. Integration is a key component of compliance.

Low-Cost Manufacturing

The drive to develop low-cost manufacturing sites has led many domestic companies into the international arena in search of low-cost labor. The initial trend toward standalone operations has given way to increased integration with upstream organizations to minimize lead time and reduce overhead. The effectiveness of low- cost manufacturing sites can be improved by:

- Integrated material requirement planning and procurement systems.
- Computerized engineering drawing and transfer.
- Numerical control equipment program download capability.
- Purchase order, sales orders, and schedule releases using EDI.
- Financial consolidation and cost control.

Time to Market

Another factor driving change and the need for integration involves the time-to-market concept. Time to market refers to firms' abilities to develop new products quickly as well as deliver existing products effectively. Compressing time to market requires increased integration of activities across functional departments, such as design, engineering, manufacturing, purchasing, distribution, and service. Time compression has always driven developments in IT. The communications infrastructure in particular would not have developed at its present pace if it were not for time, the competitive edge. Many companies, realizing the Japanese threat, have been forced to cut product-introduction lead time by over 50%. Studies indicate that products delivered six months late generate significantly less profit over a five-year period than products delivered on time but 50% over budget.

Customer Service

Also dependent on the competitive variable time is customer service. Delivery of goods and services to customers within a contractual or perceived time frame is critical to long-term business viability. Faster service, once a competitive edge, has become a universal global expectation. In addition to faster time to market, the ability to make quick modifications in the event of poor product acceptance can affect customer satisfaction significantly. Integrated information systems can effectively support product differentiation. Product customization, including tailoring for an individual customer, becomes increasingly affordable as the cost of information processing decreases. In markets in which customer tastes change rapidly, rapid product turnover is a matter of survival.

Employee Empowerment

Facilitated by IT, employee empowerment takes many forms. Whether completing individual tasks or working on teams, workers increasingly turn to computers as an indispensable tool for communications, data acquisition, computational capability, document imaging and, of course, training. Driven by extensive downsizing, organizational flattening, and time compression, individuals dubbed "knowledge workers" expand their control through increased information visibility. For instance, an empowered purchasing administrator can use integrated systems to look back into time-phased material requirements to qualify for a minimum price break and scan forward to evaluate recent supplier incoming part-inspection results while on the phone. Historically, PCs have been used to bolster personal performance. Today, collaborative work groups have driven requirements for software that allows interactive dialogues with team members, whether they sit in the next cubicle or around the world. A growing body of applications, generically termed *groupware*, supports this need. Groupware not only further integrates desktop software components, such as word processing and spreadsheet, but also allows extensive collaborative sharing between co-workers. Generic workflow procedures or requirements specific to a particular project can be embedded into these applications to facilitate team efficiency, regardless of personnel location.

Mergers and Acquisitions

Most mergers and acquisitions are justified in part through the elimination of redundancy after combining two entities. Thus the cost of the sum of two integrated organizations is less than the two standalone. Whereas there is nothing magic about this type of synergy, the managers faced with the job of integrating two diverse organizations have their work cut out for them. Back-office systems are usually the first to go. Most important are the cash-control applications, such as payroll and accounts receivable, followed by front-end application integration, affecting sales and customer support. Systems integration takes many forms, including various combinations of the following:

- *Reporting only.* Financials, capital plans, and budgets are delivered on paper or in electronic medium.
- *Shadow files.* The acquired business retains its own systems and provides file copies formatted to merge with primary systems files.
- *Interfaced systems.* Both entities' systems are electronically connected to batch process transactions, such as journal entries, purchase orders, and material requisitions.
- *Converted data.* All or most data from the acquired firm is converted to the owner or primary system, and the acquired firm's systems are dismantled.

Maintenance Repair and Overhaul (MRO)

This represents a welcome growth segment in many flat or declining manufacturing businesses. Many manufacturing companies have dealt with MRO manually or using off-line systems separate from their new business MRP applications. As their MRO business grows, the need to track and manage this function better has forced companies to look at existing systems, only to find that homegrown and packaged systems may not address MRO adequately. Companies and software vendors alike are scrambling to integrate MRO into their mainstream business processes.

IT TRENDS

Trends in IT often drive system-integration activity. IT is pulled into the organizations by the company functional area executives in response to trade press, direct marketing by vendors, consultants, and a growing awareness of the dependence on systems for revenue enhancement or cost reduction. IT is in turn pushed into an organization by the IT organization employed to manage data processing. Whereas there is a general lag in the time between trend acknowledgement and ultimate implementation, integration efforts in many firms is driven by trends, which include:

- Open systems.
- Client/server computing.
- Global communications.
- Data center consolidation.

Open Systems

The drive toward open systems encourages the use of standard operating environments and hardware, source languages, GUIs, and communication protocols. An open systems computing architecture minimizes islands of automation and supports connectivity and interoperability within a heterogeneous computer environment. Ultimately, any user anywhere in the organization worldwide should be able to access information transparently, regardless of the equipment. Key components of this trend include:

- Standard operating systems, which enable an application to be ported to any compliant hardware to take advantage of the best price performance.
- Common source languages, which permit recompilation for execution within various operating system and hardware combinations.
- Standard data base management systems (DBMS) and access interfaces, which allow use of multiple DBMS vendors without major program modification. The major trend today dictates relational data base management systems using Standard Query Language (SQL).
- Common user interfaces, which imply a common look and feel relative to end-user screen interaction. This eliminates the need for users to learn a different set of rules and instructions every time they move to a different application. The trend today is to provide the same screen interface for core business applications as users have come to expect from popular windows-based PC software.
- Standard communications protocols, which allow multiple vendors to communicate with one another. Although there is no one industrywide communications standard today, users have a choice of widely used protocols, in particular, TCP/IP. Standards bodies are at work to develop a truly international protocol, called Open System Interconnection (OSI).

Client/Server Computing

Client/server architecture splits processing between two systems: a client workstation, with which the user interfaces directly, and a server, which processes client requests and returns information to the client.

There are several types of client/server implementations. Early implementations served to "front end" large, host-based applications by putting the user presentation portion of the system on the client only. This application-server approach left the application processing logic and data on the host. Screen information is literally "scraped" off older generation, host, and online displays and reformatted for the user using a graphic user interface.

The data base-server approach moves both the presentation and the application logic to the client, leaving only the data on the host server. In this case, the application running locally on the client can achieve better response times.

The distributed client/server approach moves the data off the host as well as onto one or more servers to take advantage of lower-cost computing hardware in support of overall downsizing.

Even though the trend toward the client/server processing model is pervasive, multiple infrastructure issues must be taken into consideration. Unwary organizations that rush into client/server technology with religious zeal find that the same controls that have been perfected in the mainframe environment over the last 30 years are now being reinvented in the even more complex client/server world. In fact, many of these infrastructure issues represent trends themselves and by nature increase the difficulty of business systems integration within a data-processing organization. For global and virtual organizations, much of the hidden expense of client/server processing is embedded in the underlying support structures. The challenge for IT management is to justify this expense through improved business integration. Step one is to identify these cost components, including:

- *Server Administration.* This includes physical-disk space management, maintaining users, diagnosing errors, and upgrading hardware and operating systems on server hardware. Unlike a centralized mainframe, companies may support a variety of servers that are graphically dispersed. It is interesting to note that the trend toward extensive server decentralization has reverted to a more centralized scheme to take advantage of economies of scale. Server clusters or very large servers, including symmetric multiprocessor (SMP) or massive multiple processor (MMP) machines, have grown in popularity to meet this need. Centralization of servers does, however, place a premium on network bandwidth and reliability.

- *Network Administration.* This has been driven into the forefront, as organizations move more applications onto local and wide area networks (LANs/WANs). Client/server computing is the paradigm by which the IT function moves systems onto networks originally "hodge podged" together to support office automation requirements. Most LANs are not ready to support client/server technology. Production application may require 24-hour, seven-day-a-week support. Bandwidth, or the ability to move large amounts of traffic, may be insufficient. Consistent response times, a primary objective of mainframes, may be difficult to maintain under typical random LAN loads. As the complexity of the traffic on the wire increases, sophisticated LAN monitors and analysis software may be required to resolve problems and prevent bottlenecks.

- *Data Base Administration.* This takes on a new twist as data originated from a host system may now be replicated across servers around the globe. Data and the parameters that define data, called *metadata,* must be kept in sync across multiple platforms. Data base administrators design data bases, assist application developers, perform backups, recover corrupt data, monitor data bases, and tune for access performance.

- *Security.* These features vary widely in server and LAN operating systems. Unfortunately, complex multiserver environments require staff support, in some cases, more than the corresponding host mainframes. To limit overhead costs, the ratio of machines managed per administrator may climb, and security is usually the victim. Unique to the network are modern-day viruses and applications that run on wide-open PCs. Unlike mainframes protected in the glass house, PCs have disk drives and other peripheral devices that allow users to load personally acquired applications from a multitude of sources. In addition, security features tend to be embedded in a variety of sources. Security features are typically included as part of in-house or packaged business applications. Network operating systems and data base management systems have security features bundled off-the-shelf and, of course, separate security packages and firewalls are available to further augment security capability. Security integration has become a pressing concern in any organization with a substantial investment in hardware, software, and data resources.

- *Software Distribution.* This represents the timely, coordinated migration of changes and new releases of software to end users. Difficult enough in the single-program-source, mainframe environment, distribution of software updates across LANs and desktops presents an enormous challenge. First, IT organizations do not fully control the desktop. Second, programs are duplicated over an unlimited number of servers and workstations. Automated tools are beginning to emerge to control this problem effectively. One solution removes hard drives altogether from user PCs and places all programs and data on tightly controlled servers. Users "boot" off these servers— start their PCs off the servers—from a single copy of the software but suffer a lower grade of performance.

- *Capacity Planning.* Usually nonexistent in the client/server world, capacity planning presents a problem. Although a new application may be initially sized correctly, even overpowered, few mechanisms are in place to monitor and forecast load due to follow on applications, new users, growing ad hoc query load and proliferating files. Adding more servers to solve problems must be frequently ruled out because of the difficulty involved in partitioning application data across more than one machine.

- *Data Backup and Recovery.* Taken for granted on mainframes, recovery of systems across multiple servers is a state-of-the-art problem. Many organizations compensate by only distributing across a network read-only data that can be easily restored from a common source. As soon as updates are introduced as a by-product of business integration, the company is at risk. Most medium to large mainframe installations also have a disaster recovery plan, including a "hot site" or alternative computer ready to activate in the event of a sitewide, catastrophic problem. For some companies, the loss of their business systems for even a day can represent millions of dollars in lost revenue. Replacing one mission-critical server is one matter; providing an alternative network of servers and user workstations fully loaded with current data is a new challenge.

- *Production Operations.* In particular, the scheduling of batch processes versus interactive user sessions, production operations is a strength of the IT organization in mainframe environments. Batch does not disappear in the client/server arena—it just gets more complicated. Software designed to administer, schedule, and control production batch work on servers is beginning to mature, however. Additionally, most server operating systems, including UNIX, do not deal well with traditional batch jobs. Concurrent batch and online processing is a major problem when response time is a concern. Jobs that ran in an expected time frame may run considerably longer on a server not tuned for input/output.

- *Standards and Procedures.* These are often ignored; in fact, one of the attractions for many young IT professionals migrating to the client/server environment is the lack of controls, which—roughly translated—equates to more fun. Mainframe environments are typically locked down tight in layers of bureaucracy designed to enforce operational consistency. Client/server architecture must impose similar levels of standards and controls that run the business on the host computer. After all, the applications are basically serving the same objectives.

Global Communications

Communications technology, having evolved from the first standalone PCs, has dramatically improved an organization's ability to integrate workers at all levels. Business people today have come to expect a utilitylike capability from a company's communications infrastructure. Global organizations in particular have exploded communications technology to the fullest to reduce duplication and improve overall coordination. Some of the more significant aspects of the systems integration process and its impact on the organization include:

- LANs and WANs form the pipeline for computer traffic flows between strategic business units. Previously the domain of interoffice mail and the telephone, LANs and WANs connect individual workstations to each other, to peripheral devices, such as printers and fax machines, to local and corporate data bases, including mainframe computers, and to customers and suppliers. By definition, a degree of integration is achieved merely by implementing a network. LANs can serve to enforce standards both at the network operating system level and the protocol level. This is essential to minimize technical complexity and corresponding staffing overhead.

- E-mail is an enabling technology that can revolutionize interpersonal and group communications throughout a worldwide organization. Essentially a store-and-forward, message-based technology, E-mail can be implemented at relatively low cost using standard dial-up modems and PC equipment. In addition to moving previous voice traffic to an electronic message medium, E-mail is capable of moving other objects as attachments, including drawings, spreadsheets, word processing documents, and forms. E-mail can be used to integrate applications by incorporating automatic message-based interfaces, such as action queues and triggers, which execute stored procedures.

- *Groupware* is a generic name for a body of software designed to support collaborative work groups. Groupware often provides calendaring, resources scheduling, group editing, managing electronic messages, and tracking progress on project specific tasks. "Virtual" corporations—those that employ a dynamic, mobile work force— have come to rely on groupware as the glue that holds the company together. Reduced travel, timely communications, and iterative decision making are a few of the many benefits.

- Video conferencing is an emerging communications technology as implementation costs become increasingly more affordable. Voice and video capability has evolved beyond the classic meetings held in a special room with a two-panel, interactive session, to geographically dispersed workstations enabled with voice-aid video cameras for casual, real-time communications on demand. A poor-man's version of video conferencing includes group white-board capability and voice without the video.

- Decision-support and expert-systems benefits drive integration efforts in many companies. Decision-support systems typically rely on aggregate data bases supporting ad hoc queries to assist professional and executive data analysis. Executive information systems (EIS) represent a particular class of decision support targeted at top management. Systems must be integrated to provide consistent views as an executive navigates through financial or operational statistics across functional boundaries. Expert systems focus on the automation of specific business rules to reduce reliance on manual intervention, minimize errors, and produce repeatable and consistent results. Examples include group technology, airline reservations systems, and even voice mail menu-selection services.

Data Center Consolidation

As a final consideration, data center consolidation, including outsourcing, is supported by systematic integration efforts. Integration tends to reduce the number of software systems, hardware devices, and support personnel required to run an operation. The fewer the pieces, the more cost-effective it becomes to centralize processing or subcontract to third parties. Using off-the-shelf software packages is a form of outsourcing. Inherently integrated are packages such as enterprisewide systems, including MRP II and supply chain management. In addition to the software's purchase price, maintenance is charged annually to qualify for telephone support and, more important, software upgrades. The vendor, rather than in-house staff, is responsible for system enhancements and user liaisons. Many companies are struggling with a decentralized and geographically dispersed group of business units that trade internally. They are finding that to remain competitive, transaction cost and duration must be reduced. This is particularly difficult if each unit is operating its own set of application systems. Merely speeding up transactions, such as purchase orders, with EDI does not necessarily solve the problem. One common solution is to consolidate data centers and run operating units off one common set of applications. This eliminates most arm's-length transactions and provides common visibility into inventories and work-in-process status. As network communications become more reliable and increasingly inexpensive, this option is becoming more attractive, in lieu of the costs associated with the required conversions. An alternative methodology calls for the creation of global applications that sit above applications running locally. For instance, a common forecasting, order fulfillment, and MRP application supports centralized procurement and optimization of orders placed against remote feeder plants. The feeder plants would, in turn, operate local execution systems to manage the shop floor and feed financial statistics into a common consolidation package. This hybrid approach requires extensive integration but may provide for a level of individual unit autonomy essential for success.

SUMMARY

Changes in the global business environment and organizations in general are driving the need for new technologies and a high degree of systems integration. The functional and technical differentiators reviewed tend to polarize individual business units and must be overcome to minimize cost and maximize revenue in the global marketplace. No longer can organizations and their internal IT functions be content with multiple systems, incompatible operating environments, and fragile interfaces. An assessment of current IT across all business units serves to highlight the scope of change necessary to produce the desired results. Technology infrastructure upgrades, core business systems integration, and decision-support capabilities can be driven in conjunction with any number of global business initiatives used to create the parallel organizational changes necessary for success. Systems integration provides the technology and architectural framework for global repositioning.

2-3

The Business Promise of Hub-and-Spoke Systems

JOHAN VINCKIER

For many managers, the stories of information technology (IT) accomplishments seem like wish fulfillment. Within nine months, a retail bank overcomes the limitations of 30-year-old legacy information technology (IT) systems to launch a high-tech operation that offers its customers a complete range of banking services over the telephone. Another bank is able to roll out, within three months of an acquisition, information systems that deliver the same level of service to all customers at all branches, new and old. An industrial goods company succeeds in building the IT support it needs for order processing in the course of a nine-month reengineering project.

For these companies, IT has genuinely become an enabler of change that boosts competitive performance. How they did it, as well as the lessons they learned along the way, can be copied by others. The secret: a "hub-and-spoke" approach to building the kind of information systems that lend themselves to rapid, continual improvement.

THE AIRLINE ANALOGY

At the simplest level, an IT hub-and-spoke-based approach is much like the strategies adopted by US airlines in the 1970s and 1980s after deregulation intensified domestic competition. At that time, most airline networks looked like spaghetti, with cities haphazardly connected as local market opportunities emerged. Airlines trying to expand or improve their services quickly found the complexity of their networks becoming unmanageable. Changing a single service or adding even one flight to a preset schedule created such a domino effect on connecting flights that real optimization was impossible.

Airlines then discovered they could manage their networks much more effectively if passengers were routed through a central airport "hub," which offered fast connection times to flights along individual route "spokes" to other cities. As a result, airlines could offer a truly integrated network between as many city pairs as it had planes—a dramatic increase in customer service without heavy spending on additional planes. They could also add a new city to their network more easily; all they had to do was establish a connection between it and the hub.

Today, many corporations face a similar problem with their computer systems: they have a spaghetti-like mess of interconnections between legacy systems and applications, inherited from decades of IT investments and scattered across dozens of different computer technologies. To assemble the information required to support a single new business initiative, these corporations need to be able to integrate IT support across business processes, preferably using existing systems and without spending heavily on new development projects. They need to be able to add new applications from time to time and integrate them quickly with existing systems. Finally, they need to be able to change systems step by step, without always being confronted by endless interdependencies.

Current approaches to IT seldom meet these needs. Managers are trapped between unwelcome alternatives. Green-field solutions—scrapping old systems and rebuilding from scratch—usually take too long and cost too much. Worse, a frightening number of projects are canceled part way through, after huge time and cost overruns, leaving nothing to show for millions of dollars of investment. More focused attempts to upgrade existing systems piece by piece usually get bogged down in the infinite complexities of connecting old and new systems together. The IT analogue of the airlines' hub-and-spoke approach offers a way out of this trap.

The logic is simple. The IT initiative must start with a piece of "middleware" software, preferably one that is available off the shelf, to serve as the IT equivalent of an airline to this software hub without modifying them in any way. (This allows companies to move data to and from current applications that serve crucial business functions.) New capabilities are plugged into the hub, or existing systems are upgraded as needed. This gives new applications access to all the data and functionality of the existing systems. And that is it. Thanks to the capabilities of the hub and the overall simplicity of the approach, each of these steps can be accomplished by small teams of people in months rather than years.

INVENTING HUB-AND-SPOKE FOR IT

To understand how IT hub-and-spoke works in practice, its origins in the competitive cauldron of direct marketing can be examined. The hub-and-spoke approach first emerged in the early 1980s as companies set up "call centers" to create the new business processes needed to sell products and provide services over the telephone. In the course of a phone call lasting only a few minutes, these call centers had to have access to all the information and capabilities needed to give customers whatever they wanted. The IT challenges in meeting this business imperative are immense. One example that illustrates why is the progress of a hypothetical call through the call center of a bank.

When the call arrives, the caller's telephone number is automatically compared with a customer data base. If it matches a known customer, all relevant information on that customer immediately comes up on the sales agent's screen. When the customer makes a request—say, for a new loan approval—the system prompts the agent to gather all relevant information. The loan application is automatically sent to a loan-approval system to see if it can be approved on the spot.

At the same time, a marketing data base mulls over the customer's recent pattern of purchases and requests. As a result, it may prompt the agent to offer the customer a product that it estimates he or she may want to buy. Any new transactions are then recorded in product processing systems, and the relevant paperwork is automatically printed and mailed. Finally, a workflow system reminds the appropriate people in the bank to ensure that any necessary follow-up work is executed properly and on time.

Rightly sensing that building call centers could become a booming market in several industries, a number of innovative companies began developing software tools and products that would enable businesses to create their own call centers within three to nine months, even when their existing systems were messy. To be successful, this software had to improve continually on six measures of performance.

1. *A quick and efficient connection with any and all other programs, new or old.* Call center providers understood that they could not expect a company to change or replace its legacy systems to deliver a new phone-based service. They therefore invented a variety of technological tricks to enable legacy systems to work together, connected by a software hub, even if all that was known about them were the instructions that went in at the keyboard and the responses that came back on the screen.

    With a technique called screen scraping, for example, the legacy application sends data to the hub as if it were sending it to a terminal screen. The hub then "reads" this virtual screen and sends data back to the legacy application by "typing" it on a virtual keyboard. Of course, the spoke between hub and legacy application does not actually have a physical screen or keyboard, but the legacy application works perfectly as long as it "thinks" they are there.

2. *Fast response times.* Customers usually expect quicker, more definite responses when they do business over the phone than when

meeting face-to-face. A five-minute telephone call can easily involve a few dozen unrelated applications and data bases spread across a company. To keep the information moving quickly, call center hubs must be able to manage the simultaneous execution of different tasks on the different systems connected by its spokes.

3. *Access to new applications.* Given the fleeting nature of their contact with customers, call center operators are eager for applications that make them smarter about their customers' needs. This means integrating new applications, such as expert systems, using interfacing techniques available on the market, such as message-based communication with modern object-oriented applications.

4. *Seamless integration.* Working under intense time pressure, call center employees require seamless access to everything that their systems can provide. They have neither the time nor the inclination to learn different techniques for accessing each of the systems they use. An effective hub hides from its users the technical differences between systems, so they see only a single, coherent user interface. The hub also protects them from the problem of keeping track of which data is located where. If legacy systems do not contain a central customer data base, for example, the software hub might be extended with a new data base that remembers that customer A in legacy system X is really the same as customer B in legacy system Y.

5. *Seamless guidance.* Call center employees also need a system to help them manage the overall complexity of their work. Thus, workflow tools on one of the system's spokes keep track of incomplete tasks, send messages to all involved to ensure that deadlines are met, follow jobs that are executed in batch systems, and collect productivity statistics.

6. *Fast implementation and low cost.* Their 10 to 15 years of experience have helped call center providers to streamline their approach to setting up operations in three to nine months with small teams, no matter how complex the systems their clients possess. Competitive pressure between providers has ensured continuous improvement in implementation time and cost

Once perfected, call center technology—the ability to link and integrate computer systems—quickly proved applicable across a whole range of businesses, not just in direct marketing. After all, any business process, whether front or back office, can use at least some of the information support provided in call centers. In today's information economy, is not any office really a call center? Vendors have thus started to offer software hubs as standalone middleware products, to which companies can attach a variety of legacy systems, data bases, and user interfaces, such as remote PC networks and the laptops of a traveling sales force. One US bank has even moved to a hub-and-spoke system, with more than 10,000 internal and external user interfaces, as its overall architectural choice.

KEY HUB-AND-SPOKE ATTRIBUTES

The software hub—the central component of the hub-and-spoke approach—brings together the capabilities needed to get data from one application to another, no matter how different the underlying computer systems. The least possible processing and transformation of information, however, is done within the hub. By pushing such work out to the spokes, the hub itself stays uncommitted to any particular task and thus remains relatively "future proof" even in the face of significant business and technological change.

The essence of hub-and-spoke can best be understood by looking at the tasks that are executed when an application fires a request to the hub and expects a reply back. To answer a query, the hub must know where to get the relevant information, how to get it quickly, and how to put it into the form expected by the application that asked the question in the first place. To do all this requires:

- *Knowledge about information location.* For every type of request, the hub must contain some logic that acts as a master script specifying which applications on which spokes have relevant information and so must be consulted to formulate an answer.

- *Knowledge about the fastest method of access to information.* The master script must also know which information has to be requested in sequence from different systems and which tasks can be executed in parallel. For example, some queries might entail translating a name and address into a customer number before further information about the customer can be gathered from other systems. If the customer does not have a number, the query stops there. But if he or she does, then requests can be fired off simultaneously to the necessary systems.

- *The ability to communicate with applications.* To gain access to information, the hub contains a software module for each spoke that enables it to talk to the application on that spoke. If this application can understand any of the IT industry's emerging message standards, this software can be simple. If, however, the hub must do screen scraping (i.e., if it has to be able to mimic the activity of a human user by first logging onto a legacy system, and then navigating through various menu screens and other procedures to get the data it wants), the software must be more complex.

- *The ability to synthesize an answer.* Finally, the hub's master script

67

must pick the relevant pieces of data from the material it receives and then combine them into a single reply message to the requesting application. It must also be ready to pass on error messages or incomplete answers, in case one of the applications was not able to reply or was simply out of order. It must also complete bookkeeping tasks, provide security, and gather performance statistics.

These capabilities distinguish the software hub of a hub-and-spoke system from a network hub. The software hub knows about data, where it is located, and how to get it. The network hub knows only about communicating bits and bytes; it leaves the user to deal with the complexity of managing data. In addition, workflow software for managing business tasks is strictly an add-on to a hub-and-spoke system. The hub manages only the dependencies between data sets; workflow software manages dependencies within business processes, such as to-do lists, deadlines, and authorization procedures. Thus, to keep a hub lean, quick, and efficient, workflow software should be added only on a spoke.

Software hubs have been constructed for all kinds of computers, including UNIX servers and mainframes. Different hubs handle varying volumes of work, offer different features, and provide many approximations of the promise of universal connectivity. Not surprisingly, hubs have been built to cater primarily to the most widespread technological standards. Although the list of technologies supported by commercially available hub-and-spoke products is already impressive, no software hub can provide instant connectivity to all computer platforms. Obviously, managers contemplating buying a software hub need to ensure that it caters to the specific technologies they need. Given the capabilities of currently available hubs, however, management attention can finally shift away from the technical arcana of network protocols and data structures to the intelligent design and implementation of new business processes.

PUTTING HUB-AND-SPOKE TO WORK

As leading-edge companies have found, moving to hub-and-spoke usually requires four steps.

Step 1: Assessing the Current Setup to Define the Required Information Support

Forging working combinations of human and machine decision making that temper human creativity with computer reliability remains one of the most challenging tasks facing today's managers. A hub-and-spoke approach cannot make a science of this art. But it can simplify the art in two ways.

First, hub-and-spoke technology, when used effectively, can support a wide range of strategies and decision-making styles—that is, it helps reconcile the apparent conflict between a company's desire to maintain its own unique way of doing things and the pressure to save money by buying standardized computer systems off the shelf.

In personal insurance, for example, a hub-and-spoke approach has been successfully adopted by firms with radically different decision-making styles. At some companies, it provides raw information to the desks of highly educated salespeople, who decide precisely how to tailor products for individual customers. At others, it enables a workflow system to prompt a salesperson step by step through a dialogue with a customer. It then feeds the customer's answers to a statistical marketing data base, which determines which standardized products are best suited to that customer.

Second, and far more important, hub-and-spoke encourages IT systems to evolve—with relatively little pain. It does not require businesses to freeze their information requirements for years to allow IT departments to implement clearly defined specifications. Nor does it place obstacles in the way of establishing connections with any piece of corporate data or any computer system. In fact, hub-and-spoke encourages an exploratory process of design for new applications. If something goes wrong, it also makes it easy to fall back on the old system. The conventional approach of modifying the old system often takes away this safety net.

Step 2: Accessing Legacy Systems over Hub-and-Spoke

The hub-and-spoke method enables companies to treat legacy systems as assets rather than liabilities. By contrast, traditional approaches to systems development usually call for disposing of the old system as the first step in building the new. Often, the main reason for doing this is the difficulty of getting information to pass smoothly between the two. By allowing legacy systems to communicate with new ones, hub-and-spoke lets companies build on past IT systems rather than perennially striving to recreate them at ever higher levels of complexity.

Building on past success has obvious advantages. One, certainly, is that companies can avoid the need to devote time and money to rebuilding capabilities that a company has already developed. Another is the help it gives to managers who find it difficult to think about information needs in terms of the abstract data and process models used by IT specialists. Hub-and-spoke allows them to express their needs for information in familiar, business-relevant terms (e.g., how information should be integrated, what sequence transactions should follow, and what new functionalities should be added).

Step 3: Hooking Up Breakthrough Capabilities on New Spokes

The most visible payoff of hub-and-spoke comes from the speed with which new capabilities can be added to existing systems to create entirely new business processes. The broad connectivity that it provides allows firms to take advantage of innovations made by commercial software companies, instead of forcing firms to recreate new applications in-house to get them to work with existing systems. Consider, for example, such innovations as:

- *Event-driven marketing.* A pioneer in direct telephone banking boosted its cross-selling when it connected to its hub a new marketing package to help anticipate customer needs. The system enables the bank to define a list of significant events, like a recurring overdraft or a lost credit card. Each event is linked to a product that the customer might want—an approved loan, for example, or a credit card insurance scheme. When an event occurs, the system checks to see if the customer is a possible candidate for the product. If so, it prompts the salesperson through an appropriate sales dialogue. By offering new products when they are most likely to be wanted, this approach improves sales effectiveness. Better still, by tracking the success or failure of each sales pitch, it keeps the firm abreast of changes in the demand for its products.

- *Trying out new prices and product terms.* A successful direct car insurer uses a similar system, also purchased from a commercial supplier, to try out new prices, terms, and conditions on its prospects. Most are offered car insurance in line with the existing pricing strategy. A few, however, are chosen randomly by the system to be offered the same insurance but with slightly different prices or terms. By comparing these sales experiments, the company discovers how price sensitive its prospects are and can adjust its prices accordingly. This not only enables the company to improve its product offerings continually but also informs the planning staff on how to respond to competitive threats.

- *Electronic data interchange (EDI).* The vision of paperless buying and selling is appealing, but its realization has not come as fast as many had hoped. The challenge for EDI is related to legacy systems: when a customer wants to order a product electronically, a number of existing systems must be consulted, such as creditworthiness, available inventory, and transportation. For one major computer manufacturer, this means forging links with 13 separate product processing systems. A hub-and-spoke package makes EDI's access to these systems feasible.

Step 4: Upgrading or Replacing Legacy Systems When Needed

The flip side of the coin is the ease with which hub-and-spoke technology enables companies to replace their legacy systems when they are ready to do so. Because it can deal with both old and new systems. decisions on when to upgrade old systems can be taken simply on the merits of the technology, rather than on the need to extract data from the grip of a legacy system. Indeed, one of the major advantages of hub-and-spoke is that it allows old systems to be upgraded with little impact on users. So long as the user interface can still send queries to the hub and receive appropriate replies, the user need not know—or care—whether the application dealing with the messages is old or new. As a consequence, the IT department can manage its applications portfolio without constantly retraining or otherwise disturbing users.

SUMMARY

It is worth noting that hub-and-spoke as described above is different from client/server computing, which strives toward many of the same goals as hub-and-spoke. Client/server technology divides work between various machines. Most often the client provides a friendly user interface and personal applications on the desktop; the server manages data and corporate applications. Client/server computing stresses clean interfaces so server or client applications can be removed and new ones plugged in with a minimum of effort and disruption. It enables rapid development and change. Client/server technology, however, often overlooks the easy, near-universal connectivity between old and new systems, as stressed by hub-and-spoke.

Few client/server technologies make it a simple matter to connect with legacy systems. As a result, many companies pass up the benefits of client/server because their analyses tell them that it would be prohibitively expensive to connect, for example, their proposed new client/server marketing information system with the marketing data on their old mainframe. Worse, some client applications communicate poorly with any servers that do not come from the same provider.

In an ideal world, where there would be no legacy systems and any client would be able to talk with any server, hub-and-spoke would be unnecessary. In practice, however, hub-and-spoke technology fills the gap, supplying the missing piece that easily and inexpensively links clients with servers, legacy systems with new technologies and, most important, the vast capabilities of information technology with the urgent need for redesigned business processes.

Section 3
Business Process Reengineering

F ew management processes have generated the interest that business process reengineering has. Reengineering has become the avant garde term of the 1990s. As organizations strive to be more competitive and cultivate worldwide markets, they are critically examining all aspects of their operations, organization structure, and their staffs' competencies. In doing so, many are trying to reinvent themselves through reengineering or engineering new business processes. Their quest is to create a phenomenally agile organization, one that is quick to spot opportunities and respond to customer needs and competitor initiatives while increasing the stakeholder and economic value of their companies. Process innovation, creativity, and technology are at the center of reinventing process. Information technology (IT) can be an enabler, inhibitor, or driver of the process.

This section explores the need, implications, and methods for reinventing processes and organizations and the role of integration in doing so. The authors contributing to this section provide a broad and balanced perspective of what is and is not reengineering, how systems integration enables the reengineering process, and how IT must be used to support such systemic initiatives as reengineering and organizational transformation.

Section 3
Business Process Reengineering

3-1

Defining Business Process Reengineering

MICHAEL A. MISCHE

WARREN BENNIS

The avant-garde terms for the 1990s are *reinventing the organization* and *business process reengineering*. Both are concepts that are highly touted, often doubted, and generally misinterpreted, misunderstood, and misused by management, consultants, and the public alike. Nonetheless, there is a growing phenomenon of organizations stampeding headlong, sometimes blindly, into organizational reinvention, restructuring, and reengineering projects. At the center of this frenzy is reengineering.

According to the media and many consulting firm studies, virtually every one of the Fortune 500 companies is performing some type of reengineering. Even the federal government is touting reengineering as a means of reinventing itself and improving taxpayer value. However, although there is tremendous excitement and a rush to embrace reengineering as the management savior of the 1990s, there is also significant confusion as to what honestly constitutes business process reengineering (BPR).

Unfortunately, this is an age of quick- fix management fads, short-lived business trends, and glitzy slogans. Reengineering is not immune, and the word has become a catch-all term to mean almost anything from downsizing to true reinvention. Many organizations and consultants can claim to know what reengineering is, but few can actually produce a tangible methodology developed specifically for reengineering. Although they may be able to point to purportedly successful reengineering projects, even fewer can demonstrate that the projects actually began with a clear objective and intent to reengineer. The route followed and processes used were surreptitious, with the term *reengineering* usually applied only as an afterthought because it was catchy or had some subsequent internal appeal or external marketing value.

Many organizations that proudly proclaim themselves as having reengineered generally did so in a rather inadvertent manner or out of embarrassment, dire necessity, or sheer fear of extinction. As experience demonstrates, some type of reengineering, albeit limited, almost always occurs as a result of a major project; something in the way work was performed and resources were managed was changed by the effort. In this sense, reengineering is not new. In fact, it has been around for quite some time. Many organizations have actually performed elements of reengineering without ever calling their efforts reengineering, and most of the examples cited, such as Hallmark and Ford, occurred long before the term gained any widespread recognition. Years before it became fashionable, the term *reengineering* was used to describe a variety of projects. For example, in 1981, a team of management consultants was working at one of the Fortune 500 companies when one of the team members proudly proclaimed that the consultants and the project were "reengineering business practices and transforming the company." The statement provoked the wrath of both the client and consulting partner responsible for the project. The client said that the project was about system implementation—not process and organizational change. The consulting partner said that only engineers could reengineer, and because the staff member was not an engineer, he could not reengineer. Nonetheless, the company experienced profound change when it implemented automated systems and procedures that replaced predominantly manual processes.

There are, however, a number of companies that have reengineered out of vision and passion. General Electric reengineered out of a vision to be world class with a relentless competitive drive and internal challenge. Motorola reengineered out of a passion for excellence and the need to constantly challenge itself to improve earnings. Bausch & Lomb performed reengineering because it knew the time was right, and it wanted to establish new standards of excellence and performance. These organizations were already doing well when the decision to reengineer was made. Disney, on the other hand, set out to quietly and purposely reengineer itself back to dominance by reinventing its entertainment operations. It was driven by an insatiable creative culture and a passion not to just provide entertainment but to be entertainment. The Home Depot reengineered the way many people shop when it launched retail warehousing concepts: high percentages of in-stock items, everyday low prices, and highly paid knowledge workers. In doing so, it not only reengineered the traditional hardware and building supply industries but created a completely new industry. So it is clear that something is happening—and that can be called reengineering.

Perhaps the most obvious example of early reengineering can be found in manufacturing. When manufacturers rushed to implement material requirements planning (MRP) and MRP II in the 1960s and 1970s, they reengineered their materials handling, production scheduling, inventory planning, and labor-allocation practices. The changes associated with these efforts were dramatic and permanent. The use of electronic data interchange (EDI) and electronic funds transfer systems (EFTS) to support more cost-effective transactions, which began in the mid-1970s and continues today, is another example of early reengineering. These technologies pioneered reengineering and forever changed the administrative practices of retailers, freight carriers, manufacturers, and distributors—long before the term came to the forefront of management's attention. Finally, the advent of the personal computer (PC) and its related technologies reengineered the way people work, interact, and communicate more than a decade before business and the consulting profession embraced the term.

The many examples of early reengineering all have one thing in common. They are situational, that is, specific to a unique project, problem, or activity. Therein lies the difference between what has been the historical context of reengineering and what is today called reengineering. In its truest sense, reengineering has far-reaching organizational and management implications and is not restricted to just an activity, function, or department.

Unlike the accounting and legal professions, reengineering has no codified rules, educational requirements, preprinted forms, or professional standards. There are no examinations, certifications, professional associations, or business majors in reengineering. Reengineering is still evolving as a management doctrine, and there are very few definitive sources and little supporting research available on reengineering as a management tool. Most important, very little information and direction are available on how to actually perform reengineering and what it truly means to the enterprise and its people. Although much has been written about reengineering, the actual methodologies, techniques, work products, and results are not described at a level that can be useful to the manager or executive.

Consultants and managers who make their claims are quick to tout the benefits and hype the brilliance of reengineering, but they usually stop far short of explaining exactly how to reengineer, and what it really means to reinvent the enterprise through reengineering. Nonetheless, it is a catchy term, and the consulting firms and enterprising managers are capitalizing on a certain "herd mentality."

So what does reinventing the organization really mean, and exactly what is BPR? That is the hundred-million-dollar—perhaps billion-dollar— question of the 1990s. Those who have a vague understanding or naive appreciation of reengineering tend to attach a singular definition to reengineering by associating it with traditional techniques or situational management concepts, such as productivity improvement, total quality management (TQM), continuous quality improvement (CQI), systems integration, or downsizing. Reengineering, however, takes on a variety of forms and has a number of unique behavioral and philosophical characteristics, which differentiate today's meaning and application of reengineering from its predecessors and the traditional situational methods that are used to address a specific problem or issue.

This chapter refutes the 10 myths about reengineering and presents a definition by building a new context for reengineering.

THE 10 REENGINEERING MYTHS ... AND THE REALITIES

This chapter dispenses with the 10 myths of reengineering by exploring its realities.

Myth 1. The move to implement reengineering means that it is assumed that the organization has made all the wrong moves, all along.

The reality is that reengineering recognizes that the organization has been successful, and it has done a number of things correctly—maybe not to perfection, but correctly. Reinvention through reengineering strives to identify and capitalize on opportunities for improvement and innovation and reinvent processes and organizations.

Myth 2. Reengineering involves information technology, systems integration, application development, client/server technology, and the migration away from mainframes.

The reality is that reengineering recognizes that information technology is an enabling agent of change and is essential to any reengineering effort. However, reengineering is not, in and of itself, about information technology and cannot, by itself, reinvent the enterprise.

Myth 3. Reengineering means downsizing and personnel reductions.

The reality is that nothing could be further from the truth! Reinventing the enterprise through reengineering means doing things differently and more effectively, with or without existing resource levels. Reengineering involves process innovation and creating more agile and adaptive organizational structures. Reinventing requires leadership and team work that only the human element can provide.

Myth 4. Reengineering means doing more work with fewer resources.

The reality is that reengineering means creating greater leverage and efficiency through process innovation, seamless and harmonious work flows, more agile organization structures, improved use of information technology, and optimal and empowered human resources. Reinvention means increasing the value content of work and the contribution to the shareholders.

Myth 5. Reengineering can be used to fix any problem and address any issue.

The reality is that reengineering is not a short-term, quick-fix management tool. Reengineering is systemic and facilitates tremendous cultural and institutional changes, which create new processes, new systems, new organizations, and new measures. Reinvention through reengineering is not a situational cure for chronically mismanaged organizations and isolated problems.

Myth 6. Reengineering can be performed by anybody and managed by anyone.

The reality is that reinventing the organization through reengineering requires mature business judgment, extensive and broad-based experience, a bold vision, and a refined methodology. Reengineering leads to a permanent and pervasive change in the way organizations are structured, managed, and operated. Leadership, vision, knowledge of the process, and the commitment to change are absolutely essential to the reinvention process.

Myth 7. Reengineering can be sponsored by anyone in the organization.

The reality is that reengineering must be sponsored and supported by the highest levels of leadership in the organization, or it will fail.

Myth 8. Reengineering can occur without significant organizational change and transformation.

The reality is that reinvention through reengineering is transformation. Reengineering creates a new organization, different organizational structures, and new leadership mentality. It also creates new performance measures and compensation metrics.

Myth 9. Reengineering creates chaos and anxieties and can be disruptive and detrimental to the organization.

The reality is that reengineering causes change, and change for most organizations and their managers can be excruciatingly difficult. If reengineering is not managed and performed properly, chaos and lasting scars will result.

Myth 10. Reengineering is scientific.

The reality is that reengineering is not a physical or natural science. Reengineering is not about design of experiments, two-pronged hypothesis testing, statistical process control, Pareto charts, or object-oriented analytical techniques. These concepts and techniques could very well find their way into a reengineering process, but they are not reinvention through reengineering.

REENGINEERING DEFINITIONS

Reengineering definitions span the spectrum from the mundane to the revolutionary. At one end of the spectrum are the more traditional definitions, which characterize reengineering in a narrow, relatively myopic context that emphasizes incremental improvement, changes to fundamental processes, head-count reductions, self- funding projects, and continuous TQI or productivity-improvement programs. At the opposite end of the spectrum are definitions that describe reengineering as revolutionary in nature and substance. These place reengineering in the realm of reinvention and transformation, and that is what this chapter defines it to be.

There are four principles of reengineering:

1. Process innovation and harmonization.
2. Information technology (IT).
3. Organizational transformation.
4. Leadership.

All four of these principles must be present and addressed for reinvention to occur through reengineering. Anything short of using these four principles in reengineering is simply not reengineering.

Process Innovation and Harmonization

The first principle of reinventing the enterprise through reengineering involves the flow of work and its value content. Whereas traditional projects and situational methods are confined to individual activities within a specific function, such as customer order entry, reengineering addresses an entire process and its supporting organizations.

Reengineering strives to harmonize the various activities and tasks performed randomly in the organization by arranging activities into processes that have a continuous flow, accelerated velocity, consolidated functionality, and a common system of management practices and measurements.

Information Technology (IT)

Although IT may be one of the most important pieces of the reengineering puzzle, approaching the reengineering effort solely from an IT perspective only produces tremendous fees for the consultant and technology vendors and job security for the IT staff. Quite simply, reengineering and organizational reinvention is not about IT, application development, new software, client servers, mainframe migrations, outsourcing, systems integration, or office automation.

Organizational Transformation

The third and perhaps most important principle regarding reengineering involves organizational transformation. Quite simply, reengineering is about reinventing the enterprise, and the two concepts are inseparable. To reinvent the enterprise, one must reengineer, and to reengineer, one must reinvent. Many management consultants and managers are quick to claim that reengineering means doing more work, producing more product, and supporting more volume with less resources. This usually implies fewer people—meaning downsizing—and it is misleading. There is nothing strategic or unique about downsizing. In fact, downsizing represents a relatively easy process.

With the repositioning of many companies in the 1990s, outsourcing, corporate downsizings, and rightsizings became popular methods for quick cost reductions and short-term profit improvement. An uncertain economy and fear for earnings created a "lemming mentality" to downsize, and many organizations, regardless of industry, followed suit. Unfortunately, some organizations went too far in their penchant for overhead reductions and cut deep into their core cultures and value systems. In doing so, they exorcised many of their values and culture leaders, losing not only knowledge but role models as well. For those companies that voluntarily forfeited their rich cadre of talent, downsizing became "suicisizing." These organizations may have forever lost the essence of their cultural fiber.

Much of this activity was misrepresented as reengineering. Reinventing the enterprise through reengineering does not start from the premise of eliminating employees or jobs and does not approach innovation opportunities from the perspective of greater activity through fewer resources and personnel. Reengineering seeks to optimize resources, value, volume, and content. Reengineering means leadership, leverage, and empowerment. The organization's operations and values are critically evaluated, boundaries and hierarchies are carefully scrutinized, and work value is assessed, all from a cross-functional process perspective and in the context of the company's vision. The key objective in reengineering is to create higher value and greater synergy through innovation and organizational reinvention—not through employee terminations, massive layoffs, and lower morale. Reengineering is not about downsizing, and it certainly avoids "suicisizing."

Leadership

The fourth and final principle of reinvention through reengineering is leadership, the most elusive of qualities. Reinvention cannot occur without effective and visible leadership; it is that simple and direct. Reinvention through reengineering requires visionary skills, creativity, and broad conceptual abilities. The reinvention and reengineering process is compromised and falls woefully short of its potential if the leader does not posess these seven attributes:

1. Visionary influence and creativity.
2. Knowledge of the process and business.
3. Exceptional interpersonal skills.
4. Ability to select and coach the right people.
5. Credibility through experience.
6. Character.
7. Judgment.

THE TENETS OF REINVENTION THROUGH REENGINEERING

As described previously in this chapter, reengineering is the profound reinvention and transformation of the organization's leadership practices, business processes, measurement and reward systems, self-image and awareness, culture, and value system. Reengineering is a systemic and permanent change in the way organizations are led, structured, managed, staffed, and measured. In addition, there are five tenets that are essential to reinvention through reengineering:

1. *Increasing throughput and productivity.* Reengineering seeks to optimize activities through the creation of seamless and harmonious processes that have an uninterrupted flow and occur in a natural order with a natural velocity.

2. *Optimizing quality and shareholder value.* Reengineering strives to optimize shareholder value and create competitive synergy through innovation designed to enhance quality at multiple function points, such as product design, manufacturing, and customer service.

3. *Achieving quantum results and returns.* Reengineering is mandated with a mission and empowered with capabilities to achieve quantum

results and strategic advantage through the innovation, reinvention, and transformation of the enterprise.

4. *Compressing time and consolidating functionality.* Reengineering is designed to create processes and organizations that are leaner, flatter, and faster. Agility and the ability to rapidly assimilate innovation, market needs, enabling technology, customer trends, and competitor initiatives are trademarks of the reengineering process and reinvented organization.

5. *Eliminating unnecessary organizational levels and low-value work.* Reengineering constructively interrogates work and organizations and assesses value, purpose, and content. Work of seemingly low shareholder and competitive value, as well as responsible organizations, are repositioned to provide a greater contribution or eliminated.

The results of reengineering are measured in an upward, step-by-step progression. The reengineering process combines quantum gain with continuous improvement. This stepwise gain contrasts with the traditional incremental curves associated with productivity improvement, organizational effectiveness, and industrial engineering methods. Quantitative expectations for reengineering can be significant. Research and experience indicate the potential for productivity improvements of 25% to 100%, head count redeployments of 25% to 50%, inventory reductions of 40% to 50%, cycle time improvements of 50% to 300%, and indirect cost improvements of 25% to 50% as direct products of the reengineering effort. These percentages are not the traditional 5%, 10%, and 15% incremental improvement gains that most organizations have been content to achieve. As a general rule, reengineering should yield at least a 50% return or improvement. If the yield is not at least 50%, then it is not reengineering.

The process of reinventing the enterprise through reengineering is one of constant interrogation. Reinvention requires the continuous assessment of organizations, management practices, people, systems, customers, and the global environment within which the enterprise operates. As a result, reengineering accepts some practices, rejects many, disproves others, and casts doubt on traditions by exploring five simple but probing questions:

1. Why does the organization perform its functions the way it does?

2. What value is produced for the customers and shareholders by performing the process this way?

3. What other ways can the organization perform the process that will enhance value?

4. What results does the organization want to achieve that are innovative or breakthrough in nature?

5. What cross-organizational talents are required and who are the best people to lead them?

Reengineering endeavors are far more encompassing, inherently more complex, of much higher risk, and of significantly greater duration than traditional methods or situational practices. In reinvention, there are no one-minute solutions or quick fixes. Instead of working within the narrowly defined boundaries of a department and task and blindly accepting or acquiescing to the culture and rules of the organization, reengineering boldly challenges the traditional paradigm and established doctrines. In performing reengineering, there are no sacred stones; all tasks, organizational responsibilities, structural alignments, job titles, staffing arrangements, leadership qualities, reward structures, and performance measures are subject to constructive challenge and reinvention.

Reinvention through reengineering breaks down the traditional organizational paradigm of vertical "silos" of tasks and responsibilities and in the wake replaces them with an organization that is cross-functional, flatter, more agile, and networked. The classic top-down and span-of-control style of decision-making structures are replaced with one that is organized around core empowering processes and that is closer to the customer and execution step. Most importantly, the traditional organizational boundaries and turfs that create gaps and "pass-offs" in the work and diminish the value, speed and quality of the process are eliminated.

SUMMARY

Organizations that have successfully reinvented themselves through reengineering benefit from enhanced employee awareness of the organization, its business, and its customers. A successful reengineering project yields intrinsic value to the shareholders, organization, and customers in five areas:

1. Increased employee interest, awareness, and appreciation of the enterprise and its leadership, products, and customers.

2. Improved organizational cooperation, teamwork, and understanding of needs and requirements and communications.

3. Increased knowledge of the organization's direction, role in the marketplace, competitors, and identity.

4. Improved alignment of personnel and a better matching of skills and empowerment to responsibilities and processes.

5. New individual and group performance measures.

Having defined and placed reinvention through reengineering in perspective, the chapter has come full circle to its initial question: Is reengineering for real and can it be used to reinvent the organization? The answer is a resounding and unequivocal yes. Reengineering is not a fad. It is a distinct and permanent change in the ways that organizations can be led, managed, operated. and measured. As Frederick Taylor, Henry Ford, and Douglas McGregor defined the operating construct of the twentieth-century organization, reengineering is defining the principles of the twenty-first century enterprise. Organizations, businesses, and governments alike that desire to prosper and aspire to be leaders in the global community of the next century must reinvent themselves through reengineering. Those that do not, or that erroneously pay lip service to it, will simply run on their own inertia until the markets and competition eclipse them or the citizens that they serve revolt. Still there will be those that doggedly cling to tradition, resist the inevitable changes, or fail to discern the advantages of reengineering. These unfortunates will most assuredly become relics and, like the Iron Curtain, electric typewriter, and rotary telephone. pass as footnotes into the annals of business history.

3-2

The Role of Information Technology in the Reengineered Enterprise

MICHAEL A. MISCHE

There is little doubt that information technology (IT) is fundamental to any enterprise's operations and to process reengineering and organizational transformation efforts. The issues are how to manage the information management and technology (IM&T) resource best and what the IM&T role is in reengineering and reinventing the organization. Much has been written about IM&T's role in the reengineering process, and its critical role simply cannot be understated. In a world of real-time transactions and global electronic partnerships, IT and communications are perhaps the most important components to achieving competitive advantage through the reinvention of the enterprise. Nevertheless, the three key issues related to IM&T's role and reengineering are:

1. How do systems integration and reengineering combine to create and support a technology-enabled organization?
2. Can process reengineering and organizational transformation be achieved without a corresponding reengineering of the IM&T resource, including its organizational structure, management practices. and personnel?
3. What is the architecture and technology deployment strategy of the reengineered enterprise, and how does systems integration support its realization?

The purpose of this chapter is to discuss these key issues in the context of IT's role as a reengineering enabler.

IT AND REENGINEERING

IT supports process innovation and provides the infrastructure for the reengineered and reinvented enterprise. In addition to supporting cross-functional processes, communications, and the logical consolidation of functions and organizations, IT also provides immediate and consistent information dissemination, which, in turn, supports the physical distribution of those resources. However, as many organizations have learned, IT by itself leads neither to quantum results nor to reinvented organizations. Similarly, systems integration may cure many immediate needs but by itself seldom leads to systemic operational improvements or lasting organizational changes.

There are many reasons for this condition. First, systems integration efforts are not reengineering, because they do not possess the qualities and attributes of reengineering. Second, the base systems themselves (i.e., the point from which integration starts) represent significant technical and operational challenges.

The Current State of IT

Many systems in use evolved without the benefit of a vision, plan, or uniform architecture. Although IT clearly has moved beyond transaction processing, the majority of the technologies and systems still retain their industrial-era pedigree and hierarchical heritage. That is, they reflect design and functional concepts indigenous to traditional business practices and organizational structures and are deeply embedded in the enterprise's fabric. In effect, they are legacies of past practices and management doctrine. Rather than enabling cross-functional processes, many systems are transactions based and built to support specific functions and tasks related to the physical distribution of work as performed in the traditional hierarchical organization. In fact, in most organizations, these systems are often helping to perpetuate existing business practices, cultures, organizational paradigms, and staffing alignments. Under this structure, transactions may be processed faster, but very little meaningful customer or shareholder value is actually created. Undoubtedly a sizable amount of data is produced, but high-value analytical information is often limited and left to an employee's creativity and personal computer (PC). Thus, integrating these systems and reengineering the organization are both costly and extremely challenging endeavors.

The Data Base Explosion

Recognizing that the fissure between IT and business needs has been widening, many organizations tried to close the gap with solutions centering almost exclusively on technology. First came the rush to implement relational data base technology. Many organizations attributed the lack of connections between their systems and businesses to their systems' data and reporting capabilities rather than the applications or business processes. This led many organizations to implement sophisticated data bases with hundreds of millions of data bytes so that management reports could be generated. What soon followed was an unchecked explosion of data bases and management reports. Some of the data was rationalized and synchronized; however, much was not, thus leading to torrid management debates as to what system had the right information and correct data. For example, data that is captured and reported upon as sales data in one system is commonly entirely different from the sales data captured and reported on in another system. Data and information that were originally intended to be common to all and universally accessible ultimately gave rise to issues of who owned the data, who or what organization was responsible for the data's accuracy, who was chartered as the data's custodian, and who could view and use the data. What followed was the demarcation of data and application software based on the organization's physical structure, which generally defined how the flow of work was performed.

Another attempt at bridging the gap between systems, technologies, and users involved the mass migration to distributed systems and decentralized IM&T organizations. Unfortunately, their applications, data, capabilities, and structures began to blend with the personalities and cultures of their host organizations and developed their own individual identities, performance characteristics, and funding patterns. What soon followed was a proliferation of duplicate technologies, systems—and costs. Ultimately, the enterprisewide discontinuities between the systems and their organizations became more exaggerated and the redundancies more prevalent. Despite growing investment in decentralized systems and organizations, very little transportability of resources and knowledge has been achieved by the decentralized environment because of the management processes inherent in many organizations. Consequently, many organizations that have decentralized IM&T environments now find that their enterprisewide investment in IT is fragmented, duplicated, and generating marginal returns at best.

The Legacy Architecture's New Role

Virtually every organization has older, legacy systems that were built to satisfy a particular need or department. The reengineered organization still needs these types of systems; however, they take on a very different role. They are mainstays that support relatively stable business activities and recurring transactions, such as payroll, general ledger, accounts payable, fixed assets, and accounts receivable. They are also the least likely of systems to mandate state-of-the-art or leading-edge technology. Nonetheless, the organization has a number of compelling reasons to at least update and when appropriate, "de-mass" these systems to more flexible computing environments that are also financially and technologically advantageous.

As illustrated in Exhibit 3-2-1, there are a number of factors driving changes in technology use and deployment in the organization. Although many organizations have invested heavily in technology, studies confirm that the preponderance of PCs and client/server environments are still used to support rudimentary business needs. They yield less than 3% gains in productivity. For example, a recent Gartner Group study found that the typical business end user employed client/server and PC technology for word processing, spreadsheets, and E-mail. Undeniably, reports and graphics may be flashier than before and are produced faster, but for many organizations, the work content and job performance remain essentially unchanged. Therefore, it is apparent that IT and systems integration alone are insufficient to support or effect reengineering.

Although many organizations pride themselves on managing technology expenditures based on a percentage of annual revenues, they are learning that their actual, aggregate expenditure levels are higher. This is attributable, in part, to the decentralized management philosophies for IT practiced by many organizations that have created "shadow" technology staffs and independent islands of technology in line departments and operating units. The shadow IT organizations are almost always external to the mainstream technology organization (i.e., the senior executive responsible for technology and the enterprisewide IT budget). The projects, operational costs, personnel, applications, and systems in these shadow organizations are buried deep within various functional organizations, beyond the reach of mainstream IM&T management and cross-organizational coordination. Thus, the decentralized structure with its islands of IT usually is not integrated with other IM&T initiatives and can have serious deficiencies in the quality of personnel, system capabilities, and performance measures.

Realizing that data bases and systems integration efforts are, by themselves, largely inadequate to support any significant reengineering initiative, organizations are embracing open architecture concepts and client/server technologies as alternatives to traditional mainframe computers and proprietary technologies. This movement is characterized as a de-massification and decoupling of applications, data, and users from the traditional computing platforms to highly integrated midrange and client/server environments. The drive for de-massification and decoupling is fueled by many factors, some of which include:

- Continued technological innovation and advancement.
- The need to reduce hardware and software acquisition and long-term maintenance costs.

The Industrial-Era Organization	The Twenty-First Century Organization
• Traditional Organization and Hierarchy	• Networked Diamonds and Circles
• Mass Production and Homogeneous Markets	• Mass Customization; Directed Markets
• Low Volumes, High-Order Quantities	• High Volumes, Lower Quantities
• Sell-Into Channels of Distribution	• Sell Through to End Customers
• Manufacturing- or Marketing-Driven Cycle Times	• Customer-Driven Cycle Times, Based on Event and Need (Just In Time)
• Multiple Vendors, Lease Cost Suppliers	• Global Partnerships
• Technology-Supported Operations	• Technology Enabled
• Build to Forecast/Push Demand	• Pull Demand; Demand Flow; Kanban
• Competitive Advantage Achieved by Products, Technology, Price, and Availability	• Competitive Advantage Achieved by Products, Service, and Experience; Price, Availability, and Quality Are Givens
• North American Marketplace	• Global
• Exclusive Brand-Name Franchise	• Multiple Choices, Copy-Cats, Private Label
• Limited Logistics (i.e., Warehouse It and Ship It)	• Integrated Logistics, Final Assembly
• High Command and Control of the Employee; Vertical Development of the Human Resources	• Empowered Employees; Cross-Functional Development; Decisions Made at Point of Execution; Knowledge Workers
• Employee Dependent on IT	• Employees Dependent on IT
• Limited by Technology	• Limitless Technology

(Now →)

SOURCE: M. Mische, *The Consultant's Guide to Reengineering* (San Diego CA: Pfeiffer & Co., 1995)

Exhibit 3-2-1. Factors Driving Changes in How Technology Is Used

- Improved computing capabilities, ease of use, and the need for scalable and portable applications and computing platforms.
- Accelerated development of application and system delivery capabilities required to deliver business-enabling solutions—today, not in three years.

- Economies of scale and performance resulting from scalability, portability, connectivity, and increasingly greater computer power.

The implications of systems integration processes using demassification and decoupling are obvious: expenditures for IT will continue to increase, investment return will remain low, and the disconnection between IT, application systems and the IT organizations, and the enterprise's strategic and competitive needs will persist, albeit on a smaller scale. This conclusion is supported by five factors:

1. Prevailing management doctrines and traditional organizational structures perpetuate the notion that technology be used to support point-specific missions and departments. The mandate is to use technology to automate how processes are performed today, not how they should be performed. Thus, without reengineering, systemic change cannot be accomplished.

2. Many IT organizations lack a strategic vision for how technology should be used to support the enterprise. The IM&T organization is, itself, not fully aligned with the organization and its strategic direction. Consequently, technology is acquired and applications evolve as a patchwork of individual set-point solutions, rather than as part of an enterprisewide vision supported by a computing solution.

3. The majority of IT organizations are managed as an administrative overhead function and as part of the accounting and finance department. This has led to a misdirected and misplaced emphasis on cost control, automating the obvious transactions and point-specific solutions.

4. Many internal IT organizations continue to maintain a "fortress" mentality. The all-too-familiar management attitude and operating styles of many information organizations has been to concentrate solely on applying technology to immediate needs and to the symptoms of operational problems rather than focusing on reengineering processes and curing the problem. Even in today's competitive arena, with a preponderance of research supporting change, most IT organizations continue to display an inward focus and are reactive to user needs. Only a few IT organizations, such as those found at Motorola, Federal Express, Chrysler, Saturn, and Toyota, are known for aggressively partnering with the end user to identify and enable change through technology.

5. Many organizations have used IT to try to treat fundamental managerial inadequacies and organizational problems rather than attacking the problem itself. Consequently, many systems progressively became more exception based and functionally specific. Every time a problem or need arose it became an exception to the basic work

flow of the organization and the processing flow of the system. What soon followed were more people. more enhancements to existing systems, and new systems.

POSITIONING IT FOR THE REENGINEERED ORGANIZATION

In reinventing the organization. IT's role is elevated to a new level of unprecedented importance. Virtually every major business and administrative process will be affected by reengineering and technology in some manner. IT provides the infrastructure and enabling capabilities that allow for empowerment and leaner, flatter, and more agile structures. Technology also enables the seamless integration of operational processes, communications. and responsibility centers by providing instantaneous access to information and decisions. IT and communication networks enhance the potential for global partnerships, worldwide business alliances, and the reality of the virtual organization. To the chagrin of many chief executive officers, their organizations are not positioned for business process reengineering, nor can they fully exploit new and emerging technology. Their legacy systems, IT management practices, and organizational structures were developed long ago and for a different type of organization. They lack the architecture capable of meeting the needs of tomorrow. In acquiring technology and developing their systems, too many organizations viewed technology in the narrow context of individual users and along organizational lines. Technology acquisitions and deployment were based on proprietary hardware and software solutions, and applications and technology strategies were formulated and implemented with a myopic sense of satisfying a specific transaction and control need. Consequently, many organizations are finding that their existing application systems and technologies are deficient and cannot support integrated processes and the cross-functional management requirements of the reengineered enterprise.

Reengineering is about profound cultural and operational change— not about having the latest and greatest technology. Technology and automation alone are not reengineering and cannot precipitate systemic change by themselves; they are enabling components of a process called reengineering. Consequently, many organizations find themselves in a dual reengineering situation. Before they can reinvent their organizations and cultures and significantly change their operational methods, they must first reengineer their IT environments.

Creating the reengineered IT environment and high-performance organization requires organizations to:

- Create a new IT architecture based on the enterprise's vision, process needs, and transformation plan.

- Develop a high-performance IT organization that is integrated with its constituency processes.
- Develop and implement seamless operational processes across the organization, enabled by scalable and portable technology.

The guiding principle in organizational reinvention and reengineering is to ensure that technology supports innovation and transformation rather than the automation of the obvious or traditional practices. Thus, there are five principles associated with the role of IT and its relationship to business transformation. IT is to:

1. Replace manual processes, paper, forms, and task-driven operations.
2. Displace and eliminate certain positions, tasks, human resources, and associated costs.
3. Enable knowledge expansion and intelligent enterprises.
4. Support the consolidation of functions and creation of cross-functional, integrated organizational responsibilities and processes.
5. Provide a common infrastructure and framework for managing shareholder resources and enhanced competitiveness.

The IT architecture for the reengineered enterprise will be dramatically different from its predecessors. Not only are systems de-massed, but users are also decoupled from the traditional IM&T organization and support structures. In the reengineered organization, users are functionally empowered through process and organizational reinvention and enabling technology, and they are less dependent on the traditional IT organization. This greater level of user independence supports the need for the rapid development and implementation of technology solutions and applications. New applications are built based on technology that is synergistic to the organization—scalable, long-lived, and based on a common operating construct. In the reengineered enterprise, IT provides for the logical consolidation of resources and the physical distribution of assets and capabilities.

The concept of logical consolidation and physical distribution is central to reengineering IT to achieve business process reengineering and organizational transformation. In the reinvented enterprise, the traditional methods of managing IT and its supporting cast (i.e., centralization, distribution, and decentralization) are not effective. The centralization of IT resources leads to further rigidity, escalating costs, and a reinforcement of inwardly focused cultures and styles. At the opposite extreme, decentralization results in unnecessary fragmentation, duplication of efforts, and the continuation of patchwork systems, each with its own support structures and duplicate costs.

In contrast, the reengineered enterprise has an IT orientation that provides for the logical consolidation and management of technology, applications, data, and personnel under a uniform architecture and operational standard, and IT provides for the physical distribution of those technologies and resources on an enterprisewide basis, irrespective of geographical location. In the reengineered organization, logical consolidation is the development and implementation of common policies, management practices, processes, procedures, performance measures, and standards necessary for the effective leadership and use of IT resources and personnel throughout the enterprise. Logically consolidating IT provides for a melding of all management activities, technologies, and resources into a common management structure using a uniform set of IM&T operating constructs and performance measures. The shadow technology organizations and islands of technology that exist in many organizations today will be assimilated into this common structure. All technologies and IT support organizations, regardless of location or purpose, would be managed under this consolidated structure using a global set of strategies, standards, and practices, as well as a uniform set of performance metrics.

The physical distribution of technology requires that the key supporting IM&T structures and personnel be located with their constituent processes and function as a service bureau to them. The supporting assets are immediately available to support process owners and cross-functional operations. Performance and service effectiveness is determined based on predefined criteria and the concept of market-based performance, which requires that the internal IM&T organization be as capable, effective, and cost-efficient as any external service provider, including outsourcing firms. Their performance is based on the same criteria that would apply to an external service provider, and users are free to choose their source of service. This orientation is important to another concept of the reengineered IT organization, called the high-performance IM&T organization. Most importantly, the logically consolidated and physically distributed resource is present at the empowerment level of the process. The reengineered IT organization has a number of distinguishing characteristics:

- The majority of applications and their computing platforms are demassed and decoupled from the traditional mainframe and scaled for cross-platform processing on midrange computers and client/ servers.
- Enterprisewide data is rationalized (i.e., consolidated into accessible warehouses with single sources and uniform definitions).
- Generic application shells that are provided by third-party vendors are used by the organization as common frameworks for the creation of new systems and porting of existing systems across platforms and functions.

- Process-enabling application systems, built to support reengineered processes, are highly structured and constructed to common standards using a variety of third-party tools.
- Applications and computing platforms are shared among various process constituents.

One of the keys to supporting reengineering with IT is the rapid delivery of application systems and services. The use of rapid prototyping, joint application design, and object-oriented development methods facilitate accelerated creation and delivery of enabling software. Organizations using these methods can create the potential to leapfrog those employing traditional development practices and create windows of competitive opportunity through technology.

The Need for a Real-Time Environment

In the reengineered enterprise, access to data and the distribution of knowledge and information are integral to a leaner, faster, and empowered organization. Whereas once organizations competed on product offerings and marketing slogans, the keys to successfully creating the competitive advantage in the reinvented enterprise are technology, knowledge, empowerment, and intelligence. Organizations that have access to the most appropriate customer information and competitor intelligence and are able to translate that information into intelligent decisions are, undoubtedly, best positioned to be the most competitive. This positioning is accomplished through the real-time exchange of information. In many organizations, the technical and operational infrastructure for the real-time exchange of knowledge is virtually in place through the use of PCs, local area networks, and the advent of the information superhighway. In part, the creation of mutual work groups, less hierarchical organizations, and employee empowerment is facilitated by the evolving use of such sophisticated software products as groupware.

Groupware is software designed for multiple users and collective work groups. Unlike traditional PC software products that are predominantly designed for a single user or access on a network by a number of single users, groupware provides the capability for multiple access and use by multiple employees. Communications and access to the same working files and information are provided to many users on a many-to-many, rather than one-to-one or one-to-many, basis.

Such technology as groupware and E-mail enables the access to and dissemination of information by virtually any level of the organization. This helps break down the traditional chain of command and departmental barriers in the organization. Deployed correctly, these technologies support the simultaneous use and creation of information by any number of individuals or working groups. For example, groupware and E-mail make it possible for many project team members, regardless of location, to contribute to a project or document in a real-time environment. This ability accelerates productivity and improves the quality of the effort. Collective technologies also have a significant influence on the way information is managed in the enterprise. Because accessibility to technology and information will be far greater and significantly easier, organizations must learn new ways of managing their information assets and the staff who access and use that information.

Integrating Management and Reporting

In the reengineered enterprise, it is no longer effective for the IT organization to be managed as an overhead function within the purview of an administrative executive, such as the chief financial officer or controller. The management and reporting relationships of the IT organization must be tightly integrated with business processes and elevated to an enterprisewide leadership level, transforming the IT organization into a market-based service provider and business process enabler. IT performance is measured by value generated and contribution to an operational process or unique strategic needs rather than by the number of projects managed, lines of code written, and CPU response time. Reinvented and reengineered enterprises show a surge in IT investment. Budgets for such items as hardware, communications, applications software, and professional services expand, whereas processing costs decline. The reengineered IM&T organization possesses the following behavioral qualities. The organization is:

- Driven by a cross-process orientation.
- Displaying a broad background and expertise in multiple technologies and business processes.
- Cross-functional.
- Highly proactive as an advocate of reengineering and enabling technology.
- Organized around centers of service excellence and core technology and process competency values.

SUMMARY

The use of IT is essential to reengineering the enterprise; without IT, reengineering cannot occur. However, to be successful, the IT organization and its management practices must first be reinvented. In the process of reengineering the enterprise and its operations, management is striving for quantum results through innovation and organizational transformation rather than through incremental gain. However, achieving these results requires not only business process reengineering but a reinvention of the IT infrastructure, organization, and management practices.

3-3

Systems Integration and the Information Technology Architecture

MICHAEL A. MISCHE

The complexities and efforts associated with integrated computing environments and systems integration projects necessitate a responsive and comprehensive information technology (IT) architecture. However, an integrated computing environment and a reengineered enterprise require a new approach to developing that IT architecture. Based on different assumptions, designed to satisfy a set of needs that has since changed, and developed to satisfy an outdated organizational model, historical development methods are simply inadequate to address the reengineered enterprise's systems integration needs.

All too often, systems integration projects are performed as single, standalone efforts. As noted in Chapter 2-1, an effective IT architecture is central to the ultimate success of the integration effort and to generating a high return on using IT, and architecture is especially important to integrating systems using a client/server strategy. This chapter presents a new approach for developing the information management and technology (IM&T) systems architecture. This new approach links many of the concepts associated with business process reengineering (BPR) and organizational transformation with systems integration.

DEFINING THE IT ARCHITECTURE

A new IT architecture is necessary to support reengineering and the resulting organizational transformation. This architecture is being driven by new businessrequirements and more seamless work flows; these driving factors ultimately require more integrated technologies and accessibility to data. Developing the IT architecture for the reengineered enterprise requires that a number of issues and strategic decisions be addressed:

- What are the reengineering imperatives, business driving factors, and enabling technology that will provide customer value, shareholder value, and competitive advantage?
- What enabling technologies, capabilities, application systems, and staffing investments are required to support the reengineered enterprise's new imperatives, organizational initiatives, and operational processes?
- What are the most appropriate enterprisewide architecture and management structure for IT and the realization of reengineering objectives?
- What IM&T organizational structure and management qualities are necessary to support the logical consolidation of IT capabilities most effectively while providing for the physical distribution of those resources?
- What IT and business initiatives are competitors pursuing and why?
- What, if any, windows of competitive opportunity are created by systems integration and reengineering the organization's IT technology resource?
- How can IT and its resources be leveraged to optimize the total enterprisewide use of and investment in technology?

Significant research and literature have been dedicated to the need for and elements of the IT architecture. The investment base in and reliance on technology by most organizations is far too high to have a casual or informal definition of how to deploy technology and what technology should be used. Operationally, process reengineering and organizational changes are placing many organizations in a position of greater dependency on technology. The need for enabling technology to support new processes and leaner organizational structures has increased the need for integration, and organizations can ill afford an IT policy or direction that is not carefully choreographed with business requirements.

Despite these needs and the work performed, however, much more needs to be done. The technologies and dynamics of changing user needs are constant forces that necessitate a flexible and responsive IT architecture. Exhibit 3-3-1 lists some of the factors that are driving the need for a fresh perspective and the need for the IT architecture.

Agreeing on the definition of what an IT architecture means is usually as difficult as crafting one. To better clarify the concept of architecture, it is best to understand what it is not. The IT architecture is not:

- A budget.
- A strategic plan.
- A schematic depicting disk drives, controllers, and CPUs.

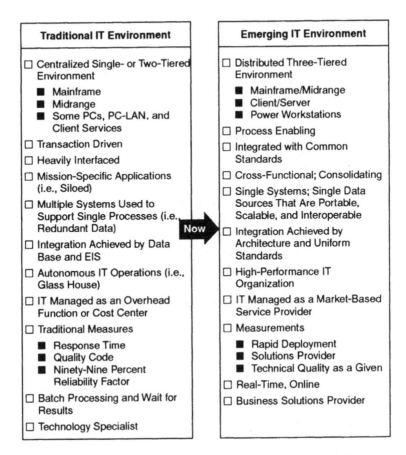

Traditional IT Environment	Emerging IT Environment
☐ Centralized Single- or Two-Tiered Environment 　■ Mainframe 　■ Midrange 　■ Some PCs, PC-LAN, and Client Services ☐ Transaction Driven ☐ Heavily Interfaced ☐ Mission-Specific Applications (i.e., Siloed) ☐ Multiple Systems Used to Support Single Processes (i.e., Redundant Data) ☐ Integration Achieved by Data Base and EIS ☐ Autonomous IT Operations (i.e., Glass House) ☐ IT Managed as an Overhead Function or Cost Center ☐ Traditional Measures 　■ Response Time 　■ Quality Code 　■ Ninety-Nine Percent Reliability Factor ☐ Batch Processing and Wait for Results ☐ Technology Specialist	☐ Distributed Three-Tiered Environment 　■ Mainframe/Midrange 　■ Client/Server 　■ Power Workstations ☐ Process Enabling ☐ Integrated with Common Standards ☐ Cross-Functional; Consolidating ☐ Single Systems; Single Data Sources That Are Portable, Scalable, and Interoperable ☐ Integration Achieved by Architecture and Uniform Standards ☐ High-Performance IT Organization ☐ IT Managed as a Market-Based Service Provider ☐ Measurements 　■ Rapid Deployment 　■ Solutions Provider 　■ Technical Quality as a Given ☐ Real-Time, Online ☐ Business Solutions Provider

SOURCE: M. Mische. The Consultant's Guide to Reengineering (San Diego CA: Pfeiffer & Co., 1995)

Exhibit 3-3-1. Business Driving Factors for the Information Technology Architecture

---

- An operating plan, a manual, or an inventory list of hardware and software.
- The latest technology, application software, object-oriented code, or data bases.

All of these are components of an IT architecture, yet by themselves they do not constitute an architecture. An IT architecture does, however, provide a blueprint. Thus, this definition serves the purposes of developing a unifying architecture: an IT architecture is a conceptual and logical representation of how application systems, data, technologies, and supporting resources are constructed and integrated into a harmonious environment that supports definable operational, organizational, and information requirements.

As discussed in Chapter 3-2, many of the systems in use today evolved without the benefit of a guiding architecture. In many organizations, they are a patchwork of applications and data bases, most of which are interfaced but few of which are integrated under a prevailing architectural standard and plan. This patchwork of systems is generally the result of the organization's having satisfied a specific need (e.g., accounts payable or order entry) and possesses a distinct departmental-ownership orientation. This design may have been adequate in an environment of work fragmentation and hierarchical organizational structures, but with the change to a borderless economy and the need to create organizational agility through reengineering, many organizations are finding that their existing systems cannot support cross-functional and consolidated business processes and the key imperatives associated with reengineering. Therefore, integration represents a much larger and complex issue, as it involves legacy systems, structures, and technologies. The combination of all of these factors creates a challenge and an enhanced need for an IT architecture.

Reengineered enterprises are quite different from those of their ancestors. For example, the reengineered organization is compressed and networked, not hierarchical. In the reengineered enterprise, information is shared and rapidly disseminated, as opposed to moving through strict protocols based on titles and organizational levels. Finally, reengineered organizations leverage employees by empowering them.

Exhibit 3-3-2 compares the reengineered organization to the traditional model. The lists of these characteristics highlight the dichotomies between the traditional uses and deployment of technologies and those that are required by the reengineered enterprise. Thus, many organizations are faced with a dual reengineering effort: not only must they reengineer their organizations and processes but they must also reengineer their IT organizations.

Many existing IT structures have been based solely on technology or a specific suite of vendor product offerings. This orientation, however, can force an inward focus that is technology driven and constrained rather than business enabling. Traditional approaches to systems integration and developing the IT architecture run the risk of viewing application software and data in a narrowly defined, transaction-based, mission-specific context rather than in a cross-functional and enterprisewide manner. Basing the architecture solely on technology and being mission specific can rob an organization of higher returns on investments for technology and improved competitive position. Usually the organiza-

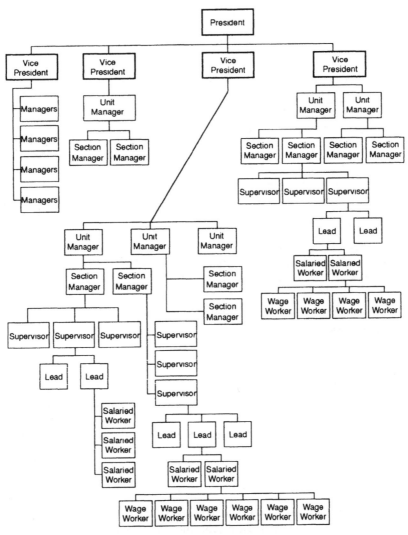

Traditional Organization

Key Characteristics:
- Hierarchical
- Strong Command and Control
- Boxes and Lines
- Siloed Skills
- Defined Boundaries
- At Least Eight Levels Deep
- Organized Around Functions

SOURCE: M. Mische, The Consultant's Guide to Reengineering (San Diego CA: Pfeiffer & Co., 1995)

Exhibit 3-3-2. Comparative Profile of Organizational Structures

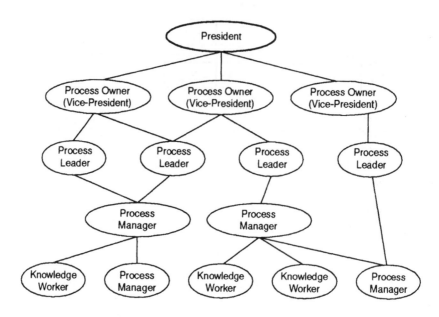

Reengineered Organization

Key Characteristics:
• Highly Networked and Cross-Functional
• Transcends Boundaries
• Fewer Levels, Three to Five Layers Deep
• Diamond-Shaped Structure
• Organized Around Core Processes and Outcomes

SOURCE: M. Mische, The Consultant's Guide to Reengineering (San Diego CA: Pfeiffer & Co., 1995

Exhibit 3-3-2. (*Cont.*)

---

tion's internal IT group leads the architecture effort to migrate onto new platforms or into new operating systems. The result is that functionality remains unchanged or is only marginally improved.

To be an effective tool, the IT architecture for reengineering and systems integration must represent five key values:

1. A long-range vision for the use of technology by the enterprise.

2. A blueprint for ensuring technology applications that best support the organization and optimize resources.

3. A direct and seamless link to the strategic and operational goals of the organization.

4. A rigorous process that involves all significant operational pro-

cesses, organizational levels, external benchmarking, and consensus.

5. A description of the long-term technical, application, data, and organizational structures necessary to support the enterprise, including priorities, needs, IT industry trends and directions, and relationships and linkages.

CREATING THE IT ARCHITECTURE: A NEW MODEL

The integrated IT architecture of the reengineered business has five important qualities. The architecture:

1. Is designed to provide a long-range vision for systems integration in supporting the enterprise.
2. Is a blueprint for systems integration that depicts routes, dependencies, and relationships among systems and the integration effort.
3. Provides a seamless linkage between the organization's strategic and reengineering goals.
4. Reflects the interactive process between systems integration and various organizational levels and users.
5. Describes the technical, application, data, and organizational structures necessary to support the systems integration effort, including priorities, needs, industry trends and directions, and relationships and linkages to internal and external business constituents.

Four Ingredients for the IT Architecture

In addressing the IT architecture needs of the organization, four distinct components, called the four cornerstones, are required. The technologies, systems, and personnel may change, but the cornerstones remain constants. Yet, the four—technology and communications, application systems, information and knowledge, and organization and supporting structures—are adaptable and evolutionary.

Technology and Communications. This component of the architecture is the most fundamental. Including the logical and physical integration capability that enables a common structure and uniform IM&T framework, the technology and communications cornerstone encompasses all mainframes, midrange machines, personal computers (PCs), workstations, client/server processors, data base management systems (DBMSs), communication networks, and electronic linkages. Without the proper IT, supporting organization, and communications infrastructure, reengineering cannot occur. Therefore, it is essential that this cornerstone be developed and stable before any significant investment and effort in operational reengineering and organizational transformation is incurred.

Application Systems. The second cornerstone includes the common and unique application systems supporting specific operational processes and can be either internally developed systems or those obtained from third-party vendors. In the reengineered organization, application systems acquire a different mission and enhanced sensitivity to cross-functional business needs. In contrast to the transactions-based and user-specific traditional software, the application system portfolio of the reengineered organization is less proprietary, far more open, portable across various platforms, and shared among many interactive users. Group-oriented applications support a working arrangement of many-to-many configurations, in which users all access common data and projects and perform common operations concurrently. In the reengineered enterprise, the traditional individual or mission-specific application is no longer considered strategic. Rather, mission- based systems, such as those that support order entry, are tactical assets, because almost all organizations can acquire and or develop the same applications with no capabilities that differentiate one company's from another's. Many of the capabilities that are provided in third-party application solutions are replicated over time, and eventually neutralize one another. Therefore, the strategic advantages of mission-specific and third-party-provided application systems quickly dissipate. However, how an organization chooses to deploy its application systems is strategic, and this is when differentiation and competitive advantages are gained.

Information and Knowledge. The third cornerstone of the reengineered IT architecture is the dissemination of information and knowledge, which are very different but inseparable concepts. Information means data that is used to initiate, support, and complete a process and is the result of any process.

In business today, many systems provide data; however, very few produce value-added information, because traditional systems were originally designed to support and process homogeneous transactions rather than operational integration and organizational cross- functionalization.

In a reengineered enterprise, data is normalized, synchronized, and accessible. Common and specialty data bases provide for the enterprise-wide distribution of information and dissemination of knowledge, which are necessary for employee empowerment and the use of common practices. Because data is critical to the enterprise, the roles of data modeling and data administration become increasingly important in reengineering.

Organization and Supporting Structures

The organization and supporting structures for IT are often overlooked in the effort to migrate and convert systems and data onto newer technologies. The reengineering of the IT organization is necessary to ensure a proper alignment of resources and staffing skills to the cross-functional and process-integrated qualities of the reengineered enterprise. In doing so, a new organization and management structure are created. Correspondingly, new performance metrics are developed, and human resources are deployed differently. To support the reengineered organization and its empowered employees, the IT staff's skills must change from predominantly technical to a more business-process and service-provider orientation.

INCORPORATING THE FOUR CORNERSTONES: THE METHODOLOGY

The enterprisewide IT architecture demands a dynamically balanced and methodical approach. Architectures that focus solely on a particular area, such as data modeling or application portfolio, may miss more meaningful opportunities to leverage technology for breakthrough results. Most often, it is the infrastructure component—the organization, and the IM&T human resource capabilities—that is overlooked. Information and data models may be built, but they can be inadequate without complementary processes and IT organizational transformation. Using the cornerstones as a basis, the development process for a new IT architecture can be represented as six phases (see Exhibit 3-3-3). The objective is to create an architecture that is a multidimensional component of the enterprise's customer, organizational, and operational structures. The phases arrange the four cornerstones of the architecture into a process that ensures that an outwardly driven perspective is developed to address the key issues driving reengineering and systems integration. The phases ensure that common requirements and emerging needs are identified, ranked according to priority, and formulated in an enterprisewide business and technology context. The systems integration process and individual projects are natural extensions and components of the architecture.

The direct coupling of the architecture to systems integration is one of the differentiating and significant attributes of this model. Developing the IT architecture under this approach requires the successful completion of 18 key processes, shown in Exhibit 3-3-4. A sample plan for devel-

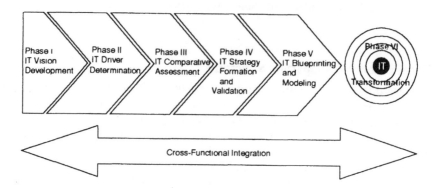

Exhibit 3-3-3. Developing the Reengineering Information Technology Architecture

oping the IT architecture is outlined in Exhibit 3-3-5. Each of the phases are described in the following sections.

Phase 1: Developing a Vision

Developing a future vision of the enterprise is a leadership process. Often the vision is lost among operating plans, budgets, immediate urgencies, the current period's financial performance. and novelty management slogans. Nonetheless, enterprise leadership must have a clear definition of what the organization is. what it wants to be. and how it is going to get there. This vision serves as the rallying point for identity and direction. Knowledge of and commitment to the vision are critical to the enterprise and its ability to create breakthrough opportunities through reengineering and enabling technology. Vision can be defined as the articulation of concepts and ideas that depict the enterprise's direction, goals, and philosophy. Vision has a sense of purpose and urgency that moves people.

The vision for how IM&T resources are to be used and deployed in the organization must be linked to and aligned with the overall vision for the enterprise. Failure to have an adequately articulated vision ultimately results in an incomplete or inadequate IT architecture and a disconnection between the business's architecture and needs. Thus, if the four cornerstones are used, appropriate linkages will be obtained.

Phase II: Determining the Business Driving Factors

Identifying the key trends, cultural influences, technology vulnerabilities. economic indicators, and best practices that drive change in the

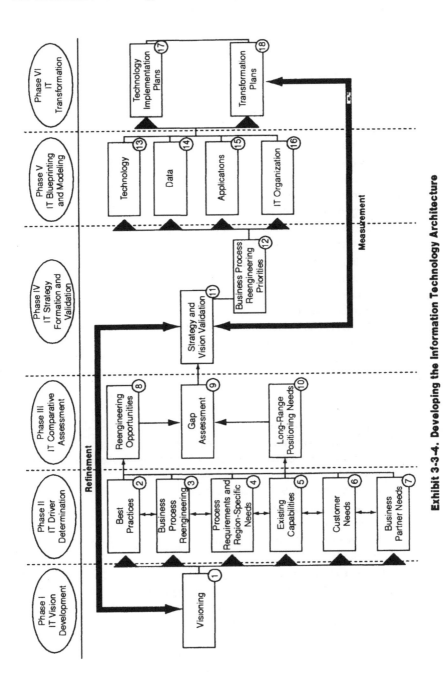

Exhibit 3-3-4. Developing the Information Technology Architecture

1. Establish the vision and direction for IT.
2. Perform IT assessment using the technology continuum and determine the organization's stage.
3. Develop the IT architecture for the reengineered enterprise.
   - Confirm the architecture's business drivers.
   - Confirm the linkages to reengineered processes.
   - Develop IT architecture and blueprints for:
     — Technology.
     — Application systems.
     — Data and information.
     — Communications network(s).
     — IT organizational structure.
   - Develop technology blueprint and architecture:
     — Types of technologies used.
     — Distribution of technologies.
     — Commonality of technology.
     — Standards.
     — Assumptions.
   - Develop application blueprints and architecture:
     — Assumptions and acquisition methods.
     — Processes.
     — Platform.
     — Integration and interfacing capabilities.
     — Standards.
     — Sources (buy, build, modify).
     — Capabilities, features and functions provided.
     — Distribution of applications by process and organizations supported.
   - Develop data blueprints and architecture:
     — Assumptions and acquisition methods.
     — Processes supported/enabled.
     — Platform supported on.
     — Integration and interfacing capabilities.
     — Standards.
     — Process requirements.
     — Rationalization and normalization requirements.
     — Data administration standards and policies.
     — Security and control requirements.
     — Standards and technologies used to support data bases.
     — Organizations and applications supported.
     — Data ownership and access standards.
     — Data definition and rationalization standards.

Exhibit 3-3-5. A Sample Plan

4. Develop communications blueprints and architecture:
   — Assumptions and acquisition methods.
   — Processes supported/enabled.
   — Back-up and recovery.
   — Integration and interfacing capabilities.
   — Standards.
   — Access requirements.
   — Volumes supported.
   — Topology.
   • Develop IT organization blueprints and architecture:
   — Assumptions used.
   — Processes supported/services provided.
   — Key measures for performance.
   — Management practices.
   — Standards of management.
   — Staffing skills and qualifications.
   — Structural alignments.
   — Behavioral qualities.
5. Complete formal architecture and integrate with reengineering project.
6. Develop IT Transformation Plan.
   • Develop requirements definition.
      — Hardware.
      — Software.
      — Personnel.
      — Data.
   • Evaluate hardware, software and IT service providers and alternatives.
      — Internal.
      — External.
      — Outsourced.
      — Make, buy or combine.
   • Select service providers.
   • Develop plan and link to reengineering effort.
7. Implement IT transformation process.
   • Technology.
   • Application systems.
   • Data and information.
   • Communications network(s).
   • IT organizational structure.
   • Culture.

Exhibit 3-3-5. (*Cont*)

enterprise is another fundamental process in developing the IT architecture. This phase involves:

- Ascertaining and understanding best practices and their appropriateness to a specific enterprise.
- Identifying business process reengineering opportunities.
- Identifying specific process requirements and region-specific needs.
- Evaluating existing capabilities and resources.
- Defining customer needs.
- Understanding business partner needs.

Enterprises usually find that up to 15 major influences drive the need for a new IT architecture. Some of the more common influences are customer demands, competitor and host-country business practices, internal operational requirements, shareholder mandates, and changes in technology. These driving factors determine the planning assumptions, strategic tenets, principles, and imperatives intrinsic to the IT architecture.

Phase III: Comparative Assessment

The comparative assessment process evaluates the results of identifying driving factors and determines their influence on information technology. During this process, gaps are identified; reengineering opportunities are evaluated and selectively initiated. Most organizations find that there are many gaps and redundant projects—opportunities for consolidation. They also learn of the major disparities that exist in the caliber and management of technology across multiple organizations, operating units, and divisions.

The Technology Continuum. A major factor in determining priorities and the enterprise's ability to transform itself through business process reengineering is its location on the technology continuum. IT capabilities and organizations have distinct life cycles and characteristics associated with each phase of the technology continuum. They are:

- Where the organization is in its IT evolution.
- Where the organization's IT should be to support strategic and operational imperatives.
- Whether the enterprise can evolve naturally to the next level or leapfrog to another level.

Research on IT organizations' growth stages started with Richard

Nolan's landmark 1979 *Harvard Business Review* article, "Managing the Crisis in Data Processing." Although the original studies focused on the management and behavioral aspects of the data processing organization and not necessarily on how technology was integrated, the research and conclusions provided a definitive view of IT and its probable evolutionary path in most settings.

Many of the basic premises and forecasts contained in the article have proven true and are used as fundamental theories today. There are five distinct phases of evolution that are a natural outgrowth of the original research: this chapter calls these phases the technology continuum, indicating growth, management, level of sophistication, degree of integration of information technology, and the organization and leadership of the IM&T asset relative to industry direction. Exhibit 3-3-6 provides an example of the technology continuum.

Within the framework of the technology continuum, there are five characteristics that help define the qualities and characteristics of each of the phases. Although the characteristics are common to each phase, each has its own attributes that define specific behavioral properties. In total, there are more than 100 individual attributes and properties that define the organization's position in the technology continuum. The characteristics that help to define an organization's IT status include:

- The state of technology.
- The application portfolio and degree of integration.
- Data bases and accessibility.
- IT organization and culture.
- Organizational culture, attitude, and views toward technology.

Understanding where the IT organization and the enterprise's use of technology are in the continuum is fundamental to developing the architecture and crafting the path for transformation and greater return on the IM&T investment.

Phase IV: Formulating the Strategy and Validating the Vision

During this process, a cross-functional, forward-looking strategy for IT is developed. This strategy is based on the four cornerstones of the IT architecture and is directly calibrated and refined against the five values for the architecture, visions, and driving factors. The validation process is critical to ensuring that the vision is attainable, realistic, and properly aligned with the strategy, operational goals, organizational structure, and financial parameters of the enterprise.

	Stage 1 Data Processing	Stage 2 Management Information Systems	Stage 3 Integrated Systems and Technology	Stage 4 End-User-Enabled Integrated Technology	Stage 5 Integrated Process and End-User Environment
Technology	Centralized Large Mainframes Data Center Based Single Vendor Proprietary Closed Systems Captive Network	Centralized/ Decentralized Mainframes Midranges PC Terminal Emulation Networked Some Standalone PCs	Centralized/Distributed Mainframes/Midranges PC-LAN Some Client/Server Multiple Vendor Networked	Centralized/Distributed Multiple Tier Client/Server Multiple Vendor Open Architecture Networked	Logically Consolidated Physically Distributed Three-Tiered Architecture Optimized for Return on Investment Optimized for Flexibility Open Environment (DCE)
Applications	Standalone Batch Oriented Transaction Specific Single Function Based Limited Interfaces Complex Code Structures	Standalone Batch/Some Online Transaction Consolidation Multifunctional Many Interfaces Complex Code Structures	Integrated Online Batch Function Linkage/Transaction Consolidation Distributed Application Portfolio EIS	Integrated/Graphical User Interface Cross-Functional Scalable Portable Distributed Analytical Based	De-Massed Decoupled Scalable Portable Interoperative Process Automation Reengineering Driven
Data	Single Source/Multiple Application Specific Nonrationalized Single Dimensioned IT Ownership IT Custodianship	Multiple Source Application Specific Some Rationalization Limited DBMS (Hierarchical) IT Ownership IT Custodianship	Data Repository DBMS Distributed Data Rationalized Data Formal DBMS Administration User Ownership	Data Warehouse DBMS Distributed DBMS Rationalized Multi-DBMS Administration User Ownership	Multiple Warehouses Enterprisewide Access Single Sourced Uniform Rationalization Formal Standards Multi-DBMS Administration User Accountability

Exhibit 3-3-6. The Five Stages of Information Technology Evolution

	Stage 1 Data Processing	Stage 2 Management Information Systems	Stage 3 Integrated Systems and Technology	Stage 4 End-User-Enabled Integrated Technology	Stage 5 Integrated Process and End-User Environment
IT Culture	Inward Focused High Control Reactive Labor Intensive Large Backlogs No CASE Do-It-Anyway Development Practices	Inward Focused Control Oriented Reactive Some Tactical Plans Dedicated Specialties No CASE Diverse Development Practices	Outward Focused Standard and Control End-User Responsive Formal Strategic Plan Dedicated Specialties Some CASE, Some Common Development Practices	Business Partner Proactive Leadership Value-Added Services Rapid Development Tools Dynamic Planning Process Cross-Functional Perspective Some CASE, Some Common Development	Enterprisewide Resource Organized Around COEs Multiple Reporting Relationships Rapid Delivery of Systems Formal BOD Focused Application Factories Extensive CASE
Organizational Attitude	IT Viewed as Cost Center Formal Charge Out IT Glass House Administrative Center Reports to Overhead Function	Viewed as Cost Center Formal Charge Out Inhibitor Cost Center Reports to CFO	Corporate Asset Charge Out Facilitator Cost Center Reports to CFO	Strategic Asset Supporter Supplier Cost Center Reports to CFO	Strategic Resource Reengineering Enabler Market-Based Service Provider Equal Business Partner Reports to COO, CEO

Source: M. Mische, *The Consultant's Guide to Reengineering* (San Diego CA: Pfeiffer & Co., 1995)

Exhibit 3-3-6. (*Continued*)

Phase V: Blueprinting and Modeling

Blueprinting and modeling are the actual process of developing the IT architecture. As conceptual representations and narratives of each of the four cornerstones of the IT architecture, the blueprints are directly linked and aligned to business driving factors, influences, trends, and reengineering efforts. Particular emphasis is directed to downsizing application systems and providing cross-functional applications.

The applications blueprint must be constructed in a manner that demonstrates the clear linkages and relationships among users, business processes, data warehouses, supporting technology, and relative business and shareholder value. Under this new model, the resulting application architecture becomes:

- De-massed and scalable to region- and process-specific needs.
- Decoupled from the mainframe and cross-platform functional with processing on mainframes, midrange computers, and client/server machines.
- Consolidated, using data warehousing and single sources of data.
- Cross-functional and shared among various business processes and constituencies.
- Transportable across geographical and organizational boundaries.
- Analytically oriented rather than transactions based.
- Built to common standards with emphasis on replication and reuse.

During the process of blueprinting, the types of technologies, vendors, utilities, support products, productivity tools, and longevity of technologies are assessed to determine how the technology can be leveraged to provide competitive advantage. Technology is selected to enable cross-functional business processes and application systems. The technical architecture, network topology, and application systems need to be distributed but not necessarily decentralized, as decentralization frequently leads to redundancy, inconsistent and often conflicting standards, and a diminished return on investment. The goal is the logical consolidation of assets and capabilities while providing for the physical distribution of computing resources.

For the reengineered enterprise seeking application and data integration, there is no single technical solution. The needs of the reengineered organization are too complex, varied, and dynamic to assume that a single hardware and application solution can address all the needs on an enterprisewide basis. Rather, there are a number of technologies that must be melded to create the integrated IT architecture. Based on this need and the tenets that drive reengineering and systems integration (see Chapters 3-1 and 5-3), the IT architecture for the reengineered organization has a diamond-shaped design. as opposed to the traditional hierarchical design and is composed of three interconnected and networked tiers. The tiers provide for the use and deployment of IM&T assets and resources in a logically consolidated and physically distributed manner. Some of the key attributes of this multitiered architecture include:

- Consolidation processors in the form of mainframes and midrange computers.
- Client/servers and local area networks.
- Individual workstations.
- Common data repositories and warehouses.
- Common and consistent application presentations and navigation tools.
- Common networks and communication protocols.

In the reengineered IT architecture, common and mission-defined data warehouses are integrated within the technology infrastructure to ensure compatibility among and accessibility to information, common tools, and key applications. Common graphical user interfaces and standard applications and presentations are used to support connectivity, scalability, portability, maintainability, and adaptability of the architecture to the organization. Exhibit 3-3-7 provides an illustration of the multitiered structure for the architecture of the reengineered enterprise.

Within the IT architecture of the reengineered organization. systems that are integrated and cross-functional demand IT support and management skills, with a distinct emphasis on understanding the processes and organization of the enduser, rather than just the traditional measurements that concentrated on technical proficiency. In the architecture of the reengineered enterprise, the performance of the IM&T organization and professional is measured by business value generated, not just by projects managed, lines of code written, CPU reliability, and response time. Hence, there is a cultural change associated with the IT organization.

In developing the IT architecture, the personnel, performance measures, and business skills needed to support the integrated IM&T environment of the reengineered enterprise must be identified and evaluated against existing resources. Personnel reeducation, retraining, and replacement must be addressed when developing the architecture. Critical personnel must be identified and used to ensure cross-functional connectivity and resource optimization.

Highly specialized work units can include such functions as applications development or end-user computing and are organized to provide economies of scale by maintaining extensive resources.

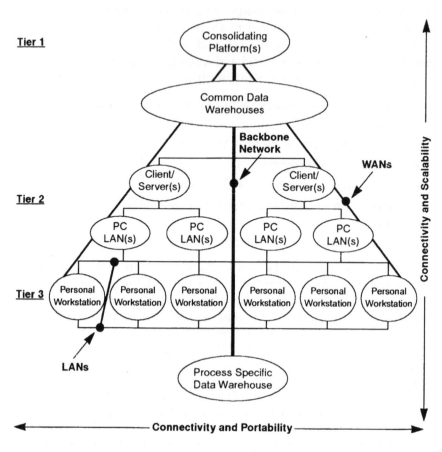

Exhibit 3-3-7. The Multitiered Information Technology Architecture

Phase VI: Transforming the Organization

The transformation phase of this new model for developing the IT architecture provides for the formalization of the investment, organizational
change, and implementation plans, schedules, and processes necessary to
realize the vision and blueprints. Structural alignments, staffing needs,
skill-set requirements, and personnel placement are all addressed and
woven into the architecture. Financial requirements, internal and external resources, migration paths, and conversion issues are integrated into
the architecture.

SUMMARY

Building a new IT architecture to support the dynamics of reengineered organizations that are dependent on integrated systems solutions is an immensely challenging and costly process. Rather than attempting to force-fit an existing architecture to a set of business requirements, this new model seeks to create a cross- functional process development process. The approach is driven by business opportunities, reengineering, and the need to provide an integrated set of enabling solutions that are built to a common standard, transportable, and scalable. Emphasis is placed on optimizing the technology investment and use through the elimination of duplicate IT environments, redundant systems, nonsynchronized data, and unnecessary layers of organization and personnel. Using this approach, organizations and their IM&T managers can define the business, organizational, and technology priorities that guide architecture decisions.

3-4

The Evolving Role of the IT Organization: A CIO's Perspective

MARK T. TURNBULL

Lewis Carroll wrote in *Alice in Wonderland* that "if you don't know where you're going, then any road will get you there." Thus spoke the Mad Hatter—an additional twist to an observation that may actually describe very closely the information systems support functions in many organizations. Most, if not all, businesses are involved in some stage of reengineering, and no matter what the definition of systems reengineering, the information technology (IT) professional is being seriously challenged.

Confronted with an increasingly demanding user community, IT professionals must avoid the "any road will do" syndrome. How they choose to do that is critical to their personal success and professional contributions. Even more important is the effect the chosen road has on organizations as they try to adapt to the massive changes that they must make today if they are to be successful in the future. A great deal has been written in recent years, and even more information has been coming out in recent months regarding the role of the IT organization and professionals in reengineering and systems integration. This chapter explores the evolving role of the IT organization in an environment that is constantly changing and being defined by the realities of reengineering.

REENGINEERING'S IMPACT ON IT

The reality is that what has been written and said about reengineering is accurate—and the message conveyed is an absolute necessity. All organizations must take a closer look at how they do business and how they meet their customers' needs. A major concern is that many organizations are looking at the reengineering phenomenon with the same skepticism that they have had with other management styles that came along in the last 20 years—namely, zero-based budgeting, management by objectives, management by walking around, and more recently, total quality management. Any professional who lived through these cycles may be inclined to be a little hesitant about embracing all the reengineering talk, and that skepticism would be justified—but very poorly placed. Business reengineering is not only real, it is an absolute imperative for an organization if it is to survive until the year 2000 and beyond.

As IT has expanded its audience and its capabilities, it has allowed for an entirely new vision for business operations. IT enables the types of changes necessary for business growth and maintenance in the near and long term. IT's value in helping to flatten the organizational structure and to assist in the empowerment of employees must be recognized. Members of technology groups and professional organizations associated with IT face the challenge of assisting business executives in providing the vision that will allow for the reengineering of their companies.

To do that, IT professionals must establish the value of IT as an enabler in this changing business climate. They must overcome many years of failure, and many years of wanting the latest technology because they believed it would solve all the problems organizations faced. IT staff tended to oversell the value of IT in the past, and now, when it is even more essential to an organization's survival, they are working from a position that is not enviable and not easy to overcome.

Such suppliers of IT as IBM, Digital, and Hewlett-Packard have all suffered the past several years, and the credibility that they once had may not be as strong. Relative newcomers, such as Microsoft, Lotus, and Oracle, may not have gained sufficient credibility to allow IT professionals to use the vendors' apparent strengths to their advantage. In light of the hype associated with 4GLs, CASE tools, expert systems, and object-oriented programming, many of the latest advances do not necessarily work for the average organizations. Organization leaders must choose those that can produce value for the organization and not be mesmerized by the latest new software or hardware inventions.

Changes are made so quickly that organizations must ensure that any new IT that is chosen is consistent with a long-term vision, particularly as managers move to enable the newly reengineered organization. The task is not easy; however, it is achievable, and some ideas that are explained in the following pages may help to contribute to an organization's growth as it enters the next century. Reengineering means changing what was—not change for change's sake, to promote someone, or to reward effort—but because organizations need to change. They must change not only the way they respond to customers but also the way they compensate, motivate, and most important, empower their staffs to make the transition to a new work structure.

Technology costs are dropping at a very rapid pace; more important is that the variety and functionality of this technology are changing almost daily. The complexity of older systems or legacy systems is further enhanced by the need to have even greater—but easier—access to the information that they contain. Another factor is the frequency and speed at which business is changing. Customer demands have exceeded expectations, and these demands continue to increase as the customers try to move their costs back along the supply chain. Businesses are challenged daily by customer needs with just-in-time delivery of raw materials to early customer response systems. Consumers are expecting—if not demanding, in many instances— better quality products and services, for a better price-value ratio than ever before. In addition to these two very critical challenges to business, new government regulations, laws, and global competition are growing rapidly. New competitors are cropping up at every point in the supply chain, and although they may not be after the entire business organization, they may be after specific portions—the most profitable areas. These new competitors have a green-field situation: they are not encumbered by the old ways of doing things or by the bureaucracy and legacy systems that go with the established means.

Staff Empowerment

As the business environment changes at a rapid pace, end users are coming up with new ways to cope with these changes, especially if the IT group cannot provide them with the type of timely responses they require. Users are responding by bringing in personal computers (PCs), local area networks, and client/server applications to meet specific business needs with almost a complete disregard of how they fit in with the rest of the system infrastructure. As systems and technology support providers, IT professionals must find ways to work with users to meet their systems integration needs on a timely basis while placing the required standards in place to allow for the integration of these ad hoc systems.

Information systems executives must support all of the legacy systems that have grown up over the last 25-plus years while moving forward to the new technologies of client/server and open systems. The challenge becomes one of providing value for the investment and moving these legacy systems from their current environment to a more responsive and easier-to-use open environment—which will be time-consuming and expensive. First, the IT staff must be empowered to allow them the flexibility and motivation to carry out the type of changes necessary. Without empowered staff, reengineering does not work. Without changing the processes, the staff's ability to meet the demands that they face cannot improve.

Where does the IT professional start? With education—a process that is complex and difficult. Human nature is such that people do not like to change; reality is such that they must change. Education is not only critical but essential, however difficult a task it may be. IT can enable this change and the educational process. For example, a properly set-up pilot system can be used to show how a new system will affect users and the positive effect that it will have on customer service.

COMPANIES THAT HAVE SUCCESSFULLY REENGINEERED

There are many success stories that prove that changes can realize productivity gains. Harley-Davidson Inc. came back from the brink of extinction and today is again a true powerhouse in the motorcycle world, a title it relinquished to the Japanese for a number of years. Technology was a part of Harley-Davidson's solution to the many challenges it faced.

Also well known is the story of Ford Motor Co., which changed its entire accounts-payable system through technology, increasing efficiency and reducing its staff. IBM Leasing saw a hundred-fold increase in business after it made substantial process and technology improvements and empowered the staff on leasing decisions.

One company that recognized this need for change in the late 1980s also recognized that reengineering was not a fad but an entire transformation of the company and how it did business. The company's senior management recognized that reorganizing only the office structure was not the key to success, and that ensuring the best possible customer service required the complete overhaul of all procedures. The most critical piece to the puzzle was technology; senior management embraced the need to involve IT at the beginning and planned the changes to the organization around the company's ability to find and effectively implement technology in support of its customers' needs. The impact on personnel was minimal at best and, in most cases, was positive as the jobs people did became more meaningful; their knowledge and expertise grew and were key to the success of the changes. At last count, a 10% reduction in the labor force had been achieved through attrition, and efficiency was up by 25% or more in all areas of the company. IT had added value to the procedures for this company.

IT alone did not make these changes happen; technology was only a piece of the whole, put in place to support change—it was not the reason for the change. Process changes occurred because businesses needed to change. Today's technology allows management to implement the type of changes they knew were necessary to meet customers' needs without compromising any levels of service or quality. In most instances, there were orders of magnitude improvements beyond what may have been anticipated.

Why the Reengineering Worked

How did the above company make the right choices? It empowered the employees to find solutions to some very serious business issues. The workers were challenged to find the systems and technology solutions that would allow them to solve the problems they were encountering while providing a long-term solution for areas that were yet to be dealt with. The employees, organized into teams, determined what the best solutions would be and then presented management with the preferred solutions. The solutions came with several caveats, without which the solutions would fail.

The key to the project's success, however, was senior management involvement—from the president and his direct reports to all of their senior staff. The president listened to the explanation of the proposed solutions, understood their costs, and had the foresight to see what the ultimate results would be. As he later explained, he was not sure they would work; however, he trusted his staff and supported them because he knew the proposed solutions were necessary to achieve the company's strategic business needs. He showed faith in his staff's knowledge and commitment to get the job done. The key was in understanding that this was not an IT solution but a business/user solution that was using IT to enable the business process reengineering to take place.

ALIGNING IT WITH THE BUSINESS FUNCTION

What follows is common sense, but then everything organizations tried to do in the IT world for the past 40 years was based on common sense. Traditionally, customers have called the customer service desk with their questions, and businesses have done an exceptional job of passing them around the organization, depending on the information requested. Businesses had all the traditional silos of information. Unless the customer knew exactly what silo to get into, the customer was in for a rough ride, as the customer service representative politely but firmly said, "I am sorry, but for that information you will need to talk to Mr. Anybody in the XYZ Department. I'll transfer you." As time went on, businesses got better at dealing with their customers by implementing voice mail customer response systems, better known to the customer as "Voice Mail Jail."

The difference is that the "glass-house gurus" must understand now, more than ever, that without a broad-based business knowledge and an appreciation of the consequences that IT can have, they cannot be successful. Another key factor in the successful use of technology to reengineer the business and to integrate the necessary systems into the business processes is the level of IT in the organization. The organization's vision must include a place for IT and how it will support the achievement of the organization's mission and the objectives that are put in place to drive the business toward its ultimate goal. IT must understand all the objectives and the processes that support the achievement of those objectives, as well as how technology can be used to enable the meeting of these objectives. IT cannot be viewed solely as a cost to the business, and it cannot be buried within an administrative department—i.e., finance. IT must be a key component of the executive committee, with a voice to assist in the integration of an IT strategy that will support the organization's mission and the strategic imperatives that make the objectives happen.

There are many questions that could be asked as these projects are approached. Some of these questions now involve a greater understanding of the business and its needs.

- How critical is the need for specific raw materials? The urgency of the order may dictate the type of method used to create the supplier order in the first place. IT professionals may have to understand the production schedule and the volatility of this schedule.
- How does the supplier normally fulfill the business's orders as compared with other customer's orders they may receive?
- Is the company the supplier's primary customer?
- Does the company have other suppliers?
- Is electronic data interchange (EDI) an option with some suppliers? Can they accept orders? Send promotions? Notify of inventory shortages?
- Are suppliers allowed access to inventory levels? To production schedules?
- Did the company partner with them? Will the suppliers benefit if the business does well?

These are only a few of the questions. Simply matching invoices to purchase order numbers has been done; however, it is not what customers want, nor is it what the business needs. The business needs a fully integrated solution to the very complex challenges it faces.

Information specialists who are not successful cause serious damage to the viability of the organization. Information professionals have a great responsibility, and they must look hard at what has been done in the past and what must be done in the future. Today IT is viewed as a way to enable the business to achieve its goals and not just as a vehicle to automate what was being done. IT managers must break the old paradigms and provide the enabling capabilities of technology. In sum, IT professionals need to understand the business and what IT can do to assist it. It is not necessarily business's responsibility to understand IT.

CUSTOMER NEEDS MUST BE MET

Information technologists must establish themselves more as business generalists than ever before. They must understand the company's products, its customers, its suppliers, and most important, its employees and what they do. It is no longer sufficient to deliver only products that will solve the users' problems: IT professionals must rise to the next level and understand how suppliers, employees, and customers mix to provide the products that are needed by the ultimate consumer. The competitive environment that organizations face is far more complex than ever before. Technology must be used to enable the organization to become nimble and responsive to competitive pressures.

No matter what the organization does, the products produced must be acceptable and meet the needs of the consumer. IT professionals can no longer view the organization in its functional roles; they must see the processes that these functions support and begin to think laterally.

The development cycle must aggressively pursue solutions to business needs that can address these needs in a matter of days and weeks, versus the traditional months and years. Staff must be empowered to achieve a rapid development goal and make decisions that can help users without having to refer to any committees. Standards that can be followed by the implementors and are flexible enough to meet their needs must be established. These standards can be called consensus standards. These standards can be established by centers of excellence that bring the best and the brightest in the organization together to decide on the most effective use of technology within the company.

IT allows the organization to flatten, by placing information—not just data—into the hands of those who can best use it. IT, with the correct software support, can assist in the cycle-time reduction so necessary in allowing a company to meet its customers' needs. The areas most affected by this are in the administrative and service sectors, as fully integrated systems, client/server applications, video conferencing, CD-ROM, and scanners are put in place.

Integrating systems that add value to the business processes allow for the reengineering that is so vital in today's business climate. IT must partner with the user departments to make them understand how technology enables them to remove steps from the process and to add value.

IT must be prepared to establish the standards necessary to enable the implementation of new systems that can meet customer needs. Unlike standards established 15 to 20 years ago, these standards cannot be based on one manufacturer's systems and infrastructure; they must be based on a multivendor environment that must interact and allow for data integration to create the information necessary to streamline business processes.

EDUCATING THE END USERS

Senior management must understand the need for educating the end users in the best use of these new technologies, as well as retraining existing information services and technology staffs. The days of the glass house and legacy systems are over. Some staff members who have been the traditional systems analysts and programmers may not fit into the new world of client/server computing and user-defined and managed systems if they are not reeducated.

Of the entire staff to be educated in the new technologies that will be entering the marketplace during the next five years, as many as 40% will make the grade with the new standards. Indeed, companies will be faced with a new dilemma, for example, how can the company make the transition from an out-of-date and untrainable staff to a staff knowledgeable with the new skills in client/server technologies, object-oriented coding, and rules-based systems? Will these people be difficult to find? How will this change affect the compensation and benefits structure?

The toughest part of this education is to get past the IS professionals who say, "We have always thought this way." If that were true, there would not be any specialists in manufacturing and accounting applications, for example. End users, under pressure from internal and external customers, have changed their requirements and even their demands. If organizations are to meet customer demands and successfully deliver results, they must form partnerships with their customers. The time has come to be less enamored with the technology and more concerned with the processes and how they can be enhanced.

IT professionals must understand how these new technologies will work together to deliver easily accessible and understandable systems that will have seamless interfaces for the users of these systems. IT managers can no longer afford to introduce new technology every several years and to go through massive conversions; they must learn to do this differently with minimal impact on the user.

RESTRUCTURING THE IS&T FUNCTION

Traditionally, over the years IT professionals strived to be a part of the senior or executive management of the organization. Although some have been successful, many have not made the leap. The reasons are varied, and most IT professionals can tell why they have not made it to the top.

IT has been one of the most successful empire-building environments within organizations in the last 30 years. Titles have been changed often to ensure that they reflect staff's importance and contributions to the organization. Yet, IT personnel have not always met the criteria that the titles reflected. The empire needs to be disassembled and reassembled.

Key areas of IS&T need to be focused within "centers of excellence," with the remainder of the IS staffs redistributed to work with the users in these new process-driven teams. Only then can IT make the type of contribution necessary to the organization. If IT professionals are afraid to make these changes aggressively, they will be left behind, because organizations must change now and must be in a position to change continuously to meet constantly changing needs.

Consensus standards are key to allowing the restructuring to happen. To restructure first and set standards later causes even more problems than when the change was made to solve the original problems. The explosive development cycle in new IT and software applications has made it more important to try to leverage strengths within the IS group and users. No longer can organizations expect one or two individuals to know everything about the variety of applications that now reside on computers. Establishing groups of experts to deal with several key areas, such as PCs, communications, and training, allows maximization of the IS staff productivity.

The centers of excellence are a structure that can be the core of the new IS group. Coupled with a research and development (R&D) function, the centers of excellence allow for progressive consensus standards to be established and modified to meet the new demands of both the users and the newly defined IT structure.

R&D has long been the domain of manufacturing or marketing, which determines what the next best widget or method to manufacture widgets may have been. Now R&D must be considered as a key area of an IS&T department, because it is now possible to do more with less and to be faster than ever before. Scanning applications and viewing documents on computerized systems five years ago meant an investment of tens of thousands—if not hundreds of thousands—of dollars. For this type of investment, a project team would be established and a consultant called to get the best advice. Now users can add scanning and CD-ROMs to their desktop PCs in a matter of days, if not hours, and for only a few thousand dollars. R&D activities within the IS&T group allow the organization to review the new technologies as they become available and establish guidelines and standards to ensure consistent use. Reengineering the business is easier to execute if there is a better understanding of all the different technologies that are available to support those business initiatives.

The competitive advantages established through the use of technology 10 or 15 years ago are no longer as easily obtainable or sustainable. The technology today is too easily transferred and too easily duplicated to have any kind of sustainable advantage. What R&D allows is the implementation of leading-edge technology that may be several months ahead of the competition and, therefore, keeps the company at least even. Maintaining a clear view of what is coming allows the organization to establish leadership, which on its own may be the competitive advantage. The real competitive advantage comes from how quickly the business can respond to the changes required by the users. Information systems development that takes more than 12 months is no longer acceptable, in the same way that bringing new products to market is not acceptable if it takes three or more years to accomplish. How the development cycle meets customers' needs is key to the long-term success of both IT and the organization.

SUMMARY

What truly aligns the IS&T function with the business is participation in and full comprehension of how the mission statement and objectives of the company have been established. Technology must be the enabler that allows the company to meet its goals while not becoming a cost burden to the organization. To assume their positions at the top managers' table, IT professionals must establish the value that they have added. They must bring to bear on some challenges being faced, including both their knowledge of the business and the technology. IT professionals must be aggressive and must take chances, because complacency does not help their businesses. To facilitate the process of change, the suggested action steps are:

1. IT must establish the infrastructure for the organization driven by industry standards that are currently in place and appear to have a legitimate shot at being in place for the future.
2. IT must establish a communications network that is capable of supporting today's distributed client/server environment and at the same time is capable of assimilating new information sources, such as the Internet, CD-ROM-based systems, video conferencing, scanning, interactive desktop video, computer-driven customer response

systems, and a myriad of other technology-driven information systems.

3. IT must be prepared to supply the technology support necessary to ensure that the users are not encumbered by the technology that they require to meet their customers' needs. The technology and the systems must be readily available to them when they need it.

4. The traditional systems and programming support staff must be integrated into the user community so that the technologist becomes a part of the process team, understanding the business and suggesting what technology is appropriate to enable the strategy to be implemented.

5. The IT organization must be restructured to meet the new demands on it while being responsible to the organization as a whole, so it can meet organizational business needs.

6. Users must be prepared to accept the IT group's recommendations on standards for the infrastructure. The pockets of IT that are growing in all organizations must accept the guidance from the main technology group.

7. A short- and long-term plan must be established to allow for a move from legacy systems to new systems whether they are mainframe or PC- based.

3-5
A CIO's View of Systems Integration and Reengineering

JOHN A. SANDERS

F ew business strategies have generated as much controversy and discussion as business process reengineering (BPR). Striking at the heart of years of organizational function and specialization, reengineering proclaims a new solution for improving internal and external competitive performance. The basic premise underlying BPR is that organizations must radically rethink their infrastructure and the way in which work is performed. Simply stated, it means starting over when determining work design and organizational structures.

Reengineering can be enabled by vision and technology. Top management vision drives the case for change, clarifying the mission for significant improvements in production, service, growth and reduction of costs. Technology drives BPR based on the capabilities of current and future products that enable new ways of operating a business that were unavailable in the past and provides a catalytic effect in reshaping the work design and organizational structures.

DEFINING BPR

In these times, business as usual is not the way to go. By now, all organizations are familiar with total quality management, management by objectives, zero-based budgeting, quality circles, downsizing, one-minute managing, and other management "flavors of the month." BPR is not just another flavor to add to the list.

Some BPR Success Stories

The following examples are useful when examining BPR.

A high-tech office-supply company reduced the cycle time from purchase to bill collection from 20 weeks to 2 weeks. Armed with laptop computers, the sales force processed orders out of a virtual office and sent the orders directly from the field to the central computers. This company saved hundreds of millions of dollars annually in processing costs.

A data-processing-services firm had 4,000 finance personnel spread over 38 business units. Benchmarking determined that similarly sized companies had a finance overhead expense of 2.1% of revenue. The data-processing-services firm's ratio was found to be 3.1%. This difference resulted in a $60 million target of opportunity. The firm reengineered three processes: the disbursement cycle (i.e., accounts payable), expense reporting, and financial reporting. The disbursement cycle was reinvented from scratch to eliminate paper from the process. Coupled with enabling technologies, the company reduced its accounts payable staff by 50% (i.e., down to 250 people from 500) and achieved a 75% reduction in other overhead costs associated with the prior labor-intensive processes. The expense-reporting process was reinvented to enable imaging of expense reports—roughly 30,000 per month—coupled with automated reconciliation of expense receipts to the electronic expense report. Labor costs for this reengineering effort were slashed by 70%, resulting in an annual savings of nearly $1 million. The company also reinvented the processes associated with the monthly closing of the books. They changed to a monthly "soft close," which included estimates of accruals for the next eight months. The books were "hard closed" each quarter. The company estimated that they saved two days each month for two to six finance people in each of the 38 business units.

A major metropolitan city reinvented the process for handling the parking-ticket and outstanding-fines processing. Hand-held computers were given to parking attendants, enabling them to perform on-the-spot data entry. An imaging system brought up the ticket when the case came to court. The results were impressive: a 600% increase in the collection rate in just a year and a half. In addition, 5,000 days of police time were freed up from attendance at traffic court.

One of the Big Three automotive manufacturers achieved significant results in the accounts-payable process with technology-enabled reengineering. Several indirect steps requiring a sequential flow were reengineered into shared data bases without mediation from other functions. The corporation's efforts paid off, yielding a 75% reduction in the work force, from 500 to 125.

Shared Characteristics of Success

Each of the examples shares certain characteristics, discussed in the following sections.

Radical, Outrageous Demand for Improvement Across the Board. Each of the companies developed high improvement goals in production, service, quality, and cost reduction. The companies consciously avoided merely improving the existing functions or processes and concentrated on aiming at the core business processes and essentially asking three simple questions:

- What do we do?
- Why do we do it?
- Why do we do it the way we do?

The answers to these questions and the direction for the reengineering projects required a mind set called "thinking outside the box." Everyone has a set of internalized rules and regulations, called paradigms, that influences his or her perception and behavior and tends to box him or her in. The companies avoided being blinded by their current inherited processes and paradigms (e.g., the attitude that they have always done it this way, or if it is worth doing, somebody else would have already done it). Rather, they chose to consider the existing processes and functions with the same dispassionate and objective perspective one finds in a new employee who has not yet been ingrained in the culture of the process. Freed from the constraint of simple, incremental improvement, these companies shifted their paradigms and indeed created new ones, moving "outside the box." They were able to discover radical new opportunities that simply would have been ignored without the paradigm shifts.

Top Management Leadership and Commitment. This is the unequivocal key success factor.

A Focus on End-to-End Processes Rather Than Tasks. Each company fundamentally changed the way it viewed its ability to create value. Management simply began to view the organization as a series of processes and considered what to do end-to-end to transform raw materials, information, labor, and capital into services or products that are valued by the customer. Each target process was viewed in terms of activity flow. Strong, weak, irrelevant, and non-value-adding activities were exposed.

The benefit of end-to-end analysis is that it encourages cross-functional and cross-organizational perspectives. It allows one to see beyond the current compartmentalized operations and consider ways to optimize the relationships between processes and activities.

Key Ingredients for BPR

There are five essential ingredients that emerge from the examples.

1. *The pressure for change.* Is there sufficient energy and motivation from the top to propel the reengineering effort? Without pressure for change, there simply is no action.

2. *A clear, shared vision.* Has the message been communicated, understood, and internalized by all participants? Without a clear, shared vision, there is a fast start, but the project fizzles.

3. *The capacity for change.* Do staff members have the appropriate resources and ability to conduct the project? Without the capacity for change, there is anxiety and frustration.

4. *Actionable first steps.* Do staff members have focused, realistic action steps to take right away? Without this dimension, the organization runs a risk of false starts.

5. *Supporting structures and processes.* Does the organization have meaningful communication mechanisms and an infrastructure in place to conduct the reengineering work? Without supporting structures and processes, the company is at risk for mixed signals and lack of momentum.

As represented by the examples and the shared characteristics, a working definition of BPR can be established: a formal process, energized by top-down, senior-management demand and appetite for radical improvement in quality, speed of product and service delivery, cost reduction, and value-added performance executed within a context of aggressive change management and paradigm shifts in organization structure and work design.

A BPR METHODOLOGY

The following points can assist an organization considering BPR in its approach.

A Clear Senior Management Vision and Commitment

This is the most important step in the methodology. Most work programs for system and non-system-related projects have a step like this. The need for senior management involvement is often acknowledged but pushed to the background while the design and implementation details are dealt with. Even though not all areas in the organization participate in the work to be performed, the message must be clear, unequivocal, shared across the organization, and constantly reinforced. Reengineering is about change: no one can just pay lip service to this step. The project's success depends on making the vision known.

Formalized Project Infrastructure

Early on, it is essential to identify sponsors of the shared vision in each of the business units. There are two types of sponsors: the business unit champions, who are the cheerleaders for the project and managers of change, and the gurus, who can assist in defining the details of the existing processes and functions. Working with each of the business unit champions are the facilitators that are the glue holding the project together. The facilitators are generally senior members of the management team who continue to reinforce the vision, conduct the focus group sessions, and perform overall project management. Finally, an executive steering committee should be formalized for general oversight management and control of the project; the members of the committee should include the members of top management who originally communicated the vision to the company.

Focus Group Sessions: Project Introductions

This part of the work breaks the ice and should involve discussion sessions with each business unit champion and the gurus. Led by the facilitators, the sessions should introduce change management and paradigm shifting. There are several excellent vehicles, including role playing and brainstorming games to orient the participants to the context of change within which the process work will be conducted. The objective of this step is to build a consensus and a comfort level and to communicate the details of the project work plan.

Focus Group Sessions: Process Views

This step involves identifying the current processes, activities, and their internal and external relationships. Several widely available tools can be used to construct process views. Warnier-Orr diagrams are useful for first-cut functional decomposition of processes and activities. Another powerful tool is CPM/PERT. Using readily available software packages, modeling of processes and their relationships can be completed quickly along with assignments of duration and costs of the activities. This first-cut diagramming technique is very useful in visualizing high-level exchanges and should promote substantive discussion of the value added by each exchange.

After the initial, high-level process views are completed, each of the individual exchange points can be further functionally decomposed. At the conclusion of developing the process views, each process and its corresponding exchange points can be examined in terms of the activities that promote value, the activities that are non-value added, the activities that create multiple hand-offs of information or data, and the activities that are irrelevant.

Focus Group Sessions: Brainstorming and Prototyping

Using the process views as a starting point, facilitators can lead discussion sessions to develop alternative scenarios. The process views portray the existing processes, and it is critical to think "outside the box" when conducting these sessions. The types of questions promoted in these sessions are: What if activities A, B, and C are eliminated? What effect would that have on the end-to-end cycle time of process X? What if events 1, 2, and 3 were eliminated? What if two departments that have a high degree of mediation and contact points are merged in this process? What if the two departments were eliminated altogether, activities A, B, and C and events 1, 2, and 3 were removed, and all remaining activities were reassigned to department Z? The facilitators' skill and leadership are critical in this step. The key is to continually reinforce thinking outside the box, or paradigm shifting. The work in this step is highly iterative and should result in a multiple series of new process views as prototypes for consideration and quantification. In addition, the prototypes should assist in the emergence of potentially new organization designs.

Identify Enabling Technologies

As a result of the work completed in the brainstorming and prototyping step, technology is likely to be identified at a high level to enable the new prototypes. This step formally addresses the identification of technologies required to enable the redesigned prototype processes and formalizes the alternatives and costs.

Benchmarks

There are two types of benchmarks: internal and external. Internal benchmarks are based on calculations of the costs and end-to-end cycle times of the existing processes and activities. The benchmarks should be compared with the costs and end-to-end cycle times of the new prototyped processes and the enabling technologies selected. External benchmarks are based on knowledge of industry or competitor end-to-end cycle times and cost curves. This information, if available, is obviously useful to test prototypes and to assist in the development of best practices or world-class performance.

Reengineering Opportunities

At this point the opportunities for reengineering are obvious and numerous. One of the key management objectives is choosing the appropriate opportunities to implement. There have been many failures of reengineering projects, simply because the organization attempted to implement all of the opportunities or conservatively focused on only marginal improvement. One approach might be to consider which opportunities can be implemented in time to deliver the promised benefit. Another might be to implement only one opportunity at a time to limit the chaos and change in the organization. Regardless of the approach, it is a lot tougher to implement than to design the new processes.

The Quick-Hits List

As the steps toward new, reengineered processes are completed, ideas and issues surface that do not translate into radical improvement opportunities but do reduce costs. In addition, reengineering opportunities that are not complex and do not require significant funding to implement are identified. These opportunities are called quick hits, because results can be effected in a short period of time with little investment.

INFORMATION TECHNOLOGY: THE ENABLING AGENT OF REENGINEERING

According to industry observers and research, investment in technology in the US over the last decade approximated $800 billion. Of that amount, it is estimated that 70% was directed to white-collar service and product delivery. It was also observed that productivity in the white-collar work force has essentially remained flat during that period of time. At the root of this dilemma is technology being heaped on existing, inefficient processes and functions, without a fundamental re-architecting of the processes first. At best, companies have accomplished incremental improvements in their businesses; at worst they have perpetuated inefficiency at an accelerating rate.

When reengineering is undertaken, the role of the information technology (IT) function should shift from the classic model of solution provider to one of solution seeker. IT assumes the role of enabling agent providing a catalytic effect. The primary challenge is to visualize how the business could be managed in light of the import of current and future technology. Why should employees manually complete paper expense reports and send them through interoffice mail for approval and payment? Why should paper documents be filed in steel cabinets occupying expensive floor space? Why should the end-to-end cycle time for order fulfillment be held captive by information only available in the home office, whereas the salesperson is in the field with the customer? The IT function needs to accelerate its knowledge base of product capability and transform this knowledge into enabling solutions. Reengineering is both vision and technology driven. It is simply a matter of recognizing and acting on opportunity. Why not implement a reengineering project based solely on available, enabling technology? Clearly, the opportunity might not be achievable, let alone even identifiable, without the enabling technology.

These are some of the technologies available today that could drive a reengineering effort and enable a new process vision:

- *Electronic Data Interchange (EDI).* An inexpensive way to interconnect companies for the purpose of electronically exchanging documents, EDI eliminates paper documents and speeds up the flow of transaction-level data. Supplier orders, purchase orders, and invoice payments are examples of processes that can be reengineered using EDI as an enabling technology.

- *Work flow technology.* Rather than routing paper documents throughout the organization, work flow technologies permit definition of intelligent forms and electronic paths for the forms to travel. For example, expense reports can be prepared at employees' personal computers and routed over the E-mail application to their bosses, no matter where the bosses' offices happen to be. Bosses can approve the expense report electronically, even with a signature, and route the reports electronically to accounts payable for payment.

- *Image-processing technologies.* Off-the-shelf, proven imaging solutions currently abound. From simple, standalone, personal workstation solutions to enterprisewide, networked solutions, this technology currently offers sophisticated and innovative ways to scan, store, and retrieve documents. Imaging offers major opportunities to improve work flow and records management.

- *Groupware and communications.* Teams of people can be enabled as projects and processes require. Rather than being dependent on the current hierarchical forms of organization, groupware can enable virtual offices and assemble groups as required. E-mail and video conferencing are enabling communication technologies that can compress time and space, reduce travel costs, and routinely bring people together face to face, regardless of location.

- *High-speed or broadband networks.* Modern networking technologies allow transmission of much higher volumes of data than in the recent past. Fiber optic channels, FDDI, CDDI, and ISDN data links bring high-speed, inexpensive data communications to virtually any organization. Transmissions of data, images, and full-motion video are easily enabled.

SYSTEM INTEGRATION AND REENGINEERING: CREATING BALANCE AND SYNERGY

Pulling together the working definition of reengineering and some technologies demonstrates how both process vision and technology can combine to create a reengineered process for the expense reporting process in a hypothetical company.

Expense Reporting: The Process as Is

Company ABC is a geographically dispersed organization with 10 business units in five states. The company currently employs 1,000 people across the organization. Approximately 25% of the personnel travel on company business during the year. Expense reports are completed on a weekly basis, and each division is responsible for reimbursement of expenses. Expense reports are manually completed on standard forms, and supporting expense receipts are attached to the forms. Copies of the forms and receipts are made and the forms are sent by interoffice mail to the employee's supervisor for approval. The supervisor reviews the expense report and either returns the expense report to the employee for revision or sends the approved form to accounts payable for disbursement. Accounts payable audits the form, files the report and receipts, prepares an input form for entry of the expense, and enters the transaction into the accounts payable system that then generates the check. The check is printed and sent by interoffice mail to the employee. This process is fairly typical in many businesses today: it is labor and paper intensive with multiple hand-offs, multiple reconciliations, and checks and controls. There simply must be a better way.

Expense Reporting: The Process To-Be

Company ABC's vision is to consolidate the accounts-payable function at the corporate office because management wants to take advantage of centralized cash-management services offered by a local bank. There are several strategies that can be considered.

One strategy is driven by management's vision and the knowledge of technologies that can enable that vision. How does this new process add value and differ from the as-is process?

The expense report as a paper form is eliminated. Rather, the employee completes an electronic "intelligent" form at the workstation that supports breakdowns of each category of expense (e.g.. hotel, airfare, food, and beverage). The electronic expense report is electronically routed using work flow software to the employee's supervisor over the internal E-mail connections.

The supervisor reviews the electronic expense report at his or her workstation and approves the report electronically; the work flow system forwards the expense report from the business unit via ISDN communication lines to the central accounts payable function at corporate or returns the expense report electronically to the employee for further revision and resubmission.

The accounts payable clerk reviews the expense report and forwards the expense report to the accounts payable system, which reads the data contained in the expense report and prepares an electronic entry to be processed by the next cycle of the accounts payable system.

The accounts payable system creates the appropriate journal entries to the general ledger and creates an electronic check for deposit to the employee's account by EDI with a bank clearing house. An advice of the check is routed electronically back to the employee as proof of deposit.

The employee sends expense receipts to a central imaging function with the electronic expense report number attached. The expense receipts are scanned. and the electronic expense report is retrieved. The images of the receipts are processed by a computer program which extracts the amounts from the receipts and reconciles the submitted receipts to the expense report total. The employee is informed that the expense report has been audited and that there is no change, the company owes the employee additional funds. or there is a balance due from the employee.

This is a contrived example; however, it does illustrate the synergy and balance that can be achieved between reengineering and enabling technologies.

SUMMARY

This chapter has briefly presented some basic principles that can guide the consideration of reengineering. Both vision and technology can combine to achieve significant improvements in production, service, growth, and reduction of costs and enable new ways of business operation that were not available in the past.

Section 4

Maximizing Returns, Managing Risks, and Avoiding Pitfalls

Optimizing the investment and returns for the systems integration process while reducing the risks to the organization are key objectives of the systems integrator. No longer are systems integration efforts solely within the purview of the information technology (IT) department, and the all-encompassing nature of systems and reengineering requires a broad-based perspective and multicompetency leadership and teams. The visibility of these projects is enormous and the risks high. The complexity of the process and the convergence of multiple dimensions of people, processes and technology require different methods and measures for managing the projects.

This section offers a diverse perspective on the project management issues confronting the systems integrator. The authors explore a number of different topics and present pragmatic considerations for managing the systems integration project and mitigating the risks of the process. Actual case studies as well as analytical tools are presented for consideration.

4-1

Symptoms of the Terminally Ill Systems Integration Project

MICHAEL A. MISCHE

Building responsive systems in today's demanding business environment creates many challenges, even for the most seasoned and sophisticated information management and technology (IM&T) organization. The challenges and risks become far greater when the integration of applications, data, and technologies combine with the reengineering of business processes and organizations. Although much investment has been made in project management and automated development tools, the risk of failure in systems integration remains extraordinarily high. Nonetheless, the need to integrate systems is compelling, and the systems integration industry continues to grow at a tremendous pace.

Problem projects have been a part of the technology industry since its inception. In some instances, the projects are absolute bottomless pits: the purpose, plan, process, or technology was wrong or inappropriate from the onset. In others, project leadership and commitment were at fault. The technology was fine, the implementation was not. Regardless, the propensity for failure remains high. This chapter discusses the warning signs and symptoms of a high-risk systems integration project that goes awry. This chapter dubs these problem systems terminally ill integration projects (TIIP).

IMPLICATIONS OF THE TERMINALLY ILL INTEGRATION PROJECT

The stakes for systems integration are indeed high. No longer are errant integration projects just an IM&T technical issue or an internally embarrassing problem. Errant systems integration efforts and problem projects can find their way onto the front pages of newspapers and into professional journals. For example, *Information Week* has reported on projects in a number of companies, including AMR Corp.'s failed $125 million consortium effort for a new reservation system. The effort, which was discontinued, allegedly cost AMR over $60 million in out-of-court settlements, in addition to what had already been incurred in executing the project.

The *San Jose Mercury News* reports that California's Department of Motor Vehicles (DMV) and Office of Information Technology disregarded dissenting analysis about the appropriateness of a selected hardware and software solution for a new system. The project's costs escalated from $38 million to more than $155 million. As a result, California taxpayers are out $44 million in payments for hardware, software, and services as state officials write off the effort and begin again with an entirely new approach and different hardware and software.

The federal government is not without its problem systems, either. According to *Information Week,* the Federal Aviation Administration's new Advanced Automation Systems project is so far behind schedule and over budget that it is difficult to fully measure. Originally budgeted as a $3.6 billion project, congressional reports are citing project delays in excess of five years and estimated cost overruns of more than $2.6 billion. This is in addition to the $2.3 billion already spent. Total estimated costs for the project: at least $6 billion.

Another example involves a project originally budgeted at $8 million, this integration and development effort became a five-years-past-due project costing over $100 million. Another involves a major company's migration to a client/server computing environment that has grown from $5 million, to $10 million, to somewhere between $40 and $70 million to complete.

These are but a few examples of integration projects gone awry. The list is extensive, populated with integration projects that started with the best of intentions only to find failure. Undoubtedly, countless others occur every day and are kept secret by those involved. Chapter 4-2 provides additional insights and examples into errant systems.

The effect of these terminally ill integration efforts can be devastating to the organization and its IM&T group. Some of the consequences of a TIIP include diminished shareholder value, diluted asset value, reduced return on investment, loss of earnings, reduced consumer confidence, outright interruption of business, and the potential for noncompliance with regulatory requirements and contractual obligations. Finally, there is the effect on people. All too often, a TIIP results in employee terminations or the emotional decision by management to stop projects or outsource IM&T.

SYMPTOMS OF THE TERMINALLY ILL INTEGRATION PROJECT

After decades of experience and hundreds of millions of dollars invested in technology, virtually every major organization continues to experience difficulties in the delivery of integrated systems and reengineered business processes. Companies may be reluctant to acknowledge them, but the problems are there.

There are three essential issues related to the management and performance of business systems integration projects:

- Why are the traditional project management methods seemingly so ineffective in predicting and correcting problem integration projects?
- How can the risks of systems integration projects be minimized?
- How can systems integration and reengineering projects be better managed to achieve superior results?

An understanding of some of the influences on the integration process provides a framework for addressing these issues. First, systems integration projects usually involve large and complex technologies and applications, including legacy systems that have been in the company for years, if not decades. For example, one financial institution's demand-deposit system that was subject to an integration process had executable code structures, data structures and language dating back to 1967. The existence of these obsolete structures increased the level of effort as well as the risk of the integration project. Second, traditional project management methods and practices tend to concentrate on what has been accomplished, not necessarily on how it was accomplished and its relationship to formal work plans and measures. Third, systems integration not only requires changes in technology but some combination of process, cultural, operational, and organizational change, and these types of changes are difficult for many organizations.

Complex Technologies and Applications

The first major influence on the systems integration process relates to the diverse technologies and application systems that are the targets of the integration effort. Found in virtually every enterprise, these applications include standalone, specialty applications that are shared in a network or work-group environment and legacy systems that were built for a single purpose and user. The legacy systems represent the core of applications for many organizations, as they support common business processes. Developed over a period of years, the legacy systems have become part of the organization's culture and fiber, and users commonly exercise a high level of ownership over them. By today's standards, legacy systems are cumbersome and usually involve highly fragmented applications and data bases built over the years to satisfy specific transaction and information needs of the owners. These systems are fragile and are inherently risky to integrate because of their extensive interfaces, fragile

code structures, and numerous processing dependencies. Further complicating the issue of legacy systems and systems integration is that many
are processed on mainframe computers with proprietary-operating-
system software requirements. In certain instances, application and system software support may no longer be available from the source vendor.
Thus, the older applications and legacy systems are, by their very nature
and composition, complex, large, and difficult to integrate. They represent high-risk endeavors.

The target "go-to" technologies for integration also provide new challenges for the systems integrator. The migration paths from traditional
mainframes, aged applications, and rigid data structures to new client/
server technologies and distributed midrange computing environments
are, for the most part, uncharted or only partially mapped. Still others
are unproved, both technically and functionally. Regardless, there are no
truly convenient or fail-safe methods. As many organizations are learning, migration and integration costs are quite high. In many instances,
organizations attempting to migrate do not have sufficient resources to
assign to or support the integration process adequately. Therefore, they
must place a greater dependency on external systems integrators and
contractors.

Traditional Project Management Techniques

The second major influence on the integration process centers on the
basic assumptions and characteristics of traditional project management
methods. IM&T professionals have relied on a variety of project management techniques and automated tools to track the effort of a project.
Some of these methods have been helpful in the management of traditional systems development and implementation projects, but few have
been successful in significantly reducing the risk of integration project
failure. They are not designed to withstand the systemic changes associated with systems integration, nor are they designed to improve the
quality of results delivered. Traditional project management methods
concentrate on a predetermined work plan and do an excellent job of
explaining what has happened; however, they provide very little insight
into what will happen. In this respect, traditional project management
techniques are neither diagnostic nor preventive. They are largely reactionary in design and use.

Organizational Change

The third influence involves both organizational and operational
changes. As integrated systems are implemented, the organization must
also change to best optimize its use of technology and integration. Many

organizations have duplicate procedures and deeply entrenched management practices that the systems integration changes. Acceptance of the integration process and the related operational changes can be slow and, in many cases, management openly opposes the changes. Very often, the systems integrator hears, "This is how the system does it, and this is how I want it." The process can almost be self-defeating. Thus, the systems integration project can be compromised by the very operational processes and organization that it is designed to support. Successful systems integration almost always requires the management of the design of new processes, new strategies, and organizational change, and therein lies one of the main reasons for the failure of many integration projects.

IDENTIFYING THE TIIP

TIIPs have very distinct qualities and many early warning signals. Unfortunately, many of these signals are subtle; if viewed individually, each on its own may not necessarily provide any alarms. However, when viewed collectively and with confirming indicators, the collective qualities of the TIIP can be quite revealing. There are 10 qualities that distinguish and differentiate the TIIP as a problem integration effort. The integration project:

1. Exceeds its financial budget, estimated work effort, and forecasted completion schedule by least 25%.
2. Is poorly documented on changes in project objectives, scope, requirements, assumptions, work plan, and staff.
3. Is not linked to any meaningful and measurable strategy, business objective, or organizational need.
4. Is failing, will fail, or has failed to meet its intended objectives, stated benefits and established scope.
5. And project team are in a constant state of turmoil, marked by date slippages, inconsistent plans, and numerous changes in team composition, requirements, and users.
6. And its work products are considered by independent parties to be inadequate for the purposes of the project or to be of poor or inferior quality.
7. Is not structured with the appropriate skills, project management techniques, time commitments, project sponsorship, and leadership.
8. The project appears to lack sufficient organizational commitment and support.

9. The effort's objectives are not clearly aligned with tangible goals and value derivation targets.

10. The human implications of the effort, as related to changes in performance measures, cultures, organizational alignments, and behaviors have not been addressed.

The majority of terminally ill systems integration and reengineering projects can be avoided. As defined earlier, a TIIP has distinguishing characteristics that relate to personnel, processes, technologies, and project management practices. The 10 most prevalent warning signs of a TIIP include:

1. Improper or inadequate project leadership.

2. Inappropriate or inadequate project sponsorship and commitment.

3. Inappropriate or inadequate project partnering and shared responsibility.

4. Unclear strategy and vision and ineffective project planning that is linked to the vision.

5. Unrealistic or mismanaged user, sponsor, or senior management expectations.

6. Inadequate or ineffective project management practices.

7. Inappropriate target go-to technology and applications.

8. Ineffective or inadequate project measurements, communications, and status reporting.

9. Inappropriate or inadequate project team staffing, organization, and management structure.

10. Unidentified or underestimated project risk and operational exposure.

11. Unclear linkage to new business processes and related organizations.

Each of these symptoms influences the dynamics of the systems integration and reengineering process differently. The level of influence that any symptom can exert on a project depends on the organization's culture and the integration's type and complexity. The symptoms can occur and reappear at any time throughout the life of the integration and reengineering effort. Although there are many more factors that can influence the process and outcome of the integration process, these symptoms are pervasive as signals of the status and pending outcome of the project.

Improper or Inappropriate Project Leadership

The proper type and level of project leadership are essential to any integration and reengineering effort. Integration projects must have leadership that provides for the effective management and coordination of resources and activities within the enterprise. Effective project leadership ensures that responsible personnel demonstrate management over the entire spectrum of integration and reengineering activities, while facilitating process changes and the removal of organizational barriers. Effective project leadership and ownership extends beyond the traditional boundaries of title, organizational lines, and budgets to encompass a personal commitment to the success of systems integration and the reengineering process. Some of the key indicators of improper or ineffective project leadership and management are listed here. The project leader:

- Lacks a thorough understanding of the project, its objectives, issues, processes, and team dynamics.
- Demonstrates apathy toward the project or off-loads the responsibility for project leadership to the project team or users.
- Fails to provide a sufficient amount of quality time and attention to the project and members of the project team.
- Fails to understand or practice sound project-management standards, such as time reporting, quality reviews, and team-building sessions.
- Lacks the organizational credibility or support for the project, fails to influence other departments, or fails to remove organizational and operational barriers.
- Fails to provide the project team with effective direction on team assignments, quality reviews and morale building and fails to serve as a protector for the team in the organization.
- Fails to communicate the project effectively to the appropriate areas of the organization.
- Fails to forge synergistic relationships with key users, external contractors, and organizational peers regarding the integration effort.
- Underestimates the difficulty of change or barriers to change.

Inappropriate and ineffective project leadership and management occurs when the wrong people are trying to manage the integration process. In the context of a TIIP, project leadership is inappropriate when the leaders do not possess the necessary knowledge, empowerment, or technical and business skills to manage the integration process and resources assigned successfully. An all-too-familiar example of inappropriate organizational sponsorship occurs when the IM&T organization becomes the proponent for process and organizational change. In these situations, the integration process is driven by IM&T and new technology that may be superimposed on an unwilling user community. It may also be unnecessary. This type of leadership virtually guarantees an unsuccessful effort and does not result in any significant systemic changes. Usually, this approach results in the complete organizational rejection of the project.

Inappropriate or Inadequate Project Sponsorship

Inappropriate or inadequate project sponsorship is similar to ineffective project leadership. However, project sponsorship usually occurs at a higher level in the organization and often involves cross-departmental processes and managers. Inappropriate sponsorship occurs when the wrong person or group of people attempts to sponsor and support the integration effort. In these instances, the sponsor is usually beyond the managerial purview and organizational empowerment necessary to effect the process, cultural, and organizational changes related to successful business and systems integration. When this occurs, the integration project sponsor simply does not have the authority for effecting change external to his or her immediate area of influence. Key indicators of inappropriate sponsorship include:

- Lack of organizational acceptance for the project.
- Increased organizational resistance and tensions surrounding the project.
- Diminishing political support for the project sponsor and for the project in general.
- Failing to align the integration project with the organization's strategies, goals, priorities, and business initiatives.

Quite simply, for any systems integration effort to be successful, it must be sponsored by the appropriate senior executive and aligned with the organization's requirements and priorities. For example, in the early 1980s, a major financial institution was funding a $5 million project designed to provide a new worldwide financial reporting system. The project was about one-third complete when a senior-level officer thought to conduct a briefing for the executives in charge of that portion of the business. The project team prepared and delivered a spectacular status report. The most senior executive complimented the project manager and then informed him of the organization's decision to de-emphasize its international operations! The major question is why the project was allowed to move forward in the first place.

Inappropriate or Inadequate Project Partnering and Shared Responsibility

In integrating systems and business processes it is imperative to have a constructive and active partnership with all organizational elements involved. The likelihood of a failed integration effort significantly increases when owners, sponsors, users, and those empowered with the integration effort fail to forge synergistic and constructive working relationships. In managing and executing the integration project, cultivation and continuous evaluation of the interaction between project partners and their shared responsibilities is essential. Warning signs of inappropriate partnering include:

- Failure of project sponsors, partners, and leaders to adhere to formal strategies, goals, and objectives.
- Failure of project sponsors, partners, and leaders to participate actively in project meetings, working sessions, and team-development sessions.
- Reluctance of project participants to assume equal responsibility for the outcome of project and collective responsibilities for the results achieved.
- Reluctance of partners to contribute the necessary time, resources and skills to the project or to express a personal interest in the success of the project.
- Failure to facilitate organizational and operational change on the part of project sponsors, partners, and leaders, who may exhibit a reluctance to abandon proprietary systems and processes.

Unclear Vision and Ineffective Project Planning

The creation and sharing of a common vision for the integration project is absolutely essential to the success of any integration and reengineering effort. A vision is important to establishing expectations for the effort's success and for implementing effective quality and progress measures. The vision also helps to establish the parameters, scope, objectives, and expectations for the project. The vision is also necessary for understanding the interrelationships between reengineering, integration, and the deployment of enabling technology. Key indicators of a poorly defined vision and disconnection between the vision and integration effort are when the vision is:

- Unstated, under-developed, or unclear.
- Not shared among participants and those who have a stake in the project.

- Not linked to any meaningful business strategy, goal, or need.
- Inconsistent or is subjected to many changes.
- Not understood by the project team or organization.
- Not clearly and consistently communicated.

Effective project planning is fundamental to the success of any systems integration effort. Projects with poorly developed work plans, task lists, milestones, activity dependencies, staffing assignments, timing, requirements, and project management standards are destined to become a TIIP. A responsive systems integration project plan provides for the description of the project, its intended approach, the assumptions used, major benefits and risks, contingency plans, project team(s) and staffing assumptions, and the relationship to key prerequisites and requirements. Some of the more important warning signs of poor project planning exhibit themselves when the project work plan:

- Is not fully developed at the outset of the effort or is being developed on the go.
- Does not integrate the corresponding process and organizational changes within an overall effort.
- Does not adequately address the timing requirements of the project and is not linked to any significant objective or schedule.
- Does not stipulate the project's staffing requirements, assignments, and management responsibilities.
- Does not stipulate the milestones for the effort.
- Does not define the deliverables and work products to be produced or does not link them to specific tasks and assignments.
- Does not identify the essential activities, their sensitivities to change, and the project's critical path items and their potential level of importance to the overall integration process.
- Does not consider or provide a contingency plan for recalibrating the project.
- Is not fully communicated to and understood by the project team or organization attempting to perform integration.
- Is not linked to tangible business goals and objectives.

Another consideration that affects the outcome of the integration process is the duration of the effort. Projects that have a long completion horizon are extremely vulnerable to disruption and represent a far greater risk for failure than those that have a short and tightly defined completion schedule. Protracted systems integration efforts are susceptible to any number of threats, including management, personnel, and organizational changes, changes in vendor product offerings, economic downturns, and user impatience. In general, systems integration and reengineering projects that require more than 12 to 14 months to complete are at significant risk for failure or major financial and schedule overruns. Some of the key warning signs of inordinately long or delayed projects are when:

- The project milestone dates and relationships among milestones are constantly changing.
- The project schedule is protracted because of personnel shortage, insufficient time dedicated to the process, organizational inertia, changing business priorities, or for the accommodation of other priorities.
- The project goes more than two months without any meaningful results.
- The project is subject to delays caused by vendors, suppliers, internal organizational changes, or contractors.
- The project is delayed because of changing sponsorship, scope, or objectives, unstable user requirements, or inappropriate technical design.

Unrealistic or Mismanaged User, Sponsor, or Senior Management Expectations

Establishing the proper level of organizational expectations is critical to the project management process for systems integration. Expectations that are distorted or miscommunicated adversely affect the systems integration project and the efforts of the project team. Some of the more common examples of mismanaged expectations include:

- Commitment to unrealistic objectives, deliverables, and completion dates.
- Oversimplification of the complexity of the project and its requirements, issues, and organizational implications.
- Failure to understand fully the risks associated with the effort.
- Reliance on technical solutions to cure fundamental operational problems or organizational and managerial inadequacies.
- Failure to establish the essential criteria for measuring both success and progress.

Inadequate or Ineffective Project Management Practices

There is a direct correlation between the quality of the project management processes and the results of the systems integration process. Systems integration mandates that the project is completed on time and within budget, satisfies sponsor expectations and objectives, and minimizes the risks of and possibilities for failure and, most importantly, delivers value. Effective systems integration project management requires the constant tracking and monitoring of results, quality, and effort. Specific warning signals of ineffective project management and progress tracking include:

- The project managers and project team members are not adequately trained in the concepts, principles, and practices of project management.

- The project has inadequate or inappropriate change-management procedures.

- The project undergoes constant reshuffling of priorities and leapfrogging of design and requirements.

- The project team does not fully comprehend the project's work plan, schedule, design constraints, or staffing assignments.

- Project management is inconsistent in updating the project work plan, staffing assignments, and status of project deliverables.

- The project team members are not tracking, evaluating, and explaining variances for the project's budget, labor, tasks, staffing assignments, milestones, schedules, and time estimates.

- The project team is not practicing or does not have the ability to perform dynamic task and responsibility allocation for resources, work products, and personnel.

- The project team is not documenting their work results.

- Project management practices and the individual capabilities of the project managers are of paramount importance to the systems-integration effort and the avoidance of terminally ill projects. Effective project management must anticipate the direction, issues, and potential pitfalls of the effort. Failure to create a forward-looking project management process leads to a terminally ill systems integration project.

- Adherence to a realistic time and milestone schedule is also fundamental to the success of the integration process. Significant slippages in time lines, labor estimates, task assignments, work products, and deliverables are all indicative of a more severe problem with the overall project and its management structure. As a rule, any unplanned variance greater than 25% has an adverse effect on the project and its budget.

Inappropriate Target Go-To Technology and Applications

The selection of the target go-to technology and the originating technologies for the integration process influence the project's success. The use of new or unproven go-to technologies creates additional risks and complications for the integration effort. The new go-to technologies for integration often involve additional labor, incremental training, and higher learning curves for those involved—all of which contribute to a higher propensity for failure. Many organizations that are attempting integration, for example, through the use of client/server technologies, are learning that they lack a full understanding of that environment's technical requirements; therefore, their effort is greater than anticipated. Some of the more common signals of using inappropriate technology to integrate include:

- Using technology to treat, but not cure, an organizational problem or fundamentally deficient business and management process.
- Selecting the wrong go-to technology.
- Using go-to technology that cannot support user requirements, processing loads, and required response times.
- Using go-to technology that is unproven in performance, vendor support, and acceptance in the marketplace.
- Choosing go-to hardware or software that is not mainstream technology or that is difficult to support.
- Custom developing essential applications using unproved methods or unproved external integrators.
- Failing to link the integration effort with an overall organizational architecture and strategic and reengineering plan.

Ineffective or Inadequate Project Measurements, Communications, and Reporting

Failure to measure and communicate adequately the project's status contributes to many integration projects' demise. The project's status, direction, needs, issues, and results need to be constantly and consistently communicated to the appropriate elements of the organization. Key signs of dysfunctional measurements and communications include:

- Lack of standard project management practices, tools, and reporting formats.
- Lack of predetermined measurements for quality, consistency, progress, and milestones.
- Lack of periodic reporting intervals.

- An absence of milestone variance tracking for individual team members.
- Absence of monitoring the quality of project work products and deliverables.
- Lack of substantive participation and feedback from users, project sponsors, and leadership.
- Lack of tracking and managing against the appropriate criteria.

In systems integration projects, effective project communications possess four essential qualities:

1. Communications and presentations are consistently formatted.
2. Communications occur at regular and frequent intervals and are delivered to the same audience.
3. Communications are unfiltered, realistic, and based on published metrics, work plans, schedules, and milestones.
4. Communications and status reports are distributed to and reviewed by the proper audiences.

Each of these must be present to ensure proper and responsive communications.

Inappropriate or Inadequate Project Team Staffing, Organization, and Management Structure

The proper mix and use of staff together with the correct distribution of management responsibilities are significant determinants to the systems-integration process and the quality of results achieved. Clues to a poorly organized or managed project team include:

- The project team has insufficient time commitments to perform the project and has conflicts in assignments, performance measurements, and time.
- The project team is composed of an inappropriate mix of staff skills, experiences, and organizational levels.
- Project team members are inadequately trained or technically deficient in the areas assigned or are placing too much reliance on on-the-job-training and outside consultants.
- The project team members do not understand their roles and responsibilities or are not personally committed to those roles.
- Individual project team members and project managers lack the discipline and structure to support the necessary project-management practices.

- The project team receives inconsistent direction or incorrect leadership signals.
- The project team does not have adequate cross-process representation or organizational participation or is perceived by users and management as an IM&T project only.
- The project team has undergone constant changes and turnover in its composition, management, membership, and structure.
- The project team and integration effort lack a protector who functions as an organizational insulator from distractions, disruptions, and adverse managers.
- The project team has poor rapport and morale among its members.
- Project team members are not documenting their work and reporting their efforts and accomplishments in a structured and consistent manner.

In assembling the staff for the systems-integration effort, the project team members should be selected and assigned to the effort based on the project's technical, functional, and organizational requirements and on the members' individual ability to contribute to the effort. Some of the key factors to be considered in assigning personnel to the integration team include:

- The professional qualifications and personal attributes of individual team members.
- The level of working rapport, relationships, and camaraderie among team members.
- The leadership qualities of individual team members.
- The individual morale and enthusiasm of team members.
- The specific project knowledge, awareness, and cross-functional business perspectives of the individual team members.

Unidentified or Underestimated Project Risk and Operational Exposure

Understanding, estimating, and managing risks are all crucial to the success of any integration and reengineering process. Often these risks are overlooked or improperly evaluated. Risk is a relative measure and concept for any organization. Evaluating the risk of the integration effort requires the identification of three types of risk: technical, financial, and business.

Technical risk includes factors that are related to the technologies and applications that are the target of the integration process. Technical risk includes assessing:

- The level of technical integration risk of hardware, communications, software, and data bases.
- The degree of risk surrounding compatibility issues.
- The communication network risk.
- The risk of vendor stability and ongoing maintenance support.
- The risk of not developing the technical competencies for sustaining the integration process.

Financial risk refers to the sensitivity of the project's budget to variances for hardware, software, personnel, and training. It also includes a measurement for return on investment in the project and the ability of the integration effort to achieve its financial goals. Financial risk involves the assessment of:

- The total project budget.
- The cost of hardware, software, and contractors.
- The ongoing support and maintenance.
- The total cost of process and organizational changes.
- The probability of succeeding at generating the anticipated direct and intrinsic value.

Business risk involves the measurement of the project's impact on current and future operations. The key measure and concern of business risk is determining the organization's sensitivity to any missed time lines or shortfalls in the integration effort. In its most severe sense, a late integration project can completely disrupt an organization and cause a cessation of operations. Business risk requires the assessment of:

- The impact of the project on the operations and organization.
- The organization's sensitivity to delays in the project and its potential impact on customers and business partners.
- The risk associated with using internal and external resources on the project.

THE TIIP ASSESSMENT METHODOLOGY

There are a number of statistical methods available to assist in profiling the behavior of a systems integration and reengineering project and its potential outcome. The 10 symptoms described in this chapter can be applied to virtually any systems integration and reengineering effort using the TIIP Assessment Methodology. The methodology includes a simple intuitive algorithm that gives the 10 symptoms subjective weightings and evaluation criteria. The methodology facilitates the identification of potential problems and helps in developing corrective action steps for projects representing a risk of becoming a TIIP. The specific weighting criteria are:

- Improper or Inadequate Project Leadership and Management:.15.
- Inappropriate or Inadequate Project Sponsorship:.05.
- Inappropriate or Inadequate Project Partnering and Shared Responsibility:.10.
- Unclear Vision and Ineffective Project Planning:.10.
- Unrealistic or Mismanaged User, Sponsor or Senior Management Expectations:.05.
- Inadequate or Ineffective Project Management Practices:.20.
- Inappropriate Target Go-To Technology and Applications:.10.
- Ineffective Project Communications and Status Reporting:.05.
- Inappropriate or Ineffective Project Team Organization and Management Structure:.15.
- Unidentified or Underestimated Project Risk and Operational Exposure:.05.

The total TIIP score weightings equal 1.00. The result, or TIIP score, of each symptom is based on a scale of 10 points. Application of the methodology requires the review of each of the 10 symptoms and the subjective rating of the key indicators in each symptom and experienced judgement and a high business acumen. The indicators are rated using a scale of 1 to 5. A score of 1.0 indicates the worst possible score; a score of 5.0 represents the best possible case. For example, in evaluating the Leadership and Sponsorship symptom, a score of 20 is derived as the sum of the indicators. This score is multiplied by the symptoms weighting factor of .15 to obtain a TIIP score of 3.0. A similar review and computation are performed for the remaining nine symptoms. Once the TIIP-score for each of the 10 symptoms has been calculated, the individual scores are summed to a grand total. Projects with a total TIIP score between 1 and 40 can represent significant risk with the potential for failure. In contrast, those projects that have a high numerical TIIP score range (i.e., 75 to 100) could have a high potential for success. Projects with TIIP scores of 40 to 74 are in the gray area and should be carefully evaluated. These projects can quickly become terminally ill.

The use of the TIIP score methodology is by no means fool-proof or fail safe. It can however, provide a simple and potentially effective means of evaluating the quality and tendencies of any systems integration and reengineering project. The 10 symptoms provide pragmatic insight into those items that can exert significant influence on a systems integration and reengineering project. They are tangible, measurable and effective

There are a number of statistical methods available to assist in profiling the behavior of a systems integration project and its potential outcome. The 10 symptoms can be applied to virtually any systems integration and reengineering effort using the Terminally Ill Integration Project (TIIP) Assessment Methodology. The methodology includes a simple, intuitive algorithm that provides the 10 symptoms with subjective weightings and evaluation criteria. The methodology facilitates the identification of potential problems and helps in developing corrective action steps for projects that represent a risk of becoming or are a TIIP. The specific weighting criteria are:

Symptom	Weighting Factor
1. Improper or Inadequate Project Leadership and Management	.15
2. Inappropriate or Inadequate Project Sponsorship	.05
3. Inappropriate or Inadequate Project Partnering and Shared Responsibility	.10
4. Unclear Vision and Ineffective Project Planning	.10
5. Unrealistic or Mismanaged User, Sponsor, or Senior Management Expectations	.05
6. Inadequate or Ineffective Project Management Practices	.20
7. Inappropriate Target Go-To Technology and Applications	.10
8. Ineffective Project Communications and Status Reporting	.05
9. Inappropriate or Ineffective Project Team Organization and Management Structure	.15
10. Unidentified or Underestimated Project Risk and Operational Exposure	.05
Total TIIP Score Weightings	**1.00**

Question 1

	Evaluation					Weighting	Total
	Poor			Excellent		Weighting	Total
Symptom	1	2	3	4	5	Factor	Score
Improper or Inappropriate Project Leadership and Management (Weighting Factor = 15%)							
The proper type and level of project leadership and management are essential to any integration and reengineering effort. Integration projects must have leadership that provides for the effective management and coordination of resources and activities within the enterprise. The key indicators of improper or ineffective project leadership and management include:							
• Project leadership lacks thorough understanding of the project, its objectives, issues, processes, and team dynamics.							
• The project leader demonstrates apathy for the project or off-loads the responsibility for project leadership to the project team or users.							

Exhibit 4-1-1. The Terminally Ill Integration Project Assessment Methodology

Question 1 *(Continued)*

Symptom	Evaluation Poor 1	2	3	4	Excellent 5	Weighting Factor	Total Score
• The project leader fails to provide a sufficient amount of quality time and attention to the project and members of the project team.							
• The project leaders and managers fail to understand or practice project management standards, such as time reporting, quality reviews, and team-building sessions.							
• The project leaders and managers lack the organizational credibility or support for the project and fail to influence other departments or remove organizational and operational barriers.							
• The project leader and manager fail to provide effective direction for the project team with regard to team assignments, quality reviews, morale building, and serving as protectors for the team in the organization.							
• The project leader and manager fail to communicate the project effectively to the appropriate areas of the organization.							
• The project leader and manager fail to forge synergistic relationships with key users, external contractors, and organizational peers regarding the integration effort.							
Total TIIP Score Multiplied by Weighting Factor:						.15	

Question 2

Symptom	Evaluation Poor 1	2	3	4	Excellent 5	Weighting Factor	Total Score
Inappropriate or Inadequate Project Sponsorship and Alignment (Weighting Factor = 5%)							
Inappropriate or inadequate project sponsorship is similar to ineffective project leadership. However, project sponsorship usually occurs at a higher level in the organization and often involves cross-departmental managers. Inappropriate sponsorship occurs when the wrong person or group of people attempt to sponsor and support the integration effort. Key indicators of inappropriate sponsorship include:							

Exhibit 4-1-1. *(Continued)*

Question 2 *(Continued)*

Symptom	Evaluation Poor 1	2	3	4	Excellent 5	Weighting Factor	Total Score
• There is a lack of organizational acceptance for the project.							
• Organizational resistance and tensions surrounding the project and efforts of the systems integration project team are increasing.							
• Political support for the project sponsor and project in general is diminishing.							
• Managers fail to align the integration project with the organization's goals, priorities, and business initiatives.							
Total TIIP Score Multiplied by Weighting Factor:						.05	

Question 3

Symptom	Evaluation Poor 1	2	3	4	Excellent 5	Weighting Factor	Total Score
Inappropriate or Inadequate Project Partnering and Shared Responsibility (Weighting Factor = 10%)							
In integrating systems and business processes, it is imperative to have a constructive and active partnership with all organizational elements involved. The likelihood of a failed integration effort significantly increases when owners, sponsors, users, and those empowered with the integration effort fail to forge synergistic and constructive working relationships.							
• Project sponsors, partners, and leaders fail to adhere to formal project management standards and practices, schedules, time lines, and responsibilities.							
• Project sponsors, partners, and leaders fail to actively participate in project meetings, working sessions, and team development sessions.							
• Project participants are reluctant to assume equal responsibility for the outcome of the project and collective responsibilities for the results achieved.							
• Partners are reluctant to contribute the necessary time, resources, and skills to the project or to express a personal interest in the success of the project.							

Exhibit 4-1-1. *(Continued)*

Question 3 *(Continued)*

Symptom	Evaluation Poor				Excellent	Weighting Factor	Total Score
	1	2	3	4	5		
• Project sponsors, partners, and leaders fail to facilitate change and exhibit a reluctance to abandon proprietary systems and processes.							
Total TIIP Score Multiplied by Weighting Factor:						.10	

Question 4

Symptom	Evaluation Poor				Excellent	Weighting Factor	Total Score
	1	2	3	4	5		
Unclear Vision and Ineffective Project Planning (Weighting Factor = 10%) The creation and sharing of a common vision for the integration project is absolutely essential to the success of any integration and reengineering effort. The vision also helps to establish the parameters, scope, objectives, and expectations for the project. Effective project planning is fundamental to the success of any systems integration effort. Projects with poorly developed work plans, task lists, milestones, activity dependencies, staffing assignments, timing, requirements, and project management standards will become a TIIP. Key indicators of a poorly defined vision and disconnection between the vision and integration effort include:							
• The vision is unstated, underdeveloped, or unclear.							
• The vision is not shared among participants and project stakeholders.							
• The vision is not linked to any meaningful business goal or need.							
• The vision is inconsistent or is subjected to many changes.							
• The vision is not understood by the project team or organization.							
• The project work plan is not fully developed at the outset of the effort or is being developed on the go.							
• The project work plan does not adequately address the timing requirements of the project and is not linked to any significant objective or schedule.							

Exhibit 4-1-1. *(Continued)*

Question 4 *(Continued)*

Symptom	Evaluation					Weighting Factor	Total Score
	Poor			Excellent			
	1	2	3	4	5		
• The project work plan does not stipulate the staffing requirements, assignments, and management responsibilities for the project.							
• The project work plan does not stipulate the milestones for the effort.							
• The project work plan does not define the deliverables and work products to be produced or does not link them to specific tasks and assignments.							
• The project work plan does not identify the essential activities, sensitivities to change, and critical path items of the project and their potential level of importance to the overall integration process.							
• The project work plan does not consider or provide a contingency plan for staffing and task completion based on changing project assumptions.							
• The project work plan is not fully communicated to and understood by the project team or organization attempting to perform integration.							
• The project plan is not linked to tangible business goals and objectives.							
• The key measurements for progress and quality are not defined in the project plan.							
• The project milestone dates and relationships among milestones are constantly changing.							
• The project schedule is protracted because of lack of personnel, insufficient time, organizational inertia, changing business priorities, or to accommodate other priorities.							
• The project goes more that two months without any meaningful results being produced.							
• The project is subjected to delays caused by vendors, suppliers, internal organizational changes, or contractors.							
• The project is delayed because of changing sponsorship, unstable user requirements, or inappropriate technical design.							
Total TIIP Score Multiplied by Weighting Factor:						.10	

Exhibit 4-1-1. *(Continued)*

Question 5

	Evaluation					Weighting Factor	Total Score
	Poor			Excellent			
Symptom	1	2	3	4	5		
Unrealistic or Mismanaged User, Sponsor, or Senior Management Expectations (Weighting Factor = 5%) Establishing the proper level of organizational expectations is critical to the project management process for systems integration. The more common examples of mismanaged expectations include:							
• Project members commit to unrealistic objectives, deliverables, and completion dates.							
• Project members oversimplify the complexity of the project and, its requirements, issues, and organizational implications.							
• Oversimplifying the complexity of the project, its requirements, issues, and organizational implications.							
• Project members do not fully understand or evaluate the project risks.							
• Project members rely on technical solutions to cure fundamental operational problems or organizational and managerial inadequacies.							
Total TIIP Score Multiplied by Weighting Factor:						.05	

Question 6

	Evaluation					Weighting Factor	Total Score
	Poor			Excellent			
Symptom	1	2	3	4	5		
Inadequate or Ineffective Project Management Practices (Weighting Factor = 20%) There is a direct correlation between the quality of the project management processes and the results of the systems integration process. Systems integration mandates that the project is completed on time and within budget, satisfies the expectation and objectives of the sponsors, and minimizes the risks and possibilities for failure. Specific warning signals of ineffective project management and progress tracking include:							
• The project managers and project team members are not adequately trained in the concepts, principles, and practices of project management.							

Exhibit 4-1-1. (*Continued*)

Question 6 (*Continued*)

Symptom	Evaluation					Weighting Factor	Total Score
	Poor			Excellent			
	1	2	3	4	5		
• The project has inadequate or inappropriate change management procedures.							
• The project is experiencing constant reshuffling of priorities and leapfrogging of systems design and functional requirements.							
• The project team does not fully comprehend the project's work plan, schedule, design constraints, or staffing assignments.							
• Project management does not update or is incapable of updating the project work plan, staffing assignments, and status of project deliverables.							
• The project team members do not have the ability to track or are not tracking, evaluating, and explaining variances for the project's budget, labor, tasks, staffing assignments, milestones, schedules, and time estimates.							
• The project team is not practicing or does not have the ability to perform dynamic task and responsibility allocation for resources, work products, and personnel.							
• The project team does not have the ability to document or is not documenting the results of their work.							
Total TIIP Score Multiplied by Weighting Factor:						.20	

Question 7

Symptom	Evaluation					Weighting Factor	Total Score
	Poor			Excellent			
	1	2	3	4	5		
Inappropriate Target "Go-To" Technology and Applications (Weighting Factor = 10%)							
The selection of the target "go-to" technology and the originating technologies for the integration process will influence the success of the project. The use of new or unproven "go-to" technologies creates additional risks and complications for the integration effort. Some of the more common signals of using inappropriate technology to integrate include:							
• Technology is used to treat, but not cure, an organizational problem or fundamentally deficient business and management process.							

Exhibit 4-1-1. (*Continued*)

Question 7 *(Continued)*

Symptom	Evaluation Poor			Excellent		Weighting Factor	Total Score
	1	2	3	4	5		
• The wrong "go-to" technology was used.							
• "Go-to" technology is used that cannot support user requirements, processing loads, and required response times.							
• "Go-to" technology is used that is unproven with respect to its performance, vendor support, and acceptance in the marketplace.							
• "Go-to" hardware or software is chosen that is not mainstream technology, proven, or easily supported.							
• Essential applications are custom developed using unproven methods or external integrators.							
• The integration effort has not been linked with an IT architecture and strategic plan.							
Total TIIP Score Multiplied by Weighting Factor:						.10	

Question 8

Symptom	Evaluation Poor			Excellent		Weighting Factor	Total Score
	1	2	3	4	5		
Ineffective or Inadequate Project Measurements, Communications, and Reporting (Weighting Factor = 5%) Failing to measure and communicate adequately the status of the project contributes to the demise of many integration projects. Key signs of dysfunctional measurements and communications include:							
• A lack of standard project management practices, tools, and reporting formats.							
• A lack of predetermined measurements for quality, progress, and milestones.							
• A lack of periodic reporting intervals.							
• An absence of milestone variance tracking for individual project team members.							
• An absence of monitoring the quality of project work products and deliverables.							
• A lack of substantive participation and feedback from users, project sponsors, and leadership.							

Exhibit 4-1-1. *(Continued)*

Question 8 *(Continued)*

Symptom	Evaluation Poor 1	2	3	Excellent 4	5	Weighting Factor	Total Score
• A lack of tracking and managing against the appropriate criteria.							
Total TIIP Score Multiplied by Weighting Factor:						.05	

Question 9

Symptom	Evaluation Poor 1	2	3	Excellent 4	5	Weighting Factor	Total Score
Inappropriate or Inadequate Project Team Staffing, Organization, and Management Structure (Weighting Factor = 15%) The proper mix and use of staff together with the correct distribution of management responsibilities are significant determinants to the systems integration process and the quality of results achieved. Attributes of a poorly organized or managed project team include:							
• The project team has insufficient time commitments to perform the project and has conflicts in assignments and time.							
• The project team is composed of an inappropriate mix of staff skills, experiences, and organizational levels.							
• The project team members are not adequately trained or technically competent in the areas assigned or are placing too much reliance on on-the-job training and outside consultants.							
• The project team does not understand their roles and responsibilities or are not personally committed to those roles.							
• Individual project team members and project managers lack the discipline and structure to support the necessary project management practices.							
• The project team receives inconsistent direction or incorrect leadership signals.							
• The project team does not have adequate cross-process representation or organizational participation or is perceived by users and management as an IT project only.							

Exhibit 4-1-1. *(Continued)*

Question 9 *(Continued)*

	Evaluation					Weighting Factor	Total Score
	Poor				Excellent		
Symptom	1	2	3	4	5		
• The project team has experienced constant changes and turnover in its composition, management, membership, and structure.							
• The project team and integration effort lack a protector who functions as an organizational insulator from distractions, disruptions, and adverse managers.							
• The project team has poor rapport and morale among its members.							
• Project team members are not documenting their work and reporting their time and accomplishments in a structured and consistent manner.							
• The level of working rapport, relationships, and camaraderie among team members is low.							
• Leadership qualities of individual team members is poor.							
• Individual morale and enthusiasm of team members is poor.							
• The project team has specific project knowledge, awareness, and cross-functional business perspective.							
Total TIIP Score Multiplied by Weighting Factor:						.15	

Question 10

	Evaluation					Weighting Factor	Total Score
	Poor				Excellent		
Symptom	1	2	3	4	5		
Unidentified or Underestimated Project Risk and Operational Exposure (Weighting Factor = 5%) Understanding, estimating, and managing risks are all crucial to the success of any integration and reengineering process. Often these risks are overlooked or improperly evaluated. Risk is a relative measure and concept for any organization.							
• Technical risk includes factors related to the technologies and applications that are the target of the integration process. Technical risk occurs when:							

Exhibit 4-1-1. *(Continued)*

Question 10 *(Continued)*

Symptom	Evaluation					Weighting Factor	Total Score
	Poor				Excellent		
	1	2	3	4	5		
—The level of technical integration risk of hardware, communications, software, and data bases is low.							
—The degree of risk surrounding compatibility issues is low.							
—The communication network risk is low.							
—The risk of vendor stability and ongoing maintenance support is high.							
· Financial risk refers to the sensitivity of the project's budget to variances for hardware, software, personnel, and training. It also includes a measurement for return on investment in the project and the ability of the integration effort to achieve its financial goals. Financial risk involves the assessment of the systems integration effort and occurs when:							
—The total project is high relative to other projects.							
—The cost of hardware, software, and contractors is proportionately high relative to other projects.							
—Ongoing support and maintenance is predicted to be high.							
· Business risk involves the measurement of the project's impact on current and future operations. The key measure and concern regarding business risk is determining the sensitivity of the organization to any missed time lines or shortfalls in the integration effort. In its most severe form, a late integration project can completely disrupt an organization and cause a cessation of operations. Business risk occurs when:							
—The impact of the project on the operations and organization have not been addressed.							
—The sensitivity of the organization to delays in the project and potential impact on customers and business partners have not been addressed.							
—Risk associated with using internal or external resources have not been addressed.							
Total TIIP Score Multiplied by Weighting Factor:						.05	
Total of All TIIP Scores: **(Higher TIIP Scores = Safer Projects)**						100	

Exhibit 4-1-1. *(Continued)*

indicators to monitor. Exhibit 4-1-1 provides an example of the TIIP Assessment Methodology.

SUMMARY

Systems integration remains a major area of activity for virtually every organization. Integration is necessary for many organizations, and the ability to effectively manage the integration effort is essential to the process. The use of the 10 symptoms and TIIP Assessment Methodology can be of potential value to any organization in managing and reducing the risks of these complex and important projects.

4-2

Systems Integration and Corporate Strategy: A Tale of Success, A Tale of Failure

EFFY OZ

S ystems integration, as the term suggests, is the integration of information systems into the business. This means the system contemplated must fit the business in several ways. It must fulfill its original purpose, comply with the organizational culture, be used, and perform well technically. When all of these factors are addressed in the contemplation of a new system, it is likely to succeed. When at least one of these factors is not taken into consideration before the system is delivered, chances are it will fail.

Systems integration is often thought of in terms of simply integrating hardware and software. The experts ensure that the hardware pieces are compatible with each other and the applications are compatible. However, all too often, the real challenge is how to integrate an entire information system into business processes. There are not, as yet, scientific methods to ensure the success of systems integration in this sense, but cases of success and failure provide valuable lessons.

This chapter presents two case studies: Saturn's information system (SIS), a remarkably successful incorporation of information technology (IT) into a strategic business plan, and Citicorp's MortgagePower Plus, a total failure of a would-be strategic information system. Both systems were great ideas, in the business sense. Judged in terms of providing innovation, the idea behind MortgagePower Plus was more ingenious than the former. However, Saturn's system succeeded, whereas Citicorp's failed.

SUCCESS AT SATURN

Saturn of Plymouth was opened in 1990 at the time of the introduction of the Saturn automobiles. These cars represented a new era for General Motors (GM) and the US auto industry as they attempted to regain lost market shares from Asian and European auto makers. Saturn is considered a small car in the US market and its primary competition was Toyota, Honda, and other small American cars that were in the $15,000 to $20,000 price range. It was targeted at the 25- to 48-year-old upscale professional. Women accounted for 60% of customers. In 1992, Saturn cars were selling as fast as they rolled off the assembly line, a marked contrast from the sales of other GM organizations in that year.

As the Saturn Corp. developed, GM was trying to tell its workers that the new product was totally different from the typical American car. The downturn in GM's once-massive presence had been attributed to low productivity, high absenteeism, defective products, mediocre after-sale service, and Japanese competition. GM's management set its objectives as being at the leading edge of technology in three areas: manufacturing, marketing, and information systems (IS). In the IS area, the new corporation's management decided to abandon the culture in which IS and data-communication were pursued on an individual or departmental basis. Saturn management adopted the people-technology-systems triangle. IS forms the base of the triangle, supporting people and technology equally. However, technology is only one element in the triangle; the other two factors are equally important.

In this triangle, people represent the importance management attributes to the way employees, customers, and suppliers use systems. Thus, the systems are designed to serve them. Technology is the state of the art of the means enabling the corporation to provide supporting systems. In Saturn's case, it is also the technology enabling the car maker to provide a better end-product: a superior car. And systems is the result of combining people's needs and the enabling technology into tools that make their work more efficient and effective. (Exhibit 4-2-1 illustrates this triangle.)

The Saturn Information System

Most current GM information systems are not standardized. It is therefore hard to achieve integration of multiple units in a network. As a result, many GM dealers have to purchase their own hardware and software to run their operations. Many are not connected to GM information resources. There is neither interdealership support nor any direct dealership-production relationship.

GM assigned its subsidiary, EDS (Electronic Data Systems), to de-

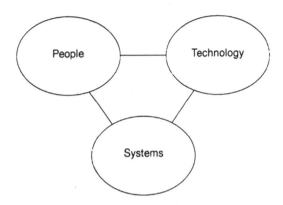

Exhibit 4-2-1. Saturn's Trinity for Success

velop the new IS and communications system for Saturn. EDS was to develop a single, integrated system to serve management, production, and the dealerships. The new system was developed on these principles:

- Applications run on a single corporate platform.
- All data bases are relational.
- Data entries are defined by a central repository.
- Rigorous standards are established regarding data modeling and presentation.

GM management decided to organize Saturn as a separate corporation rather than as another division. Management enjoys a great amount of independence. A decision was made to tailor the IS to Saturn's needs, regardless of previous practices. This gave EDS a golden opportunity to design the system from scratch. The software company established a body named the Business Systems Action Council (BSAC) that reports to the EDS member of Saturn's Strategic Action Council. BSAC ensures consistency in Saturn systems, as every decision is agreed on by members of the group, which includes all functional areas of the company, including the dealership end users.

Dealership end users are not taken lightly at either Saturn or GM. Project teams consistently include representatives from user groups. Numerically, they often constitute the majority of the team members. Study after study has shown that user involvement is an extremely important ingredient for success. In prototyping projects, systems analysts must involve the user, because the inherent factor is the iterative interplay between the developer and the user until the user is satisfied. However, when a more formal development process takes place (e.g., a formal system development cycle), users tend to be forgotten. To ensure user involvement, user representatives were included in the BSAC.

Saturn Dealer Information System

The Saturn Dealer Information System is a network of several subsystems: SALESLINE, SERVICELINE, and various transaction-processing systems that perform accounting functions. The network links the dealer sites by satellite with the production facilities and Saturn headquarters, located in Springhill, Tenn., Dealers receive current news from headquarters about new products, practices, and other topics. The dealers transmit business data, such as sales by model and color. Data on the products sold (i.e., cars and parts) and customer demographic data are transmitted to the company's headquarters nightly. This allows management to make decisions quickly and react to market trends in a timely manner. Car dealers face a common problem: a customer steps in to purchase a car with certain features. The car is unavailable, and the dealer promises to deliver the car days, or weeks, later. Often, this means a lost sale. With Saturn IS, the dealers can keep a close look at the production line of cars and parts. They have access to both the assembly facility in Springhill and the parts plant in Rochester, N.Y. SALESLINE, a subsystem of the SIS, allows the salesperson to use a microcomputer to connect to other dealerships computers and search their inventories for the desired car and features.

Another subsystem, SERVICELINE, provides the dealers with inventory management services similar to material requirement planning (MRP). The dealers enjoy the ability to carry only minimum levels of parts for serviced cars, because they can connect directly to the parts factory by satellite and order the necessary items. Thus, carrying costs are minimized, whereas customer-service time is kept low.

Studies have repeatedly shown that car buyers consider the service they receive after the purchase as important as the quality of the car itself. Saturn's harnessing of IS to improve service gave the company a competitive edge. At the time, no other car maker owned a similar system.

In addition, SERVICELINE allows customers to promptly receive information about anything that has to do with the car, such as new or used cars that the customer may want to purchase, service for the purchased vehicle, the availability of a car with specific features, and the availability of a certain part. The sales or service person processes the request and consults SIS's data base to meet the customer's demands.

For the individual dealership, SIS provides the mundane, but necessary, services of payroll, accounts payable and receivable, recording of

178

title transfers, sales and purchasing management, and analysis of customer demographics and sales trends.

Although the individual car buyers purchase the end product, the sale and service are carried out through the dealers. Thus, Saturn must keep the dealers well informed and well served. Saturn achieves several goals with SERVICELINE. First, it provides the dealer with up-to-date information that keeps inventory levels to a minimum; second, connecting the dealers to the satellite network helps the dealers communicate with each other, which improves availability of cars to the end customer and creates a closer network of dealers. And third, SERVICELINE communicates that Saturn is willing to provide the state-of-the-art information systems for their success.

A Lifelong Relationship

SIS keeps track of the history of every car from the moment it is assigned a Vehicle Identification Number (VIN) to its last service. When the production of a car begins in Springhill, it is assigned a VIN. The VIN is encoded onto a computer chip that is permanently embedded in the chassis before assembly to uniquely identify the vehicle. The salesperson can use the VIN as a primary key when searching for information on a particular automobile. VINs are also used to bring up the service record of the car when it is brought in for repairs, tune-ups, and other warranty and service maintenance. The car can be tracked throughout its life even though it may change owners.

When a customer pulls up for service, the service person can retrieve the vehicle's past service record by typing in the VIN. The computer chip activates a scanner which retrieves the record to be displayed on a microcomputer's monitor. Many customers have been amazed at a service person who knew their name and their car's service record before they rolled down the window.

Strengths and Opportunities

SIS gives Saturn and its dealers valuable information with almost no effort. SERVICELINE also allows the tracking of each part that goes into the manufacture of the car. This is accomplished by assigning each part a bin number that corresponds to the inventory address of the part. The system maintains the links between part bins and VINs and then between the VINs and their corresponding owners. Clearly, this gives Saturn a competitive advantage. For example, when product defects occur, management can use the communications network to notify every dealer what the problem is and which vehicles the problem affects. This capability has already been used. On February 12, 1991, Saturn informed more than 1,200 owners of a potential problem with a seat-recliner mechanism. Owners were asked to return their cars to a dealer of their choice to replace the mechanism. The company did this not because of government demand for safety but to keep its customers satisfied. Similarly, on May 10, 1991, Saturn voluntarily replaced 1,836 cars because of improperly formulated antifreeze fluid purchased from a supplier.

In both cases, the company used SIS to track the owners of the vehicles that were affected. EDS clearly designed the system to help with a strategic goal: keep the customer happy. The customer is contacted about a probable problem before he or she is even aware of it. The car owner feels that Saturn cares about him or her and cares about the car at least as much as the owner.

At the dealership, the customer does not have to wait long for information about service and parts. The only delay time is that of the satellite reception and transmission. The data collection and analysis modules provide both the dealers and Saturn with valuable information that helps plan production volumes and target specific population groups for marketing.

The SIS purchasing module fully adopted the Electronic Data Interchange (EDI) concept. While other GM operations, Ford, and Chrysler have implemented EDI at 60% to 70% of their suppliers, Saturn purchases only from vendors that are willing to adopt its EDI software. All purchasing activities are carried out electronically. This strategy saves the company millions of dollars annually.

In the future, SIS will help Saturn diversify its products. The advanced telecommunications are instrumental in globalizing the company's business.

Why Is the System Successful?

Until recently, the auto industry was blamed for not keeping up with information technology (IT). American auto makers were slow to adopt robots and information systems. Some experts still believe that the manufacturing sector in general, and the auto industry in particular, cannot gain from IT as does the service sector. Yet Saturn is an example of excellent technology integration into all segments of an auto maker's business, from purchasing, manufacturing, and assembly, to customer service.

SIS is a success because the major issue was not technical compatibility but business compatibility. Information systems plans were integrated into the corporate strategic plan early on. Management decided to exploit IT as the backbone of all of the company's operations and involved the employees in the integration process. In other words, there was a shared vision of IS at Saturn. Most important, corporate management envisioned how the end users (i.e., the dealers) would use the system for the benefit of the customers and designed a feasible system. Often, organizations rush to implement a great idea without such careful consideration, as in Citicorp's case.

In the context of the four conditions presented in the introduction to this chapter, Saturn's system fulfills its original purpose, complies with the organizational culture, is used enthusiastically, and performs well technically. Bingo! All of the conditions have been met.

FAILURE AT CITICORP

Ostensibly, the service sector can benefit from innovative IS more than other sectors of the economy. The automated teller machine, cash-management accounts, and airline reservation systems are just a few types of systems that brought their owners great competitive advantages. Even the greatest idea, however, may fail if not tested on a small scale and if there are loose ends. A case in point is Citicorp's MortgagePower Plus.

The scenario is: a consumer is about to purchase a house. Does he or she apply for a mortgage loan from a bank that takes a week to approve it or go to an institution that can approve the loan within 15 minutes? The provider that can guarantee a minutes-long approval process would garner a huge business, and that was the exact idea that Citicorp came up with.

Citicorp is one of the most advanced users and developers of IT. In the early 1970s, it designed its own ATMs. Its president is considered a computer guru and is said to be fascinated by technology. This largest banking corporation in America employs 4,000 programmers and spends an estimated $1.5 billion a year on IS. Apparently, it has all the ingredients necessary for successful implementation of new ideas with IT.

In 1987, Robert Horner, chairman of Citicorp's mortgage unit, conceived a great idea: a quick mortgage process, equivalent to the 10-minute oil change or one-hour photo processing. He wanted to connect real estate agents and brokers to a Citicorp electronic system through which the applicant could receive an approval or denial of a mortgage loan within 15 minutes. The attractive short procedure, he estimated, would funnel huge business to the giant institution. The company was at the time the biggest mortgage lender, originating nearly $15 billion in loans per annum. The new system, dubbed MortgagePower Plus, was to become the goose to lay the golden eggs.

Reality Checks

The launch of MortgagePower Plus was accompanied with much fanfare, but the program soon turned into a disaster. Strategically, Citicorp made the mistake of preferring quantity to quality Technologically, the failure had to do with faulty modems and software modules.

Citicorp tried a procedure that is a usual for mortgage lending. The bank did not require borrowers to obtain private mortgage insurance and under-reserved for the resulting losses. Also, to facilitate the application process, the bank made low-document and no-document loans, in which it checked a borrower's credit report and a summarized employment history but not assets or income. At worst, the executives figured, the bank could sell a foreclosed house at a profit and recoup the loan.

MortgagePower Plus was supposed to work as follows: a real estate agent using Citicorp software types up an application at a personal computer (PC). The application is then sent over telephone lines to a mainframe computer in St. Louis. The computer judges whether the applicant earns enough to afford the loan, pulls the prospective borrower's credit history from a credit bureau, and sends back a decision.

In reality, the Citicorp system was often down, the credit bureau's system was down, or there were connection problems. Static on the line caused the PCs to disconnect from the mainframe. This problem could have been easily fixed with available error-detection and -correction software, but it was not. The logic of the software was wrong, too. Many credit-worthy applicants were denied loans. In addition, some brokers committed fraud by lying to the system. In other words, the shortcuts the bank took were overly risky.

Mortgage companies often sell the mortgage (i.e., they transfer the entire contract to another lender for a fee). These lenders are referred to as the secondary market. Citicorp hoped to make huge profits by using the system to originate loans that it would later sell. But the secondary market smelled trouble. It refused to buy mortgages from Citicorp and other banks that took such risks, unless the banks guaranteed to take them back when borrowers defaulted. The secondary market lenders' fears were probably realistic. In the first year that Citicorp used MortgagePower Plus, its overall mortgage portfolio showed a delinquency rate of 7%, five times the industry average.

MortgagePower Plus rejected 70% of all applicants, well above the bank's normal 35% rate. The 15-minute objective has never been met. Brokers waited hours for an electronic reply and often did not get one at all. After several trials, they were embarrassed to use the system in front of clients. Citicorp's management reduced the size of the mortgage unit and removed its responsibility to originate loans for later sale. The bank became a second-tier mortgage lender. In a way, the technological failure was a blessing; had the system succeeded, the bank would have made more loans, and the losses might have been higher.

Why Did the System Fail?

Undoubtedly, the idea of 15-minute mortgage loan approval (or denial) was great. However, several factors brought the system down. From a business point of view, Citicorp executives should not have relaxed evaluation criteria. This decision, in itself, has nothing to do with IT. Executives should have thought of a way to receive the usual, unabridged information for an applicant's evaluation without shortcuts. If 15 minutes were not enough, the system could have taken 60 minutes or three hours; the company would still have gained a competitive advantage.

The battalions of experienced professionals should have tested the code to ensure robustness. Any IS system, in particular a financial one, must be equipped with control to prevent fraud and unintentional mishaps. Careful incorporation of controls might have caused a longer time to market but would have eliminated many of the technical problems before implementation.

It seems as if Citicorp did not have any tests planned for Mortgage-Power Plus. It is extremely important to pilot an IS before a total implementation. Piloting would have allowed the company to test the system in real-life situations and would have saved it much embarrassment. None of the technical difficulties that the system experienced seem unresolvable with the current technology, but because the system failed, it would be extremely difficult, if not impossible, to regain the agents' confidence.

Using the framework outlined in the introduction, the system has not fulfilled its original purpose. Had it succeeded technically, chances are it would very well fit Citicorp's organizational culture. The company has positioned itself on the forefront of IT, and this innovation would not have been alien to the employees. The system, however, did not perform well technically and was therefore used only until the agents got sick of its malfunction.

SUMMARY

Saturn Information System and MortgagePower Plus are two examples of strategic information systems. Each was the result of a sound idea. However, an idea alone does not always do the trick. The idea must be carefully thought out and transformed into an IS that is both businesslike and technically sound. Surprisingly, the company that had implemented IT so successfully in the past and relied on its own homegrown professionals failed to implement a new system because of lack of business and technical controls.

The many cases of successful integration and few publicized cases of failure yield an outline of several reasons for systems integration projects' failure:

1. *Lack of shared vision.* There are three parties in the integration of IS: top management, the users, and the IT professionals. All three must share a vision of how the system will be used to significantly improve business processes or gain competitive advantage. Too often, top managers believe their role ends when they have earmarked the funds for the effort. This is a grave mistake. Top managers must want the new system, they must show continued interest, and they must champion the system. If an external consulting firm is to carry out the project, the consultants must fully understand the business processes in which the system will help.

2. *Failure of the developers to understand the system's requirements, or changing the requirements after the project starts.* Lack of communication between IT professionals and end users is a major cause for problems. Experience shows that problems may be minimized by involving the users in every major phase of the development or integration process. Practically, this is achieved by nominating user representatives to the project team. Not surprisingly, some of the more successful projects have been those whose teams had a majority of user representatives. Some organizations started to experiment with an even bolder idea: user-led project management, in which a user actually serves as the project leader. This approach may yet be discarded.

3. *Unforeseen and insurmountable technical difficulties.* System incompatibility is a common cause for time and budget overruns. Some IT professionals fear that revelation of such difficulties mar their reputation. However, they are bound, professionally and morally, to inform their clients, be it another organization or end users in their own organization. The key question for management is, Are we willing to continue the effort even with these overruns? If the system is determined to be still economically feasible, financial and time overruns should not stop the effort. The colossal failure of the confirm system is one example of a failure due to technical problems. Rather than notifying the client users of the problem and stopping for reevaluation, the developers ignored the problems and wasted time and money in futile attempts to patch the system. After three and a half years and $125 million, the system was abandoned. Chapter 4-5 discusses this in a case study.

4. *Underestimation of cost and completion dates.* This is by far the most pervasive problem in systems development and integration projects. IT professionals tend to underestimate costs and milestones. Almost always, the purpose is to offer an attractive proposal and get the job. When cost and time are running out, management sometimes simply cancels the whole project. Anyone who has been involved in any development projects, not necessarily just IS, knows how difficult it is to estimate cost and lead times for these projects. Therefore, it is so important to include large enough cushions on which to fall back.

5. *Poor testing procedures.* As the MortgagePower Plus case shows, rushing the implementation of an IS may prove lethal. A carefully thought-out test scheme must be integrated into the project. Each phase must conclude with a successful test of its respective system module or component. The AMRIS Confirm project and Bank of America's TrustPlus project were continued despite clear flaws. Both systems were abandoned after millions of dollars were invested. Shortcuts are extremely risky, especially when several systems are to be interfaced. The project team must list the tests that the different modules will have to pass before contingent activities take place. The list should be prepared along with the logical design of the system. Once on the list, no test should be skipped.

6. *Poor project management.* IT professionals often lack project management skills. PERT and Gantt charts are either not used properly or not used at all. Poor project management can also be attributed, to a certain extent, to the growing popularity of prototyping, which inherently does not follow a project plan as meticulously as does a formal systems-development life cycle. Beyond knowledge of project scheduling and tracking techniques, project management includes clear and frequent communication between team leaders and their teams, and team leaders and the project manager.

Avoiding the above pitfalls does not eliminate project failures altogether, but it may minimize them. Managers must be especially sensitive in this age of outsourcing. Returning to the notion of systems integration, it is about time managers think in terms of integrating systems with people and business processes rather than integrating systems with systems. It comes with no wonder that some of the more experienced IS professionals started to speak of business systems rather than information systems.

4-3

Case Studies of Terminally Ill Integration Projects

R. JACK MUNYAN

After witnessing and participating in years of systems growth and after gaining years of experience with computer information systems, why do so many organizations continue to enter into systems integration projects that fail? To express this same question in a slightly different way, why are so many organizations critical of their information technology (IT) in its ability to deliver integrated solutions on time and within budget? As discussed in Chapter 3-3, many of today's corporate systems evolved without the benefit of a vision and blueprint of how they would be linked to either the business or other technologies. However, in spite of numerous signals given in the past, a pattern of mistakes is occurring in the development of many computer information systems today. It seems as though the only thing organizations are learning from history is that they do not learn.

There is no single answer to why computer information systems projects fail but rather a combination of many factors that span the spectrum from leadership, reporting relationships, corporate attitude, and technology to internal management practices. This chapter explores the many dimensions of systems integration projects and their reasons for failure in the context of the 10 symptoms of the Terminally Ill Integration Project (TIIP), as discussed in Chapter 4-3. However, a TIIP can also share a number of characteristics with a runaway development project.

A FEW NOTABLE EXAMPLES OF TIIPS

Twenty M.B.A. students at the University of Tampa, Fla., performed a research study of published information to examine evidence as to why computer integration and systems development projects fail. The study reports findings of more than 50 organizations, and the students quickly realized that many such failures would not be reported for obvious public relations reasons. Nevertheless, many illustrations were found to support the symptoms of TIIPs.

The following is a sample of apparent system project failures with some pertinent facts as to why they did fail.

AllState Insurance

In 1982. with software from Electronic Data Systems (EDS), AllState began to build an $8 million computer system for total automation of the company. The target date for completion was December 1987; the target cost was $8 million. Some $15 million later, AllState had a new project consultant, a new deadline, and a new cost estimate of $100 million. Although there were many explanations and reasons for this type of failure, one of the principal causes was builders of computer systems and management have not adjusted their deadlines to reflect the increased complexity of computer projects. This failure falls under two categories listed in Exhibit: unrealistic or mismanaged user, sponsor, or senior management expectations and ineffective or inadequate project management practices.

Blue Cross & Blue Shield United of Wisconsin

In late 1983, the company hired EDS to build a $200 million computer system, which was ready 18 months later than the promised delivery date and produced major errors. Some $60 million in overpayments and duplicate checks were issued erroneously, and Blue Cross lost an estimated 35,000 policyholders. The categories are inappropriate or inadequate project sponsorship, inappropriate or inadequate project partnering and shared responsibility, and inadequate or ineffective project management practices.

The system's developers never adequately used checkpoints during the project to evaluate progress of the system properly. Many of the nontechnical executives assumed a passive role in this system's development. The leaders did not fully understand the complexities and effort level necessary to make this project succeed—until the costs had spiraled out of control.

The State of Oklahoma

In 1983, the State of Oklahoma hired a consulting firm to build a $500,000 system to handle explosive growth in workers' compensation

1. Inappropriate or inadequate project leadership.
2. Inappropriate or inadequate project sponsorship.
3. Inappropriate or inadequate project partnering and shared responsibility.
4. Unclear vision and ineffective project planning that is linked to the vision.
5. Unrealistic or mismanaged user, sponsor, or senior management expectations.
6. Inadequate or ineffective project management practices.
7. Inappropriate target go­to technology and applications.
8. Ineffective or inadequate project measurements, communications, and statusreporting.
9. Inappropriate or inadequate project team staffing, organization, and managementstructure.
10. Unidentified or underestimated project risk and operational exposure.

Exhibit 4-3-1. Ten Symptoms of Project Failure

---

claims. Two years and more than $2 million later, the system still did not exist. It was finally finished at a price of nearly $4 million.

A few symptoms in this case were unclear vision and ineffective project planning and unrealistic or mismanaged user, sponsor, or senior management expectations.

The data-processing people involved said they felt they had to keep the executives happy—even by promising what they could not deliver. Suppliers promised almost anything to get the business.

PRESENT-DAY FAILURES

One would think that these examples would have taught us how to keep a project under control; however, the results do not indicate this. Here are a few examples of the same types of problems happening in the 1990s. In the first two illustrations, a detailed analysis was made for each of the 10 symptoms shown in Exhibit 4-3-1 . In the other examples, a summary of the main problems is given.

The C-17 Air Transport

The mission concept of the C-17 transport was to improve US capability to rapidly reinforce and sustain combat forces worldwide. This program began in 1981 and in 1985 the government began the full-scale development phase using the Douglas Aircraft Company, McDonnell-Douglas Corp., as its prime contractor. As of early 1992, the C-17 development program was two years behind schedule and $1.5 billion over its 1985 cost estimate of $4.1 billion. Software development has clearly been a major problem; the Government Accounting Office has stated that "the C-17 project is a good example of how not to manage software development when procuring a major weapons system." General Bernard Randolph, commander of the Air Force Systems Command, said, "We have a perfect record on software schedules—we have never made one yet, and we are always making excuses."

The following is an analysis and classification of the symptoms of the terminally ill C-17 transport project.

1. *Inappropriate or inadequate project leadership.* Leadership and users did not fully understand the complexities of this project. The Air Force assumed that software was a low-risk part of the C-17 program and did little either to manage its development or to oversee the contractor's performance. The Air Force also underestimated the software development effort's size and complexity and did not fully understand the potential software risks. As a result, a number of mistakes were made in designing contractual and management controls.

2. *Inappropriate or inadequate project sponsorship.* The Air Force awarded a contract that gave Douglas control over software development, limited the Air Force's access to software cost, schedule, and performance information, and restricted the Air Force's ability to require corrective actions, even when critical software problems became evident.

3. *Inappropriate or inadequate project partnering and shared responsibility.* There was no integrated project team with equal and shared responsibility.

4. *Unclear vision and ineffective project planning that is linked to the vision.* By not performing the preliminary analysis and risk assessments required by Department of Defense (DOD) standards, the Air Force was unaware of how extensively embedded computers and software would be on the C- 17. Douglas did not prepare an overall system engineering approach to C-17 development and could not deliver a satisfactory computer program development plan. On the other side, the Air Force did not require Douglas to prepare a risk management plan and software integration test plans. These are required by government regulations when software is to be a significant part of a new weapons system's development. However, because of the mistaken assumption that software would not be an essential part of its project, Douglas was not required to— and did not—establish the type of software quality assurance program required by military standards. As a result, Douglas developed test schedules based on meeting certain pro-

gram milestones rather than on the actual times needed to complete the various tasks.

5. *Unrealistic or mismanaged user, sponsor, or senior management expectations.* There was definitely a feeling of oversimplification, because the Air Force assumed that the C-17 software development would be routine and low risk.

6. *Inadequate or ineffective project management practices.* There was inadequate training in project management, as the Air Force either waived or ignored many of the DOD standards and guidance for managing software development— in spite of Douglas's limited software development and integration experience. Douglas lacked the technical expertise and experience to develop a coherent computer program development plan. The program office described various drafts as incomplete, inaccurate, and lacking sufficient technical content to be used as a management tool. Douglas's failure to develop an overall system engineering plan was considered by many to be the root cause for most of Douglas's software development problems. Air Force management did nothing to correct this failure; in fact, the Air Force often lacked specific knowledge about software development problems as they occurred and as a result could not ensure that Douglas took timely corrective actions. In addition, Douglas did not develop adequate systems documentation, and now the Air Force may not be able to upgrade, test, and maintain the C-17 computer systems.

7. *Inappropriate target go-to technology and applications.* The Air Force has created an inefficient and uneconomical software maintenance situation by allowing C-17 software to be developed in a diverse assortment of computer languages. Also, Douglas, with the Air Force's approval, took a number of steps to shorten the testing of the system. Douglas deleted a few preflight integration tests and concurrently conducted some other preflight tests with systems integration tests occurring, despite the fact that integration testing of the individual subsystems was still underway. They also used unproved software specifications to develop and test software. These were all done because of continued software development problems and pressure to complete the project. These shortcuts result only in long-term additional hardware and software maintenance costs.

8. *Ineffective or inadequate project measurements, communications, and status reporting.* Both the DPRO and the DOD's inspector general reported that Douglas's approach to software quality assurance was inadequate and undisciplined. Specifically, they pointed out that Douglas did not adequately staff the quality as-

surance program, and the organization lacked the independence and management clout to ensure that the problems were corrected. In spite of these reports, there was the absence of milestone tracking and inappropriate monitoring of deliverables and work products.

9. *Inappropriate or inadequate project team staffing, organization, and management structure.* Douglas did not have enough trained personnel to plan and manage properly the software development activities of its many subcontractors. Douglas seemed to lack the organizational emphasis and commitment to good software development practices.

10. *Unidentified or underestimated project risk and operational exposure.* Douglas, with the Air Force's concurrence, took a number of shortcuts that have substantially increased the risk of not successfully completing software development and testing that easily could result in higher software maintenance costs.

Del Monte

Del Monte was forced to renovate the computer system in its processing plants because the old system was creating problems with an increasing frequency. The packing season is a 24-hour operation for employees, and during the busy time, Del Monte cannot afford to slow down and handle information systems problems. However, in 1992, during a $1.4 billion sales season, the company was forced to deal with both software and hardware problems. Although the problems surfaced gradually, they became critical at a very unfortunate time, when Del Monte was being spun off from RJR Nabisco. In assessing the 10 symptoms as they relate to Del Monte, the following analysis applies.

1. *Inappropriate or inadequate project leadership.* Del Monte's leadership did not fully understand the value and complexities of the information system, especially as they related to the shoe-string budget for the IS budget.

2. *Inappropriate or inadequate project sponsorship.* Senior management did assert ownership responsibility with respect to the existing information system but failed to integrate this importance into their overall business plan.

3. *Inappropriate or inadequate project partnering and shared responsibility.* There was absolutely no partnering involved in the information system before the problems became critical. At this point, the IS group received the necessary attention.

4. *Unclear vision and ineffective project planning that is linked to the*

vision. Planning was not a symptom as much as bad timing was, which was unfortunate for a computer failure accompanied by software problems during the busiest time of the season.

5. *Unrealistic or mismanaged user, sponsor, or senior management expectations.* Again, it would be difficult to have predicted the combination of failures that Del Monte experienced, because its management did not expect too much out of the existing system.

6. *Inadequate or ineffective project management practices.* Management did not give high priority to these systems as is evidenced by their low expectations and low budget.

7. *Inappropriate target go-to technology and applications.* An analysis of the problems of the existing system determined that the existing system was incorrect for Del Monte's business needs. In addition, variances were not tracked and analyzed until the problems became critical.

8. *Ineffective or inadequate project measurements, communications, and status reporting.* There was a definite lack of leadership interest and curiosity. Also, there was an inappropriate communication forum to provide needed feedback.

9. *Inappropriate or inadequate project team staffing, organization, and management structure.* The project team for Del Monte's IS department appeared to be sufficiently staffed for the existing system; however, Del Monte decided to outsource much of the new system work.

10. *Unidentified or underestimated project risk and operational exposure.* Before the RJR spin-off, Del Monte had experienced system problems but decided against any action. The company could have weighed the potential risk of doing nothing against the risk of reengineering the information system before the major problems occurred, but management chose to do nothing without weighing the risk.

Florida Department of Health and Human Services

The State of Florida purchased a supercomputer for the Department of Health and Human Services (DHHS) in the late 1980s for $108 million. The computer was bought in an effort to use information-processing technology within the DHHS. Department executives believed that the computer would dramatically reduce human error and make department workers much more efficient. Preliminary studies pointed to estimated savings of $105 million annually through faster work performance, less paperwork, and a higher rate of child-support collections.

However, these systems never lived up to such lofty performance standards and have instead turned into an unmitigated disaster. The wake of this disaster has brought criminal probes and numerous lawsuits, high turnover rates among department workers, and massive errant payments distributed by the computer. It has been estimated that there was a computer error rate of 18.6% in the food-stamp program, amounting to over $300 million in errant payments to recipients and estimated millions of dollars in future federal fines.

The causes of the disaster seem to fall into almost every one of the 10 symptoms. Many press articles, however, point directly to mismanagement, which occurred repeatedly throughout the history of the project. State officials did not know what they wanted before ordering the system, allowed contractors to get what they wanted, failed to ensure that systems personnel were adequately trained, and ignored early signs of system problems. Management and staff began avoiding responsibility after things began to go wrong and then started to blame each other for the problems.

SimuFlite Training International's FasTrak System

SimuFlite Training International, based in Dallas, provides advanced training for pilots and mechanics operating corporate, commercial, government, and military aircraft. The company spent $10 million and three years in developing FasTrak, the computerized personal education system for pilots ground school. New management studied the problem and discovered three areas that caused the FasTrak system to fail: failure to follow proven steps in instructional design, lack of input from the user, and outdated technology. The systems designers also made a fatal flaw in assuming that pilots were comfortable with using computers. They thought that because pilots operate highly complex and technical machines, they would not resist using computers—but they did. Also, the FasTrak technology was obsolete when it was delivered, hopelessly locked into 1982 technology. The company finally had to redesign the system based on input from their customers, at an estimated cost of $2 million.

United States Army Tactical Command and Control System

The US Army has tried for two decades to develop a standard battlefield computer system. The latest attempt is the Army Tactical Command and Control System (ATCCS), three parts of which are in serious trouble. One of the control systems, the maneuver control system, has been installed, but the data that it produced was neither accurate nor timely. The cost of this program that does not work is $1.3 billion.

All three parts of the ATCCS failed for similar reasons. The developers failed to consider the users' needs, and the development phase then became the project. The project's basic premise was flawed but, as the problems occurred, they were not addressed. Instead, the tendency was to sweep them underground and push ahead with a faulty system.

Pinellas County's 911 Dispatch Center

Pinellas County, Fla., spent more than $1.2 million for EAI Systems/Bell Atlantic software and maintenance. EAI was an exclusive software designer for McDonnell-Douglas and initially operated without a county contract. Relations between the county and EAI reportedly deteriorated to the point at which the 911 center doors were locked to keep the EAI programmers out. Programmers had been coming in unannounced and dumping into the county's computers new software that had not been tested fully, and some of these programs locked up the computer system. At the end of 1993, because of seriously strained relations between the county and EAI, the county was forced to pull the plug on the EAI Systems/Bell Atlantic contract.

PacifiCare Health System

PacifiCare Health Systems, Inc., in California, is a health maintenance organization (HMO) that wanted to replace its host/terminal sales and marketing systems with what was intended to be a more sophisticated and useful system. One of the first problems PacifiCare encountered was network overload. Programmers had underestimated the complexity of this type of operation, and overloading has been a continual problem throughout the installation and testing of the new system. Another problem that proved to be much more expensive and time-consuming than anticipated was the necessity of training and education. An analysis of this situation reveals that insufficient research was done before purchasing the system, the complexity of the installation process was underestimated, and employees were insufficiently trained by both the consultants and the company. Many of these problems could have been avoided with better planning and implementation.

Texas State Bank

Texas State Bank in McAllen, Tex., decided in 1988 to switch from its existing system to an EDS system. EDS stated that it had sufficient understanding of the current system to convert the bank to the new system in three months. However, the conversion and the postconversion services were woefully inefficient. The bank now claims it missed two bank acquisitions and lost about $46 million in customer business due to downtime and inaccuracies in customer accounts.

Hillsborough County, Fla.
The county spent $2.4 million on a system that it never used. It attempted to sue a nationally known consultant for millions of dollars to save face and recoup millions of dollars lost in an embarrassing fiasco involving a computer purchase. Early in 1994, four years after the initial suit, the county attorneys were prepared to settle the case for free advice on a new system from the consultant—but not for millions of dollars.

The Government Accounting Office's 1993 Report
Although unnamed in the report, three states spent almost $30 million trying to develop computer systems that were canceled after years passed without a workable design. Another state cannot bring its $51 million system online because it failed to plan for how workers would use it. Another report says that faulty computer systems at the New York Stock Exchange may have contributed to the stock market crash on October 19, 1987.

Cleveland School Districts
Because Cleveland, Ohio, school officials lacked expertise and centralized control and relied too much on vendors, they bungled millions of dollars in wrong technology purchases.

London Ambulance Services
This organization made countless mistakes along the road to its eventual collapse. Managers were reportedly under great pressure to get a computer-aided dispatch system running in an unrealistic time frame. The computerized ambulance dispatch system collapsed primarily due to an overcomplicated system and incomplete training of the control staff and ambulance crews.

The Federal Aviation Administration (FAA)
The FAA acknowledged that its troubled $5.1 billion program to replace the country's air traffic hardware and computer system had severe problems and announced in March 1993 that there would be a 14–month delay in the project and major changes in its management.

I/S Analyzer
Organizations are realizing that their IS must have better controls to reduce their exposure to data inaccuracy, security breaches, system failures, and misalignments with business objectives. Bankers Trust Co.'s implementation of a system for its capital markets area illustrates a case in which the development team was faced with pressures to deliver immediately. This appears to be another example of mistaken expectations. Other organizations with similar computer system problems include Massachusetts Mutual Life Insurance Company, Motorola, Inc., Eastman Kodak Co., and Bank of Canada.

North American Aerospace Defense (NORAD) Command
The long-overdue upgrade of two major NORAD computer systems will be delayed seven to eight years and cost an additional $207 million due to software problems.

Exhibit 4-3-2. Other Organizations with Terminally Ill Projects

EDS used a system that it had recently acquired and with which it was not thoroughly familiar. On the other hand, Texas State Bank did not investigate sufficiently before it hired EDS and did not play a sufficient role in testing the new system before it was installed.

There are many other accounts of terminally ill projects and undoubtedly many, many others that were unreported or are suppressed by their respective organizations. Some of the other projects researched in the course of writing this chapter are shown in Exhibit 4-3-2.

SUMMARY

Even though there is no single answer to why computer information systems projects fail, there seems to be sufficient evidence found in this research to indicate common threads that are present in the vast majority of terminally ill projects. These common denominators are a lack of:

1. Top management total commitment and involvement.
2. An architecture for information technology.
3. A close team effort between users of the systems and IT personnel.
4. Communications and user involvement from the very beginning of the project.
5. Training.

When one factors in the nature of the business, the organization of the business, the human element issues, the corporate traditions and culture, and last but not least, the economics, one sees that finding the appropriate solution for successful systems integration projects is not easy, but perhaps the findings from this research will help organizations learn from history.

4-4

The Benefits, Advantages, and Maximization of Systems Integration Success

STEVE HARKOLA

PETER DAVIDSON

In most organizations, information systems were built in isolation, designed for the specific operational activities required by different areas of the organization. Often these applications were developed in different languages, using incompatible data base systems and various hardware platforms. Over time, some interfaces were developed between systems that required communications, but differences between the platforms often resulted in information inconsistency and duplication of effort.

This is the legacy that most information technology (IT) organizations must work with. The motivation to further integrate systems is driven today by the demands of the marketplace and the necessity to use IT as a competitive advantage. This chapter introduces the benefits of integration and explores the specific organizational advantages. Various methodologies to maximize the effectiveness of integration reengineering are also introduced.

THE BENEFITS OF SYSTEMS INTEGRATION

Integrating systems is both time-consuming and expensive. However, in the long run, it can be the right strategic decision. The business case to perform system integration activities must be developed to justify the capital and manpower resources necessary to accomplish tighter linkages. The extent to which integration is carried out is a product of this business case. One enlightened bank, for instance, consolidates acquired or merged banks into one data center using one common set of business applications. When other banks merge, they interface their systems by linking multiple remote data centers. The bank that performs the ultimate integration or consolidation reduces the cost structure inherent in maintaining multiple data centers, even though performing the necessary conversions may be initially more expensive. In addition to cost reduction, other integration benefits include productivity improvement and increased competitive advantage.

HOW NONINTEGRATION INCURS COSTS

Cost-reduction opportunities surface when viewing information systems that are scattered across multiple hardware and software platforms. These systems inherently have higher costs than integrated systems. First, each system requires costly maintenance and support for its hardware, software, and operating system, which require periodic upgrades and may require specially trained personnel. Backups need to be run periodically, and system performance needs to be tuned. Many of the costs are reoccurring and are not sensitive to usage, such as costs for operating systems or software site licensees. Companies that contract out for their maintenance may benefit from some economies of scale; however, they are still paying more than if they had fewer systems. If companies do the work in-house, they still need to pay the training and staffing costs for each platform. Consequently, reducing the number of systems supported directly reduces maintenance costs.

Development. Development costs are greater for scattered systems. Each operating system uses different commands and file-access methods. Even implementation of standard operating systems, such as UNIX, depends on the particular vendor. Development tools and languages are often tailored to the platform they run on, which limits portability of programs across systems and requires a more expensive MIS staff. Companies require MIS personnel experienced with the variations of each platform, often at additional costs. Many new technologies, however, have better processing capability and offer features, such as graphical user interfaces (GUIs) and open operating systems, that decrease the cost of developing new systems. Because an increasing amount of hardware is less proprietary, more suppliers exist, reducing the cost of processors, terminals, printers, and other peripherals. Pricing is more often than not based on concurrent use, better reflecting the actual utilization of products. Reduced cost allows more bang for each IT dollar.

Training and Support. Organizations pay more to train and support their end users for each platform in operation. Users require additional access, as user interfaces are not consistent across platforms. Even though some attempt can be made to keep application menus and functionality consistent, programs are still constrained by the inconsistent input/output facilities of the various platforms.

Communications. Communication costs are also greater for nonintegrated systems. Various hardware platforms often use different types of communication controllers and printers. Many older mainframe shops using a proprietary cabling topology find that introduction of open systems necessitates a completely different wiring scheme. In addition, the connection between old and new systems by definition forces interfaces rather than integration. Users may require multiple terminals or higher-cost PCs to allow them to switch between applications. Data that needs to be transferred between computers may require special communication lines, protocol converters, and software called "middleware" to provide connectivity.

HOW INTEGRATION LOWERS COSTS

Integrated systems can greatly improve productivity. First, data that is available at a central location is much easier to work with. Users do not need to waste time consolidating information from multiple sources, because they learn and use one system to get the information they need. As employees move to various departments within the organization, they find it much easier to become familiar with the information necessary to do their jobs.

Data Quality and Integrity. Today, the problem of determining which version of the truth to report has reached the boardroom in importance. Data managed on a single data base is more consistent and accurate than data spread among applications. Modern data base systems can enforce data integrity rules, and data is less prone to timing errors that arise out of information transfers across platforms. Rework resulting from inaccuracies is reduced. Decision making in the organization improves, because data is consistent, and users have more confidence in it. Users can access the information quickly and consistently. In addition, the MIS organization supports fewer data base systems, which can improve their service level. Many companies are "normalizing" their data bases in an attempt to eliminate redundant or duplicate data items.

Improved Performance. Systems that are integrated are easier to control and run. Maintenance operations, such as data backups, are easier to coordinate. Hardware performance tuning is simplified, because there are fewer system components. Processing capability is better allocated across applications. Scheduling to avoid conflicts among applications that depend on the same data can be simplified.

HOW INTEGRATION INCREASES THE COMPETITIVE ADVANTAGE

Integrating systems can lead to many competitive advantages. First, companies have more timely and higher quality information about their operations. Interface time lag is removed, and the data source is known. Integrated systems also provide greater flexibility. The greater the number of systems involved in a company's operations, the harder it becomes to create new applications or change existing ones. Integrated systems, particularly those tied directly to production technology, allow companies to develop applications faster. In many companies, systems development is often the bottleneck slowing down new- product production, particularly when the products are information. Companies that have a better understanding of the changes required to implement new products and services ultimately reduce their cost of producing new products.

Technology has advanced considerably since many companies first designed their information systems and processes. At one time, the mainframe was the only vehicle for large-scale information systems; however, not all mainframe systems were equal to all tasks. Some platforms were better suited for data management; others were better for operational control.

Today, companies have many more choices. Powerful PCs are abundant. Sophisticated networking software and hardware exists. Applications are developed using distributed technology and client/server models. Operating systems have GUIs, simplifying ease of use and learning. Technology advances have spawned a new breed of employee known as "knowledge workers." These empowered users use information to be more productive in less time and are largely responsible for the success of recent downsizing efforts.

Data. Data in an organization is increasingly being referred to as a corporate asset supporting competitive advantage. Often, the organization needs to combine data from various platforms and data bases to answer business questions. The extent to which data exists on separate systems affects the organization's ability to produce consistent, accurate, and timely information. Data on multiple systems is often redundant. This increases the organization's storage costs and reduces the overall performance of system resources, because changes to the data must be made on multiple systems.

In addition, data that has multiple locations is harder to secure. Controls must be maintained on each system, and the level of these controls differs from platform to platform. Finally, applications developed on nonintegrated systems have a strong data dependency; any changes made to the data often necessitate corresponding changes in the application's source code.

Data Bases. Data bases are an excellent candidate for integration. For example, information about manufacturing production costs is often scattered among the hardware controlling actual production. Consolidating the information provides new insights into an organization's total cost structure, helps find potential for economies of scale, decreases work in process, and helps allocate resources. Such customer information as historical purchases, product requirements, delivery requirements, and financing terms helps identify new product markets or serve existing ones more effectively. The more customer and supplier information that an organization can keep and consolidate, the better one can manage for increased profitability.

Data Warehouses. Data warehouses are growing in popularity as companies move from clerical operating systems to decision support and trend analysis. Much of the procedural transaction data locked in older systems can be extracted into a separate repository designed for fast, easy retrieval of summary statistics. Integrating core data and data transformations on a separate platform minimizes the production system impact of CPU and I/O intensive queries. Banks, for example, can analyze the customer patterns and preferences inherent in detail ATM traffic to design new financial products. This is competitive advantage through integration. Executive information systems (EIS) are an offshoot of the data warehouse. The objective is to arm senior management with critical company performance metrics to support strategic and tactical decisions.

Major competitive initiatives, such as cycle-time or lead-time reduction, supply-chain management, and just-in-time (JIT) inventory management, ultimately will drive integration efforts. The question for IT is whether to wait for the top-down direction inherent in such programs or to begin proactively the infrastructure surgery necessary to promote best practices. In many cases, pilots illustrate how systems initiatives can in fact drive business change for competitive advantage, depending on the culture of the organization and whether IT has been empowered to provide not only technical leadership but business process improvement know-how, as well.

ORGANIZATIONAL ADVANTAGES OF SYSTEM INTEGRATION

Integrating and reengineering processes have the greatest possibility for organizational improvement. Processes affect all aspects of an organization: people, technology, data, products, and customers. Processes define the business.

Today's business processes are deeply rooted in the past. When organizations were young and rapidly growing, processes were created to handle the specific tasks facing the organization. Many of these needs were short term or focused at a limited customer base. Faced with rapid expansion, many companies naturally evolved to a pyramid organizational structure, which allowed for easy growth. As organizations became more hierarchical, responsibility for existing processes split across functional areas or divisions. Over time, processes also changed. Because responsibility for processes was split across managerial areas, fewer people involved in the day-to-day operations knew the total picture. As a result, adaptations resulted less in the best interest of customers and the organization and more in the interests of the people involved in the processes.

With less overall understanding of the various processes, individuals added new tasks and procedures to optimize operations. Many of these tasks became new processes in themselves, often adding little value for the customer. Process ownership, understanding, and control diminished. Problems occurred, particularly at "hands-off" steps in the process, in which responsibility crossed managerial areas. Processes became outdated, globally inefficient, and added less value to the company. Today, a significant portion of white-collar effort adds little or no value to the product. Companies have much to gain from reengineering processes.

Target Areas for Reengineering

Process should have one major purpose: delivering an affordable, quality product to an external customer on time. Every step of a new process should be geared to adding value for the customer. Successful reengineering starts with a clear understanding of who the process is intended for and who the customers of a process are. Many of the tasks in existing processes serve as management controls. Well-engineered processes require fewer traditional controls, because workers understand the global impact of their assignments. These controls can be reduced, consolidated, or eliminated from new processes, resulting in significant cost savings.

Most reengineering efforts result in elimination of unnecessary tasks, and several jobs are combined into one, which increases the employee's individual responsibility and the need to decentralize decision making. Successfully reengineered processes give the people involved in the process much more autonomy and decision-making ability. Decision making becomes part of the job. By combining jobs, much of the costs of rework, coordination, process control, and errors are reduced. Decentralized decision making reduces process delays and overhead, making a process more responsive to the needs of a customer. Customer problem resolution improves, because employees are empowered to make local decisions.

Benefits. Reengineered processes take advantage of new automation and information technology. The easy dissemination of information about a customer and the state of the process allows for employees to make better and faster decisions. Computers themselves can make decisions and control aspects of the process. New automation tools have eliminated much manual work. However, technology is not used simply to automate the tasks of existing processes, which only increases the efficiency of unnecessary steps. Technology serves as an enabler, allowing companies to successfully reengineer poorly performing processes.

There are other benefits of reengineering processes. Redesigning the key business functions from the ground up eliminates unnecessary constraints. These constraints often evolve from problems in the existing processes but have no place in new processes. Redesigning the processes helps an organization reassess and relearn what is important to the operations and eliminate the rest. Often, employees take an active role in the design efforts, gaining a new sense of ownership and involvement in the organization. Processes redesign is an excellent opportunity for them to resolve process issues under which they have been laboring for years. Not only do new processes better reflect what is important to the organization, but the people involved feel that what they do is directly contributing to the organization's success.

DEFINING BUSINESS PROCESS REENGINEERING

Business process reengineering (BPR) is a comprehensive, IT-driven, substantial undertaking by an organization. Rather than a continuous quality improvement program whose objectives are small increases, BPR strives for dramatic improvements in performance, productivity, profitability, and expense reduction. BPR's mission is order-of-magnitude improvement within the essential functions an organization performs. In fact, one of the cornerstones of BPR is the premise that significant performance improvement cannot happen from incremental change.

BPR is undertaken by organizations that realize that the way they do business today will not work in the future. BPR programs start by identifying the essential processes of a business that deliver products and services to customers, such as production or inventory management and logistics. Next, management assesses the performance of the processes using conventional measures such as throughput, cycle time, work in process inventory, and unit costs. Through these measures and benchmarks of the performance of other companies, management identifies processes that are most in need of improvement. Finally, senior management sets specific objectives for the level of improvement they expect.

Tools That Support BPR

Business tools can both enable systems integration and substantially improve business performance. The choice of tools depends on the organization's needs and the scope of the changes undertaken. Systems reengineering is not inexpensive; therefore, before undertaking any integration or reengineering process, the organization must decide if the benefits outweigh the often substantial costs. However, depending on the company's ability to compete and the aggressiveness of its competitors, the organization may not have the option to wait.

Joint Application Design (JAD). A methodology designed to maximize user involvement in the planning and redesign effort, JAD concepts can be readily applied to BPR. A facilitator is established to lead a designated team of content experts through intensive, day-long, hands-on workshops to generate the following: mission statement, reengineering objectives, current systems and process diagrams, scope of change, alternative solutions, make-versus-buy analyses, organization changes, and critical success factors. Critical to success is top-management sponsorship, including a periodic review mechanism and metrics to evaluate progress and successes.

Rapid Application Development. Pilot or prototype implementations on a small scale are recommended to test solution concepts before major commitments in resources or capital are expended. Pilots may use RAD techniques to create systems quickly. RAD products typically generate code from higher-level process definitions, for users to visualize the solution early on in the project life cycle. New business rules and data dictionaries may then be reused as part of a more formal in-house or package system solution.

Computer-Aided Software Engineering (CASE). An entire line of software engineering tools is marketed under this generic name. Consisting of a suite of computer based tools used to automate the many tasks of system development, CASE can dramatically simplify development of new systems. The tools are based on a standard software development methodology and allow users to create outlines of the new system using such concepts as data flow diagrams or entity relational diagrams. Users then translate these into data base designs and application programs. CASE supports all aspects of system development, from conceptual design, to screen and report generation, to code development.

CASE tools have many productivity features, such as report and screen prototypers, and powerful graphical tools, such as entity-relation diagrams, that simplify diagram creation. They keep all the information about all aspects of the application in one central repository. CASE simplifies coordination of development when many people are working on the system at the same time. The tools can enforce consistency and sound application-design principles, resulting in higher quality applications. They can translate all the information in the repository into source code in such conventional languages as COBOL. They can also generate detailed system documentation.

Some CASE products even feature project management tools. Today, there are CASE tools available for various hardware architectures.

Application Software Packages. A new generation of application software packages is on the horizon. Today's packages contain not only program code but a business model and rules as well. Most provide parameters to allow non-code-base modifications and customization. The latest concept provides for the marketing of only the business model in the form of templates. Either the business model is adapted to the specific requirements of the buyer or a library of best business practice templates is purchased off the shelf to force the necessary paradigm shift. The model is then fed into a code generator to create the execution environment on whatever hardware and operating system platform is specified. No changes are made directly to the code; rather, enhancements to accommodate business change are made to the model by business analysts. Traditional programmers are necessary only to develop or maintain older legacy systems or auxiliary applications. To take advantage of new hardware advances, the code is merely regenerated from the model and ported for recompilation on the target processor. The result is nonproprietary and self-documenting and resides directly in the hands of the business users.

Relational Data Base Management Systems (RDBMSs). Designed to simplify how the data is accessed by application programs and users, RDBMSs are comprehensive tools that manage an organization's information resources. The systems create logical views of the data base, which are used by applications and users to navigate through the data. These logical views are independent of how and where data is physically stored. The RDBMS displays all the information in the data base in simple two dimensional tables called relations. Each row of a table represents information about some aspect of the business, such as an order, a production schedule, or an employee. Tables are related to each other by common information, called keys. Applications communicate with the RDBMS through fourth-generation query languages (4GLs), such as SQL (Structured Query Language). These query languages are simple to learn and quite flexible. All the work of translating the simple data requests into the steps needed to retrieve the desired data physically is handled by the RDBMS.

The systems feature data dictionaries, which give users detailed information about all the data an organization maintains. Data can be accessed through a consistent, well-documented user interface. Access to the information can be controlled on a per user and per application basis. The RDBMS provides a physical separation of data from applications. Data is no longer associated with specific applications, reducing costly redundant files, and eliminating the inconsistency of data across applications. Application programs no longer need to control how data is stored or controlled, reducing their development time. When the information needs of the organization change, modifications can be made to one central system. Any changes required in applications programs are reduced, if not eliminated.

Object-Oriented Data Base Systems. Object-oriented data base systems represent one of the latest technological innovations. They have many of the same features as traditional data base management systems, such as centralized control of information, data dictionaries and security control, but object data bases also include operations call methods that can be performed on the data.

Operations are performed on objects by sending them a message. For example, an object may include basic information about a customer. To display the information on a screen, the user sends the object a message to display itself. The data base system then executes the code needed to paint the information on the screen.

There are at least three advantages of object-oriented data bases over traditional systems. First, the data stored by these systems is much more generic than in a standard data base system. For instance, objects can be pictures, sounds, and multimedia video, as well as standard text and numbers.

Second, application code developed for object-oriented systems is much more reusable. Because all the methods for an object are stored right with it, anyone using that object in another application does not have to rewrite any of the operations. For example, the same customer record is to be used in two applications; the user wants to display the information on the screen the same way. If a method to display the screen is defined, that information stays with the object rather than the application. To display the screen the same way in another application, the user simply includes that object in the other application.

Third, objects can be derived from other objects and inherit methods. A customer record, for example, needs to be printed in two different applications. However, the user needs to include one more line of information in the second application. He or she can define a new customer record object that is based on the first one; it automatically inherits all of the methods previously defined. To add an additional line of text, the user simply creates a new method that first calls up the original print method and then adds the additional line of text. Object-oriented systems can substantially reduce application development costs if the same information and functions are common to many of your applications.

These systems not only save development time but help to provide a consistent user interface across the application. Using these tools, organizations accomplish system integration improvements in a number of ways. Ongoing maintenance in most IT organizations represents a significant portion of the total data processing budget. Users typically request application changes without regard for integration issues. They may only know that it takes an inordinate amount of time or costs an excessive amount of money to process the change.

It is important to note that the driver for systems integration is usually not user management. With the exception of customer-driven strategic improvement, IT has the difficult task of initiating and justifying integration efforts. These may be embedded in maintenance activities or identified as separate IT-driven projects. Technology solutions have a useful life that used to be measured in periods of 10 to 15 years. Today, three to five years is a long time.

IT integration projects are in full swing in many companies that have pushed legacy systems to the breakdown point. These efforts include application rewrites, data base management system replacements, operating system upgrades and packaged software implementations. All require conversions of one form or another and are disruptive to varying degrees. Package implementations are particularly attractive because they represent paradigm shifts by definition. Embedded in most packages are canned business rules or the ability to redefine business rules. Not only is the technology updated but the opportunity to straighten out the cow paths is implicit. Whereas packages may invoke vigorous internal debates at setup time, the beauty of off-the-shelf software is the ability to take advantage of best-of-breed business practices.

In addition to new development, older technology is an obvious target for integration. Up through the 1980s, mainframe technology provided considerable economies of scale. Very large organizations still depend on large hosts and will continue to for a long time. New technology, particularly distributed networks, offer ways of increasing computing power more cost-effectively. Some companies have adopted a "surround" strategy to take advantage of distributed computing while maintaining legacy applications in place.

Unfortunately, maintaining the existing high cost of the mainframe infrastructure and adding additional network applications is driving data-processing costs in the wrong direction. More important, these costs are in many cases buried in user budgets and not consolidated along with traditional IT expenses. For many, the elimination of the mainframe requires a leap of faith beyond the risk profile of many management teams. For those that do, not only is the base infrastructure capital expense magnitudes cheaper, but the incremental expansion of capacity is far more cost-effective. The major complaint of IT managers today is the apparent lack of controls that have been developed over the last 30 years in the mainframe environment as applied to the UNIX or various open system structures. This gap has been reduced to the point at which it is almost not an issue.

PRIME TARGETS FOR SYSTEMS INTEGRATION

Technology candidates that are expensive to maintain, require constant data transfers with other platforms, do not have the ability to expand cost-effectively, use outdated operating systems and languages, do not have the same level of support as other systems, and require proprietary hardware and software are particularly suited to integration. Older systems typically communicate with other applications through a myriad of flat files complicating error recovery. What these systems do contain is the accumulated knowledge and business rules developed over a significant span of time. Companies may not have the tools, resources, or time needed to create the flexible and adaptive information systems that are needed to remain competitive in their respective industries.

In general, the best opportunities for applying any of the tools of systems integration are operations that are critical to the success of the organization. They need not be the most resource intensive, but they often are. Nor must they be contained in any one functional or divisional area of the company. The bottom-line objective of systems integration is improvement in profitability, through any number of operational changes. For improvements in operation to have a significant financial impact, the process must affect company performance significantly in the first place. Some of the characteristics of systems, applications or operations that are viable candidates for integration are:

- Processes that cross managerial boundaries.
- Processes that are underachieving in terms of standard performance indicators, such as throughput and unit costs.
- Systems or operations that use outdated technology.
- Hardware platforms that are difficult to interface.
- Applications that are archaic, problematic, or costly to change.
- A high degree of offline end-user applications to augment the formal system.
- Systems, operations, or applications that are undocumented or difficult to learn.
- Applications with extensive information exchange, redundancy, and re-keying.
- Processes with a high ratio of reconciliation checks and balances.
- Processes that always have exceptions.

To maximize the effectiveness of the systems integration process there are a number of proven steps that increase the probability of success. These steps, though not specific to systems integration, provide a framework upon which to organize the teams associated with reengineering tasks. More importantly, management and project sponsors responsible for fueling change in the organization can effectively build teams and set directives. Management must:

- *Establish companywide commitment and vision.* While smaller centers of excellence can be established within an organization, it is more productive to propagate companywide programs, such as total quality management (TQM) and JIT. Under these umbrella programs sponsors can justify new initiatives and mobilize organizational resources to achieve specific deliverables.
- *Understand current processes.* Too often, management rushes out with the latest popular solution in search of a problem to solve. Before automating a bad process, the cow paths need to be straightened out. Managers must interview customers and watch for extensive interfacing, arms'-length transactions and any non-value-added activity.
- *Emphasize people.* People, not processes, are the hardest to change. Organizational issues, internal politics, and informal systems must be addressed first, for process change to succeed.
- *Build high-performance teams.* Many projects are ultimately staffed with expendable staffers peripheral to the mainstream production operation. Successful projects are staffed by key personnel most essential to the day-to-day operation. Leadership is even more prob-

lematic in that a good operational employee may make a bad task manager. Competent project leaders that understand the product, company mission and internal workings of the organization are at a premium.

- *Leverage legacy applications.* Embedded in those old, tired systems that have been around forever are a wealth of business rules that are usually undocumented. Unless the technology is prohibiting change, legacy applications can be surrounded and integrated with new processes. At a minimum, standard cost-benefit analysis is required to justify wholesale replacement. In many cases the reengineered solution may be entirely procedural requiring no system changes at all.

- *Gain early successes through pilots.* If results cannot be produced at some measurable level within one year, management should go back to the drawing board. Pilots focus on a manageable scope area and serve as a model for success in subsequent larger undertakings. Smaller projects spawn innovation and serve as technology clusters that can be packaged and migrated to other areas of the business.

- *Ensure accountability.* Employee empowerment is great but without adequate controls can result in proliferation of individual agendas and corresponding wasted resources. Metrics must be established to provide measurable status against an established plan. Deliverable expectations should be defined. Reviews can be in the form of periodic walkthroughs and status reports in addition to mandatory evaluations by the project sponsor, including top-management visibility.

SUMMARY

Systems integration can lead to many benefits, many essential to the survival of companies competing in a worldwide marketplace. These benefits include cost reduction, improved productivity, and competitive advantage through technology application. However, the process of reengineering for improved integration must be approached with a responsive methodology, a well-defined scope, and a competent project team.

The organizational implications of process reengineering represent change. Change management dictates direct involvement by the employees affected, including support by top management. There is a well-defined set of tools available to facilitate business process reengineering and corresponding process integration. The application of modern information technology has improved the level of affordability so that even smaller organizations can benefit. Finally, this chapter has reviewed where to look for systems integration opportunities and the steps required to maximize the effectiveness of the integration reengineering process.

4-5

Ethics and Failure in Systems Integration: A Case Study

EFFY OZ

On September 26, 1992, a lawsuit was filed by Marriott Information Systems, Inc. and Marriott Corp. The plaintiffs accused AMR Information Services (AMRIS), Inc. and the AMR Corp. of intentionally concealing chronic project delays and cost overruns, refusal to provide Marriott with requested information, and making misrepresentations and false assurances to induce Marriott to continue funding and participating in the joint venture of developing a multimillion-dollar information system. The AMRIS chief executive officer (CEO) acknowledged his employees' unethical behavior, and a sum of $125 million was lost in this venture. This chapter chronicles this affair and identifies the lessons to be learned.

A consortium of Hilton Hotels Corp., Marriott Corp., and Budget Rent-A-Car Corp. subcontracted a large-scale project to AMRIS, a subsidiary of American Airlines Corp. The consulting firm was to develop a new information system called CONFIRM, which was supposed to be a leading-edge, comprehensive, travel industry reservation program combining airline, rental car, and hotel information. A new organization, Intrico, was especially established for running the new system. The consortium had grand plans to market the service to other companies.

Major problems surfaced when Hilton tested the system, and malfunctions lead to Intrico announcing an 18-month delay. The problems, however, could not be resolved, and three and a half years after the project started and a total of $125 million had been invested, the project was canceled.

In a letter to employees, Max Hopper, American Airlines Information Services CEO, said, "Some people who have been part of CONFIRM management did not disclose the true status of the project in a timely manner. This has created more difficult problems—of both business ethics and finance—than would have existed if those people had come forward with accurate information. Honesty is an imperative in our business—it is an ethical and technical imperative." Apparently, the clients were misled into continuing to invest in an operation plagued with problems in data base, decision support, and integration technologies.

Undoubtedly, software developers, as experienced as they may be, may legitimately run into technical difficulties. The questions information systems (IS) professionals should ask are:

- Does a failure of this magnitude have to happen?
- If the developers realize they are facing technical problems, should they notify the client?
- How severe should the difficulties be to warrant alerting the client?
- If management does not share information with the client, are the individual members of the project team expected to blow the whistle?

These questions are supposed to be answered by professional codes of ethics and standards of conduct. The purpose of this chapter is to present the case and examine how IS codes of ethics address the issues raised. It also tries to draw practical lessons for providers of IS services and their clients.

BACKGROUND

In 1987, a potential market—centralized hotel reservations—caught AMR's attention. As the company found out, only 20% of hotel reservations are made through a centralized service, whereas in the airline business 80% of the reservations are made through a central system, such as AMR's own SABRE. The company decided to take advantage of this situation in the form of a new, comprehensive system.

CONFIRM was the name given to an information system that was supposed to be the most advanced reservation system in the combined industries of travel, lodging, and car rental. The clients relied on the professionalism of the specialists who developed the highly successful airline reservation system SABRE. SABRE was a classic example of how an information system can gain strategic advantages for its user organization.

The ease with which travelers can make reservations is vital to the travel industry. Of the more than 85 hotel companies in North America, the five major national chains are Marriott, Hilton, Hyatt, Westin, and ITT Sheraton. Over the past 16 years, each of these chains installed a computer-based reservation system that provides information to travel agents throughout the world. Some chains developed their own systems; others hired outside vendors. The systems varied in efficiency and effectiveness. For example, Marriott's MARSHA has been recognized as one of the best in the industry, whereas Hilton's NORTH, which dates from the early 1960s, is inadequate and inefficient.

Airlines, too, have acquired reservation systems. The most notable are SABRE, developed by AMR, and APOLLO, developed by United Airlines Corp. SABRE, installed in 1976 and continually upgraded since, has gained acclaim as the world's most successful airline reservation system.

In 1986, AMR formed AMRIS, the information systems arm of the corporation. AMR Chairman Crandall hired Max Hopper to head AMRIS and offered him "a chance to combine running the SABRE business . . . and expanding it into other businesses, really leveraging it." AMRIS was to exploit its success with SABRE for business in other areas. Unfortunately, the success of one system does not always guarantee the good fortune of a more advanced system. What follows is a chronicle of the events that led to the CONFIRM disaster. The information is taken from the lawsuit filed by Marriott (i.e., Suit, in the references) and media reports.

THE CONFIRM CHRONICLES

On March 13, 1987, AMRIS representatives made a presentation to Marriott executives about a new reservation system they were going to develop. The system, named CONFIRM, would be superior to any reservation system in existence in the industry. The representatives claimed it would be a state-of-the-art reservation system meeting all business needs of hotels and car rental partners in the joint venture. According to the proposal, AMRIS, as a managing partner, would be in charge of the design and development of the system, and Marriott would pay for the effort and also input the necessary data.

The partners, hotels and car rental businesses, would use the system for their daily operations. In addition, they would join AMR in an effort to market customized versions of the system to other hotel and car rental companies for profit. AMRIS was to operate the data processing center of the system.

From May through August 1987, Marriott and other potential partners met with AMRIS executives to negotiate the deal. AMRIS representatives repeatedly assured the partners that CONFIRM would be superior to any current reservation system without being more costly to use. They also promised that the project would be completed in time to outpace the competition in the hotel and car rental industries.

On September 2, 1987, Marriott, a major partner in the venture, agreed to consider the AMRIS proposal even though it already had an excellent system. The company's vice- president emphasized: "Marriott is pleased with its current reservation system .... We have one of the best reservation systems in the industry in terms of both functionality and cost." Thus, he said, his company would join the venture if "the joint venture can develop a reservation system that is functionally richer than the system we intend to operate [and that Marriott costs] will be less than the costs to operate our proposed system."

The first three partners to the joint venture were Marriott, Hilton, and Budget Rent-A-Car, and they formed a consortium named Intrico in October 1987. In late 1987 and early 1988, technical representatives from the four partners started to plan detailed performance capabilities of the new system. On May 24, 1988, AMRIS issued a press release announcing the commencement of the CONFIRM design process. In the meantime, the Intrico partners were funneling large sums of money into the project. By September of 1988, Marriott alone had spent more than $1.45 million on the preliminary design.

In September 1988, after a year of negotiations, Marriott, Hilton, and Budget signed a partnership agreement with AMRIS. According to the agreement, the objectives of the joint venture were:

- To design, develop, operate, and maintain a new state-of-the-art reservations-processing system to be marketed worldwide.
- To design and develop interfaces with airline computer reservations systems so consumers could make airline, hotel, and car rental reservations through a single computerized system.
- To market the reservations system and other communication services to customers for a profit.
- To convert each of the partner's reservations systems to the newly developed system.

AMRIS was given the title Managing Partner, Development, of CONFIRM. The agreement made the company responsible for all aspects of the design and development of the new system. The four partners undertook to pay AMRIS $55 million for the development. Each partner was to appoint a professional team that would be stationed at AMRIS headquarters, in Dallas, Tex., so that the partners would provide input as to what functions were needed and also would test and evaluate the system as it was developed.

The agreement stated two phases: the design phase, to take seven months, and the development phase, which was to be completed within 45 months after the agreement was signed. Thus, the deadline was the end of June 1992.

The contract provided that the total expenditure to develop CONFIRM would not exceed $55.7 million. AMRIS warranted that it had "no reason to believe" that the development costs would exceed this amount.

The company also undertook to develop the system so that operation costs would be limited to $1.05 per reservation.

On December 30, 1988, AMRIS presented a "base design" of the system. Marriott claimed the functional specifications were not adequate. A 1992 internal audit by AMR's SABRE personnel stated that "these documents describe the expected functionality in general terms; they do not provide sufficient detail for a developer to understand what the user is expecting."

In March 1989, AMRIS declared that the functional and technical specifications were complete. Late that month, the company circulated a preliminary development plan. The plan was unacceptable to the partners. The next six months were devoted to revision of the plan. During this time, AMRIS executives reassured the partners that the system would comply with all the requirements, and that it would be ready on time.

AMRIS completed the design phase in September 1989 and circulated a proposed development plan for the partners' review. At this time, the company increased the price of the project from $55.7 million to $72.6 million. It also stated that the cost per reservation during the first year of full operation would be $1.30, instead of the original $1.05, and that would decline to $0.72 and $0.40 in the fourth and fifth years, respectively.

According to the partnership contract, the three clients/partners could withdraw when the development plan was presented, with a penalty of $1 million. The partners had to make the decision at this point. The per-reservation cost was crucial information in their decision-making.

On August 8 and August 15 of 1989, AMRIS representatives met with those of Marriott, Hilton, and Budget to review AMRIS's pro forma financial statements. Based on these statements, the client/partners decided not to exercise their option to withdraw. To Marriott, for instance, the value of the project declined by $1 million, but still promised a net present value of more than $3 million. In September 1989, the partners accepted the development plan. The deadline was revised from June 1992 to July 1992.

In August 1991, Marriott found that the statements were false. AMRIS understated the costs of personnel and other operating costs. The company also used numbers that overstated the total number of reservations. The actual processing cost per reservation was then estimated at $2.00.

The contract outlined four major development phases: the Business Area Analysis (BAA), to develop business models, the Business System Design (BSD), to enumerate detailed descriptions of the application systems, Construction, to construct the system's code, and Testing, to test the system's code.

On October 16, 1989, AMRIS assured the partners that the project was on time and on budget. However, in January 1990, the company missed the contractual deadline for completing the terminal screen design. In February 1990, AMRIS missed the completion milestone of the BAA phase. Apparently, the developers redefined the unfinished work of this phase to become a part of the next phase.

In February 1990, AMRIS admitted it was more than 13 weeks behind schedule, but claimed it could catch up and recapture much of that lag. In March 1990, the company began a six-week "replanning" effort.

Millions of dollars kept flowing into the project. On May 15, 1990, AMRIS made a presentation to the partners, saying the project was still on time and the system would be ready by its deadline. At the same time, major players in the development effort were chastised for delays.

During the summer of 1990, both Budget and Marriott expressed concerns that the project was behind schedule and that its management was ineffective. While employees at the project office estimated CONFIRM would not be ready in time, they were instructed by management to change their revised dates so that they reflected the original project calendar. In August of that year, AMRIS declared that the first phase was complete and entered BSD, the second major phase. When Marriott representatives asked to see some deliverables of the completed phase, the developers refused to show them or explain their status. In October, the company admitted to the partners it was one year behind schedule, but company executives claimed they would still meet the deadline.

In February 1991, AMRIS presented the Re-Plan, to replace the original development plan. According to the Re-Plan, only Hilton would be using the system by June 1992, and Marriott would not receive all the features it was promised before March 1993. Marriott later claimed that AMRIS executives knew they could not meet the new schedule. The hotel company said that AMRIS forced employees to change their timetable artificially to reflect the new schedule and that those who refused either resigned, were reassigned to other projects, or were fired. The Re-Plan attached a new price tag: $92 million, far above the original $55 million and the previously revised $72 million. The AMRIS president resigned in October 1991. During the end of 1991 and the beginning of 1992, about 20 additional employees resigned.

AMRIS employees were dissatisfied with the way management handled the project. They believed that their managers kept stating unrealistic schedules and that they lied about the project status. Many realized the schedule could not be met—even with nine-hour workdays and work on weekends. By the summer of 1991, more than 180 employees, or about half of the people assigned to CONFIRM, were looking for new positions. A consultant was hired by AMRIS to evaluate the project. Dissatisfied with the consultant's findings, a vice-president "buried" the report and dismissed the consultant.

An evaluation by Marriott concluded that the developers could not complete the project. However, the hotel chain still gave them a chance: "As a partner, we hope that you will be able to perform as promised. However, as a user, we do not, based on experience to date, believe you can." (*Marriott v. AMR*) AMR, the developers' parent company responded that CONFIRM's development was on target and that the system would be fully functional. AMRIS continued to bill Marriott at a rate of more than $1 million per month.

Finally, in April 1992, AMRIS admitted it was two months to six months behind schedule. Like Marriott management, Hilton management was still hopeful that "whatever has been broken can be fixed to meet the original schedule." (*Marriott v. AMR*) But there was no basis for these hopes. That month, major problems surfaced when Hilton tried the system as CONFIRM's first beta-test user. On April 29, 1992 AMRIS chairperson wrote to the three partners:

> Unfortunately, things have not gone as planned. Specifically: (1) The individuals whom we gave responsibility for managing CONFIRM have proven to be inept. Additionally, they have apparently deliberately concealed a number of important technical and performance problems. (2) The technical staff, while skilled, has failed in the construction of the very demanding interfaces between the systems, and the extensive database, which will both be part of the completed CONFIRM system. The bottom line, gentlemen, is that in our best current judgment the system is 15 to 18 months from completion . . . .(*Marriott v. AMR*)

The company promised to repay 100% of the investment to any partner that wished to withdraw from the joint venture. A senior officer of AMRIS blamed employees for lying and accused the project management of concealing problems. The project, he said, was actually two years behind schedule.

On April 28, 1992, AMRIS fired eight top executives and replaced another 15 employees. On May 1, 1992, the company's vice-chairman circulated a letter to employees acknowledging that CONFIRM's "system interfaces and databases are inadequate to providing the necessary performance and system reliability." He explained:

> Our CONFIRM RS problem has many roots—one more serious than the other. Some people who have been part of CONFIRM RS management did not disclose the true status of the project in a timely manner. This has created more difficult problems—of both business ethics and

finance—than would have existed if those people had come forward with accurate information.

In July 1992, after three and a half years and after spending $125 million on the project, the Intrico consortium disbanded. Technically, the developers' main problem was to tie CONFIRM's transaction processing, facility-based, central reservation system with its decision-support system. AMRIS's president admitted: "We found they were not integratable." Also, it was later discovered that the data base was irrecoverable in the event of a crash.

THE LEGAL PROCEEDINGS

Apparently, some of the failure is due to bad management practices of all the four partners in the Intrico consortium. The client/partners teams met with the developer's representatives just once a month. An AMRIS executive said, "You cannot manage a development effort of this magnitude by getting together once a month. Had they allowed the president of Intrico to function as CEO in a normal sense and empowered their senior reps [to] work together with common goal and objective, it would have worked." (*Computerworld* 26, no. 41, p. 8)

AMR filed a countersuit against Marriott, Budget, and Hilton in September 1992. On May 14, 1993, AMR amended its suit to suggest that its client/partners changed an approved plan to determine specifications for the common reservation system. Instead of a single system, AMR claims, the developers were encouraged to create three individual systems under CONFIRM. The company accused its clients of being "selfish." (*Computerworld* 27, no. 21, p. 4)

By January 1994, AMRIS had reached out-of-court settlements with all of its partners for undisclosed amounts. Some sources say the firm was facing damages suits of more than $500 million, and therefore agreed to pay about $160 million.

ANALYSIS OF WHAT WENT WRONG

The CONFIRM case is likely to reverberate for a long time because of the huge investment that was lost in the effort to develop the system. However, this is not the first large-scale failure to develop an IS. In 1988, Bank of America's trust investment system, TrustPlus, was scrapped after insurmountable difficulties of the developer, a private consulting firm, and the bank's staff to iron out program errors. The bank spent $20 million on the development project and another $60 million on trials to fix it. The bank spent $60 million on the development of the system before the abandonment. Consequently, the bank sold its trust portfolios to other banks and withdrew from this lucrative business altogether.

Software development failures are not rare occurrences. According to one survey, an astonishing 75% of all system development undertaken is either never completed or the resulting systems are not used. Unfortunately, despite the great impact of such incidents on society and businesses, the topic has not been studied methodically. The reason may be the paucity of objective data. Therefore, this article relies on anecdotal information when discussing this important issue.

Clearly, journalistic accounts of these cases tend to be simplistic and to highlight the sensational. Also, lawsuits tend to emphasize the other party's failures to comply with the contract rather than to outline the subtleties of interaction between the parties. Thus, the reader should not take every claim, by either party, at face value. Also, the lack of industrywide reliable data does not allow a comparison of this case to general industry practice. In sum, there is very little reported research on the topic.

Corporations' True Social Responsibility?

The quoted words of AMRIS president raise the issue of managers' unethical conduct. Robert Jackalls and other writers on the subject of business ethics indicate that Milton Friedman's assertion that the social responsibility of corporations is to maximize the profit for their stockholders has proved itself realistic. Or, at least, that is what corporate managers think is their responsibility. Should this, however, be at the expense of professional integrity?

No one would actually accuse a respected software developer of deliberately hindering its own effort. A project may be plagued with technical problems that were not anticipated before the project began, but one suspects that unprofessional behavior might have contributed to the mishap. Before examining "professional behavior" as it relates to this and similar cases, this article examines professional status and standing.

Who is a professional? A professional is an expert whose services are required because of increasing technological or other specialty demands. Individuals and organizations seek the services of professionals because professionals are assumed to have knowledge far beyond those of lay people. Individuals and corporations trust their interests to professionals. While some people debate the label "computer professional" (especially because of the variety of occupations within this broad term), software developers usually consider themselves professionals. In fact, their responsibilities may be more comprehensive and therefore may require them to be more careful about their conduct than other, more traditional professions.

221

Mylott equates computer professionals to other professions: In the services they perform, computer professionals most resemble a combination of accountants, architects, and engineers. Like architects and engineers, computer professionals create specifications and supervise the implementation of specifications. Yet, whereas architects and engineers rarely execute the construction of the buildings they have designed, computer professionals create specification. They write computer software and propose combinations of hardware and software to purchasers. To develop computer software and to assemble configurations of computer hardware and software, computer professionals, like accountants, often perform financial and business analysis.

Deborah Johnson suggests that professional codes of ethics address these four types of obligations:

- Obligations to society.
- Obligations to employer.
- Obligations to clients.
- Obligations to colleagues and to professional organizations.

When professionals are employed by a corporation, they are faced with two sets of standards used to evaluate their behavior, and the two are not always mutually compatible. Professionals are faced on the one hand with the organization's standards that dictate success in terms of organizational goals; and on the other hand, they are faced with the standards of their professions. In cases like the CONFIRM project, at least some employees noticed that not all was well. If their superior does not act to remedy the failure or inform the client, then they are faced with a dilemma: they have to choose among the interest of the employer, the well being of the client, and their obligation to the profession.

What do professional organizations expect of computer professionals? Each professional organization has its own code of ethics. However, all, in one way or another, expect the member to honor the above four types of obligations. (The codes of the ACM, the Data Processing Management Association, the Canadian Information Processing Society, the British Computer Society, the Institute of Certified Computer Professionals, and the Information Technology Association of America and the analysis thereof are detailed in Oz, 1993; for a summary of the codes' principles, see Oz, 1992).

One important obligation to the public at large and to clients in particular is to avoid misrepresentation of information technology (IT). The Data Processing Management Association Standards of Conduct require its members "not [to] misrepresent or withhold information concerning the capabilities of equipment, software or systems." The ACM Code of Ethics and Professional Conduct mandates that the organization's members "Ensure that users and those who will be affected by a system have their needs clearly articulated during the assessment and design of requirements; later the system must be validated to meet requirements" (Clause 3.4). The Institute of Certified Computer Professionals Code of Ethics says, "One shall not make false or exaggerated statements as to the state of affairs existing or expected regarding any aspect of information technology or the use of computers" (Clause 3.4). Other IS professional organizations in the US and other countries (e.g., the Canadian Information Processing Society and British Computer Society) expect similar behavior of their members.

Most professional organizations require their members to disclose limitations of the systems they develop. For example, the ACM code mandate is: "Honesty is an essential component of trust. Without trust an organization cannot function effectively. The honest computing professional will not make deliberately false or deceptive claims about a system or system design. He or she will offer full disclosure of all pertinent system limitations and problems" (Clause 1.3). The ICCP code reads: "The personal accountability of consultants and technical experts is especially important because of the positions of unique trust inherent in their advisory roles. Consequently, they are accountable for seeing to it that known limitations of their work are fully disclosed, documented and explained" (Clause 2.7).

Typically, development ventures involve risk. Development of IS is not different. Project leaders cannot be expected to bother the client with every mishap along the development effort. Often, the developers face obstacles that are eventually overcome. How far should the developers go in their efforts to overcome a major problem? Obviously, this is a question of professional judgment. If the project leaders feel that the problem is grave enough to jeopardize time and money constraints, they should immediately inform the client. They should certainly do so if they believe that the product is downright unattainable.

In the case of employees developing the system, it is their obligation to inform the employer, i.e., management. Management, then, has to disclose the information to the client. Most professional codes of ethics in the IS field do not require the individual IS professional to make a preference between obligation to the employer and obligation to the client. In fact, come codes even mention the employer and the client in the same clause. For example, the ICCP code reads: "Certified computer professionals have an obligation to serve the interests of their employers and clients loyally, diligently and honestly," and the ACM code lumps the employer and the client together in some of its directives (e.g., clauses 2.5 and 2.6). Thus, the employed professional may face a dilemma in choosing between the interests of these two parties.

Management, on the other hand, is not faced with this dilemma. Management has an obligation only to the client. The Information Technology Association of America (ITTA), an association of companies engaged in the development of IS products and services, clearly requires the member organization and the organization's employees to refer the client to other parties: "The judgment of a professional services company and its data-processing practitioners should be exercised solely for the benefit of a client and free of compromising influences. Neither the interest nor the desires of any other party should be permitted to alter objectivity and independence when rendering recommendations in a professional situation or climate" (Basic Principle Number 1). Because AMRIS is a member of ITAA, the company is expected to follow this principle.

The code anticipates hardships in development projects and prescribes what the company (i.e., management) should do: "If unforeseen circumstances make completion unreasonable, the professional services company or data processing practitioner should be prepared to make just and appropriate compensation to the client."

What can be learned from CONFIRM and similar cases? Experience shows that one or a combination of the following occurrences are the reasons for failure to develop a satisfactory IS:

1. Unforeseen and insurmountable technical difficulties.
2. Underestimation of cost and completion dates.
3. Failure of the developers to understand the system's requirements.
4. Changing the requirements after the project started.

PRINCIPLES TO MINIMIZE THE DAMAGE IN IS DEVELOPMENT FAILURE

In its countersuit, AMRIS claims that the reason in this case was the clients' demand to make changes long after the project started. What really happened and who is culpable in the CONFIRM case will be decided by the courts. However, it seems that much of the damage in CONFIRM-like cases can be avoided with several simple principles. One could immediately see the relationship between the aforementioned codes of ethics and professional conduct, and the following principles.

Principles for Managers of the Service Provider

1. In the business of software development, the managers always know when a project is started, but they never know when it will be completed. The consultant responsible for Bank of America's Trust-Plus boasted he could complete the system by 1983. The system was abandoned, incomplete, in 1988. When outlining the project sched-

ule, managers should be realistic and include an adequate "slack" time. Technical and other problems may occur. Problems often occur when a project involves interfacing two or more IS. Trying to entice the client with an unrealistically short schedule is not only unethical but may eventually hurt the manager's own efforts.

2. When the phases must be sequential to ensure quality, managers should never start phase n before resolving all the problems of phase n-1 and should avoid shortcuts. AMRIS left bugs to be ironed out at a later time, while it went on with the next phase. Reports of other large-scale development failures point out similar practices. One former executive of Bank of America said, "There were still bugs, but the users felt they could run with it and work out the bugs as we went along." [*Computerization and Controversy: Value Conflicts and Social Choices,* eds. C. Dunlop and R. Kling (Boston: Academic Press, 1991)] Managers should view the project plan as a part of the contract with their clients, even if it is formally not. The client counts on managers to manage the project to their best professional ability. Failure to do so betrays the client's trust.

3. Executives should not make any "calming" statements about the project status before they learn the facts. Making uncorroborated statements is not only unethical to the client, it may send wrong signals to employees.

4. Managers should adopt a code of professional standards and communicate it to their employees. The code should detail what an employee is to do when experiencing a persistent problem with systems under development, to whom he or she should report the problem, and what steps he or she should take if immediate supervisors are not responsive to complaints. A clear policy ensures that both managers and their employees know what is expected of them and fosters more ethical behavior.

5. Most important, being dishonest may hurt the client, but it may also hurt the managers and their company. The financial impact of lost business because of a failure due to lies may prove much greater than the lost income from a single mishap. If it is not the monetary gain that drives the manager's judgment but the reluctance to admit professional weakness, the manager should think again. Failure to disclose the real status of the project to the client may exacerbate the damage.

Unfortunately, honesty is not always in one's economic self-interest. Often, there is economic incentive to lying, for example, when the transaction is a one-shot deal and if information of the incident does not spread. In this age of fast communication, however, especially in the IS industry, the news travels fast. Furthermore, the organization's own employees may follow the bad example by lying to their superiors.

Principles for Employees

The first to observe technical problems are, usually, the employees: systems analysts and programmers. Employees have an obligation both to their employer and the client. When employees realize there is a persistent problem, they should notify their supervisor. One wonders how long it took until the first employee stepped forward and did so in this case. However, some employees did complain about the technical problems. Several of them paid with their jobs.

Principles for Clients

1. It seems that the three client/partners kept loose vigil over AMRIS. This is surprising, because they had liaison teams that were supposed to keep track of the project progress. It is tempting to rely on a company that demonstrated success with another system, but this is not the same system. Previous success does not guarantee success with the system that is being developed for a particular client. Clients should check the status of the project periodically. If they do not have qualified personnel, they should hire an independent consultant to do that for them.

2. In a suit filed by AMRIS against Marriott, Hilton, and Budget, the plaintiff complained that the three client/partners in the Intrico partnership missed a deadline for providing a clear definition of system functionality. Clients should communicate to the developers exactly what their requirements from the new system are. Clients must realize that later modifications may result in a higher price and a later completion date.

3. Clients must pay attention to alerting signals. When executives and other employees of the developer are either massively dismissed or voluntarily look for new positions, clients should ask questions. When the rats abandon the ship, it is probably sinking.

SUMMARY

A word of caution: as pointed out several times in this article, there is no conclusive data from which to draw hard conclusions about reasons for failure in systems development. For obvious reasons, such data is extremely difficult, if not impossible, to come by. However, it seems that the

CONFIRM case contains many ingredients that are common in cases that have been exposed in the media and trade journals.

An ancient proverb says, "You are a wise person if you do not make mistakes; you are a clever person if you make a mistake but do not repeat it; you are a stupid person if you make a mistake and repeat it." It may be hard to be wise the first time around, but let the IS professional not be stupid, either. Professionals are expected to learn from their own and their colleagues' mistakes.

The CONFIRM case draws attention because of the magnitude of resources expended. It is also a case of what seems to be the result of miscommunication at best or grand deception at worst. However, there is reason to believe it is only one of many such cases. To minimize the probability of such mishaps, IS organizations have to adopt detailed codes of professional standards, which should outline to both managers and employees how to behave when projects do not proceed as expected.

Large development projects rarely proceed exactly as planned. This is true of IS development efforts as well. Management should be deeply involved in the progress of large-scale projects. If the professional team cannot overcome difficulties to comply with promised cost and timetable, it is the professionals' responsibility to duly report to management; then, it is management's responsibility to disclose the difficulties to the client and mutually outline a resolution.

Of course, there is no point in promoting a code of ethics and professional standards if executives do not demonstrate personal example. IS organizations should be honest with their clients. In the long run, honesty indeed is the best policy.

4-6

Critical Success Factors in Systems Integration

EUGENE J. ROMEO

Systems integration projects have become mainstay efforts as Information management and technology (IM&T) organizations attempt to provide business solutions while improving the return on IM&T investment. At issue are creating and sustaining an effective balance between maintaining and migrating legacy systems while supporting the introduction and implementation of more cost-effective computing solutions, such as client/server technologies. As a result, there are increasingly greater levels of pressure on IM&T organizations to deliver integration results in a timely and cost-effective manner. Consequently, the systems integration project and the overall IM&T organization have much greater visibility and vulnerability with senior management and the user community.

This chapter describes some of the critical success factors in managing the systems integration effort. There are five, as discussed in the following sections:

1. The role of the information technology (IT) architecture and strategic plan.
2. The impact and effect of project management on systems integration.
3. The role of communications in establishing and managing expectations.
4. The need for change management procedures.
5. The role of the user in integration.

THE CRITICAL SUCCESS FACTORS OF SYSTEMS INTEGRATION

There are many factors and events that can affect the outcome and experience of the systems integration effort. The experience and sophistication of the personnel assigned to the project, the time and technology

229

involved, the financial investment, and the complexity of the application systems and data bases are but a few of the many factors that influence the outcome—and the success—of the integration process.

However, there are additional factors, pervasive to any integration project, that are technology and project independent. Defined by the human elements and leadership practices of the integration process, these factors have a profound effect not only on the integration process and the results achieved but the quality of experience surrounding the process. They are what is called tonal factors, that is, they help establish the behavioral patterns, expectation levels, and context in which integration is performed and also provide a basis for measuring the success of the integration process.

CRITICAL SUCCESS FACTOR 1: THE IT ARCHITECTURE AND STRATEGIC PLAN

Systems integration projects and the rapid deployment of enabling solutions cannot be performed in a random or patchwork manner. The driving issue and incentive for integration is to avoid the dysfunctionality that accompanies standalone technologies and "islands" of technology. Integration is defined as a melding of systems, processes, and technology into a relatively seamless and cross-functional capability. In this respect, the IT architecture is absolutely essential to the process. Organizations that have attempted to perform integration without a guiding architecture have often found their investments in the processes achieved results that were far lower than anticipated. Systems integration, especially in support of reengineering and major organizational transformation, must be the product of a well- thought-out and choreographed vision and plan. It must be the result of an architectural process that involves an enterprise-wide perspective.

Choosing the Appropriate Architecture Tier

The IT architecture and strategic plan provide the context and scope for systems integration. Perhaps the most important component of the IT architecture is an understanding of how integration is to be achieved in a multitiered environment, which requires the careful assignment of data warehouses and applications to a specific tier or computing realm of the overall architecture. Not all applications and data should be resident on a mainframe or client/server computers.

The assignment of an application and data to a residence is important because it establishes expectations and a clear dimension of measuring deployment of integration. The assignment of applications and data also determines how the architecture will be translated from a logical representation of the technology to a tangible and functional system. Thus, having an architecture that clearly stipulates paths for integration and how integration is achieved is one of the most fundamental requirements of a systems integration effort.

Deploying a Strategic Information Systems Plan

Complementing the IT architecture are the development and use of a Strategic Information Systems Plan (SISP). The use of the IT strategic plan provides the rationale and linkage for integration and the IT architecture to tangible business needs and organizational requirements. It is critical that the business units and the IT organization are linked with common objectives, target the same results, and use similar assessments of those results. The SISP is a working document that provides the actual road map for realizing the IT architecture and achieving systems integration. The SISP establishes the priorities and relationships for integrating systems in a multitiered computing environment. In addition, the SISP provides a cause-and-effect matching between IT projects and business needs that is immediately discernible and definable. The IT strategic plan provides a basis for evaluating the direction and business rationale for integration and links the IT effort and use of technology to organizational needs. In this respect, the SISP provides another valuable management tool for directing and managing the integration process.

CRITICAL SUCCESS FACTOR 2: PROJECT MANAGEMENT

Much has been said and written about project management, and many organizations spend a significant amount of money on training project managers. Despite this, many projects continue to experience problems, and IM&T organizations continue to come under increasingly greater pressures to deliver not only technically correct systems but tangible and measurable business benefits. Using an effective and responsive project management method does not, in and of itself, guarantee a successful integration effort, but it can certainly help to minimize risk and facilitate the management of expectations.

Managing Expectations

Traditional project management techniques usually concentrate on reactive methods and measures and can fall short of the anticipatory needs of systems integration and reengineering projects. Adopting an approach as discussed in Chapter 5-1 clearly helps in the deployment and direction of project resources and in the management of expectations, which provides, perhaps, the greatest opportunities for creatively adopting project management.

The major issue involves user and management expectations. Even though users have become more sophisticated in their understanding of information technology, they may still have inflated expectations for new systems and technology and low appreciation for the level of effort and complexities involved. A well-prepared and scripted project plan and project management process can help level expectations and improve the overall degree of understanding of the complexities involved.

A key element of this process includes providing for the co-management and accountability for the process's results, shared equally between traditional IM&T resources, functional users and sponsors, and as appropriate, third-party vendors and consultants. The creation of co-managers and co-accountabilities is essential for ensuring the proper level of user involvement and responsibility. It is also necessary for creating synergistic project teams that are client focused.

Setting and Calibrating User Expectations

In managing the integration process, the expectations of users, senior management and the project team should be set and constantly calibrated using a combination of key techniques, including:

- The criteria for measuring the progress of the effort described in Chapter 5-2.
- Constant reference and linkage to the IT architecture and SISP.
- Comparison to other non-IT projects, such as the product design process and the implementation of new manufacturing technology.

Performed properly, project planning and management can be used to educate and communicate to the user community and senior management the complexities, issues, and key processes involved in systems integration. This educational process is essential to ensuring that users and management are in concert on the effort and have a central and common basis for measuring the success of the effort and establishing a responsive reward structure.

CRITICAL SUCCESS FACTOR 3: EFFECTIVE COMMUNICATIONS

The effective communications and presentation of the systems integration project is essential for the IM&T professional and users sponsoring the effort. Communications in the systems integration effort involve a variety of formal and informal methods. The formal methods include such common tools as status reports, periodic presentations to senior managers, written minutes of meetings, follow-ups to major discussions, and the frequent meetings of the project team with users, sponsors, and senior managers. Informal techniques include a variety of methods, such as personal communications with senior managers and user sponsors, building team rapport, and providing follow-up to conversations, as appropriate.

All of these techniques are common and are practiced in business today; however, how they are used separates successful integration projects from unsuccessful ones.

Consistency Counts

First, it is imperative to the integration process that all communications be consistent and constant. Consistency in communications means that both the formal and informal processes, such as status reports, minutes, and personal communications be performed the same way, time and time again. Consistency is imperative for educating the user and senior management about the process, for leveling expectations, and for creating credibility. Credibility is essential for any leader, especially those chartered with the larger integration efforts. In creating consistency, all formal communications should use the same formats, language, and definitions. Inconsistency compromises the communication process's effectiveness and confuses the recipient of the information and can also have a negative impact on the project, especially if faulty communications cause double work and necessary cycles for communication. Thus, it is important to communicate using consistent definitions, languages, and formats.

Constant Communications

Complementing the need for consistency in communication is the requirement for constant communications. To be effective, communications must be predictable, constant, and recurring. In the integration effort, the need for constant communications is essential to user involvement, expectation management, decision-making, and collective problem solving and consensus building. The process must be bilateral and must directly seek and engage user and senior management input and reaction. This becomes increasingly important when the systems integration effort is in support of operational reengineering and organizational transformation efforts. These types of efforts always involve systemic changes and cultural transformation, all of which require employee involvement and education. Thus, in the effective management of the integration effort, the project manager must ensure that the communication process is consistent, constant, and continuously centered on the IT architecture, SISP and stated objectives of the project.

CRITICAL SUCCESS FACTOR 4: CHANGE MANAGEMENT

As discussed in Chapter 4-1, maintaining a stable set of design parameters and functional attributes for the systems integration project and project team is essential to the effective management of the process. All too often, major integration and reengineering projects are derailed because of peripheral issues and misdirected efforts that lead not only to unfavorable project variances but to increasing frustrations of the project team. Therefore, maintaining a formal change management process is essential to managing IM&T resources, the scope of the integration effort, and the activities of the resources applied to the process.

Effective change management involves several key activities and attributes. First and perhaps most important, it must be formal and disciplined. The process must have a formal set of operating guidelines, procedures, and structures. Change management simply cannot be performed—and cannot be effective—without discipline and a formal operating construct. Change management practices must be enforced and used aggressively.

Second, change management must include criteria for quantitatively and qualitatively assessing the impact of any changes on the overall scope, budget, technology base, staffing resources, critical path, and schedule of the project. It cannot be an informal process composed of arbitrary criteria. The change management process must provide the updating of the overall work plan and must contemplate the effect of the change on the resources assigned to the project and document the decision process.

In systems integration, it is essential always to work from and toward a stable design and set of functions. In the absence of this stability, the integration process can never be accomplished. Usually what results is a terminally ill project that has no definable end-point and constantly evolving and fluid measurements. (See Chapter 4-1 for further information on the terminally ill integration project.) Under these circumstances, the effort almost assuredly ends in failure. Therefore, having a well structured and smoothly operating change management process is absolutely essential to the overall success of the integration process.

CRITICAL SUCCESS FACTOR 5: THE ROLE OF THE USER

This list concludes with perhaps the most important of all success factors, the user. Creating and sustaining user involvement is absolutely imperative to any systems integration project, regardless of technology involved. After all, the role of technology is to enable users; therefore, it is incumbent on the integrator and IM&T organization to ensure that users are actively involved not only in the integration effort but in the sponsorship and management of the integration effort.

To create this involvement, the project must be organized as a partnership, with users and sponsors functioning as co-managers who are held to the same standards as the IM&T organization. In this respect, users not only are sharing the responsibilities for the project but are co-accountable for the results and processes used to perform integration. Thus, users must be equal business partners in the integration of systems and be held to the same levels of accountability and performance as that of the IM&T professional.

Creating user accountability is difficult, as there is an inevitable tendency for the responsibility for the effort to gradually become entirely that of the IM&T organization and technical resources assigned to the project team. This is certainly the typical behavior pattern for many organizations and projects; regardless of sponsor and need, the ultimate responsibility lies with the IM&T organization and users, and senior management is somehow exonerated. However, building co-accountabilities and partnerships is essential to the systems integration process and there are several techniques that, when properly applied, ensure that users and sponsors are constantly engaged and held to the same standards and measures as the technical resources.

Joint Application Development

One of the most common and successful techniques involves the use of Joint Application Development (JAD) in defining integration requirements and in the testing and deployment of integrated systems. JAD is not new, and there have been many studies and articles written on this subject. In using JAD techniques to create accountability and shared ownership, the major objective is to leverage user and sponsor commitment into the project through direct and meaningful participation. The end result is a user-defined and user-designed system, not an IM&T system. A JAD team of both user and IM&T personnel is empowered and accountable for decisions regarding the functionality, design, documentation, testing, training, and implementation plans for the integrated system. The JAD team concept is an extremely powerful and useful technique because early consensus building is established while time lines and approaches are developed. The targets and environments for integration become stable, and consistent measures can be applied to the collective project team. Thus, JAD can create a sense of concurrency in the process that can provide a construct for mutual ownership long before the end product is delivered.

A JAD team usually has the positive result of breaking down organizational hurdles and barriers. The team creates momentum, which accelerates the system's design while improving the quality of the final product. The JAD team also assists in addressing and resolving the resistance to change that is often associated with systems integration efforts, especially when legacy applications are involved. Finally, a well-managed and well-constructed JAD team can provide synergistic relationships and positive working relationships.

Training

Training is another critical success factor which can facilitate end-user participation and accountability. Too often systems integration training efforts have narrowly focused on a training plan that emphasizes the use of the system with respect to the new screens being implemented, the use of new functions that were requested, or new technology. Unfortunately, this traditional training is not as effective as it needs to be, because it trains only for the technology and not necessarily for how operations and organizations are changed by that technology.

Thus, an effective technique to developing user accountability and co-responsibility is a new approach to training. To cultivate this sharing of ownership, training should focus on how the customer, or end user, needs to perform his or her job within the context of the integrated system. This relationship becomes significantly more important when operational and organizational reengineering is involved.

When the end user has the new process training, they can make the mental bridge between the old and new, understand the differences, and begin to assume ownership and responsibility for the entire process, including the technology integration and its ergonomics. Each end user requires a different level of training support based on his or her current knowledge of the system, comfort level with change, and overall ability to comprehend new processes and procedures. In this regard, training must include not only the technical aspects for using and maintaining integrated system solutions but also end-user operational procedures. The major emphasis in this type of training is ensuring that the users can optimize their leverage in technology through use and knowledge of the systems.

SUMMARY

Managing the systems integration process to a successful completion involves far more than melding technology and replacing legacy applications. Systems integration involves managing a process of change and transformation that involves the proactive leadership of the sponsor, user, and IM&T organization. The integration process must be managed—that is clear and simple. The major issue in the management of systems integration transcends traditional technology to include the management of user and senior-management expectations. It also involves the use of consistent measures that assess not only progress but the quality of products delivered and the quality of the experience, which provides perhaps the greatest opportunities for the IM&T executive who is chartered with providing enabling solutions. The experience is what users and senior management will remember long after integration is achieved. In the context of the 1990s, technological success is no longer a measure: it is an absolute.

4-7

An Approach to Estimating Client/Server Development Projects

ROGER KAHN

In spite of the past decade's overenthusiastic leaps into new technologies, client/server computing is winning favor as a vehicle for providing both the information and the agile systems required to compete in the 1990s' global marketplace. Whereas mistakes have been made in terms of development and implementation, client/server technology is bringing improvements.

There were many reasons for these mistakes. Perhaps the single biggest factor for projects gone awry is the IS profession's penchant for treating each new technology as the silver bullet. Beyond this generalization, however, lie the real issues, which include trivializing development complexity, underestimating the infrastructure required for successful client/server projects, and applying inappropriate estimating and project management methodologies to the client/server development process. One CIO for a Fortune 200 company likens the process of estimating client/server system development to playing Russian roulette with bullets in five of the six chambers. The odds of surviving or, in the case of estimating, the odds of being accurate, are extremely small.

Whereas a great deal has been written on estimating techniques and processes that determine project size and cost, much of that work is too impractical, academic, and complex for real-world development efforts. Furthermore, many of the approaches used for estimation are not applicable to the client/server development environment.

Developers of client/server systems need a current and practical approach to help estimate project size or to validate estimates formulated by other means. The estimating technique presented in this chapter is current and is intended to be used specifically for client/server development. It goes beyond many of the traditional approaches by factoring in the skill levels of personnel who are involved in the project's development effort, and the technique is also practical, as it has evolved over the past several years and is based on more than a dozen client/server development projects. In fact, several colleagues contributed to this process and estimating approach by using it on projects for which they were responsible. The accuracy of the approach has improved with each iteration.

The estimating process described in this chapter assumes that a prototype and data model of the target system have been developed. If these have not been completed, the process presented here cannot not be beneficial.

Users and clients are everywhere who believe that project-development estimates are a bit of witchcraft designed more for the developer's purposes than for theirs. To avoid this, the developer may find it beneficial to share this process with them to help demystify the estimation process and therefore aid in justifying the project's development budget.

A PRACTICAL APPROACH TO ESTIMATING CLIENT/SERVER DEVELOPMENT PROJECTS

The estimation process described in the remainder of this chapter is driven by two factors: proposed-system complexity and personnel level of skill. Given the existence of a prototype and data model, the technical characteristics and complexity of the proposed system are known, and the formulas can be applied successfully.

The process itself works from the inside out, so to speak. The first step calls for developing the estimate for the construction phase of work. The process then moves to estimating the size of the integration, acceptance testing, and implementation phases of a client/server development project.

The methodology anticipates that graphical user interface (GUI) and programming productivity tools, such as those available from Powersoft, Gupta, or Microsoft, are chosen for development. The process described here has been successfully applied to projects developed using tools from these organizations.

THE FIVE STEPS TO DEVELOPING AN ESTIMATE

Developing a project estimate involves five steps and the use of three tables. In the first step, each of the windows (screens) in the prototype is classified by complexity (see Exhibit 4-7-1). In the second step, the estimate derived in the first step is adjusted for the skill level of the individuals who are involved in the system's actual development. The skill-level adjustments are contained in Exhibit 4-7-2. In the third step, the

Complexity Attributes	Ext Spec	Int Des	Code	Unit Test	Help	Quality Review	Mgt. Time	Total Time
Easy	4	2	14	2	3	1	1	27
Inquiry Only								
Less Than 10 Fields								
1 Table								
1 Data Window								
Average	6	4	20	4	5	2	1	42
Inquiry Only								
10 to 25 Fields								
Simple Join (1 - 2 Tables)								
1 Data Window (Display or Update)								
Selection on 1 Field (Location, Customer, Part)								
0 to 1 Child Windows (Detail...)								
Complex	10	6	28	6	6	2	2	60
Inquiry or Update								
25+ Fields								
Complex Join (3 Tables)								
2+ Data Windows or Detail Windows								
Edits on Input Fields								
Table or Data Window Aggregates/Groups								
Selection on Multiple Fields								
Very Complex	18	10	36	10	8	4	2	88
Inquiry or Update								
Cross Tabulation/Calculation								
Table or Data Window Aggregates/Groups								
3+ Data Windows or Detail Windows								
Very Complex Joins (>3 Tables)								
Cross-Validation of Input Fields								

Exhibit 4-7-1. Complexity Factors

---

infrastructure components of the project are estimated. This step uses the results of the previous steps. In the fourth step, the additional time required to develop data base triggers and scripts are estimated. The result is the construction-phase estimate, which serves as the basis for estimating the remaining components of the project, the fifth step.

As was previously mentioned, Exhibit 4-7-1 classifies complexity factors. The attributes of each level of complexity are provided in the left portion of that exhibit, whereas the effort, in person-hours, for various

Knowledge of GUI Tool	Experience Level of Individual		
	Trainee	Average	Experienced
Detailed At Least One Life-Cycle with GUI. Knowledgeable in Development Tools' Capabilities	4.0	2.0	1.0
General Six-Months' Experience at Code Level	5.7	2.8	1.4
None GUI Trained.	8.0	4.0	2.0

Exhibit 4-7-2. Proficiency Factors

tasks is listed in the right portion. The definitions for each of the tasks are:

- *External specification.* This involves writing the external functional specification and acceptance criteria. The resulting specification must be reviewed and approved by the client.
- *Internal design.* This means the internal design of all programmable objects required to implement the external specification. The result must be reviewed with the technical team leader.
- *Code.* This refers to the creation of all programmable objects. Includes peer review for compliance with overall architecture and standards.
- *Unit test.* This means the testing of only those programmable objects related to this specification.
- *Help.* This is the writing of the window-level help text for this specification.
- *Quality review.* Testing is by an individual other than the analyst or programmer according to the acceptance criteria listed in the specification and standards compliance.
- *Management time.* This means the project management, final review, and sign-off or work performed.

Again, once the total programming estimate has been completed, the result should be adjusted to reflect the skill level of the development team. Exhibit 4-7-2 contains proficiency factors for various skill levels.

PROFICIENCY LEVELS

Applying the proper knowledge level from the table is important, so the following definitions are provided.

- *Trainee (i.e., none).* This level assumes no practical experience with systems development or with the GUI tool to be used in the development of the system.
- *Average.* This level assumes at least one year of practical systems development experience with the selected GUI tool or with an equivalent one. The individual should also have data-base experience writing structured query language (SQL) and be familiar with the process of defining requirements and in prototypes.
- *Experienced.* This level assumes three years-plus of full life-cycle development and GUI programming experience. Experience must include heavy experience with the tool to be used in the development or with a close equivalent as well as with data bases and SQL. The individual should also be experienced in developing detailed requirements and prototypes.

Knowledge of GUI Development Tools

The experience levels concerning the knowledge of GUI development tools are defined here:

- *None.* This level means the individual is GUI tool trained with little experience.
- *General.* This assumes the individual has six months-plus development experience with the GUI tool and with SQL, sound knowledge of data window development, and sound knowledge of scripting.
- *Detailed.* This assumes the individual has experience with the GUI tool in the areas of identifying common objects. Developing complex executables, and an understanding of how to use the GUI tool's features is critical.

ESTIMATING INFRASTRUCTURE COMPONENTS

After the skill-level adjustments have been calculated, the time required to develop the application's infrastructure and common objects should be calculated and added to the total to arrive at total estimated hours.

Infrastructure objects may be defined as objects the application uses that are not included within the work specified for the application's windows, such as:

- The application menu.

Complexity Attributes	Ext Spec	Int Des	Code	Unit Test	Quality Review	Mgt. Time	Total Time
Easy	2	1	3	1	1	1	9
1 Table Updated, No Editing or Balancing							
Average	3	3	6	2	1	1	16
2 - 4 Tables Updated							
Simple Editing, Limited Balancing							
Complex	5	7	10	4	2	2	30
5+ Tables Updated							
Cascading Used							
Complex Editing with Error Processing							

Exhibit 4-7-3. Complexity Factors for Data Base Triggers and Scripts

- The foreground print manager.
- The export manager.
- The background print manager.
- The security interface.
- The SQL request router.
- The work queue manager.
- The message logging.
- The common error services.
- The fax manager.

Based on applications experiences to date, infrastructure approximates 15% of the total derived in the first two steps.

ESTIMATING DATA BASE TRIGGERS AND SCRIPTS

Increasingly, client/server systems are using triggered events to handle data base activities on the server rather than tying up desktop computers. Writing triggers adds a level of effort and complexity to a project beyond that of embedded SQL. Including triggers in the application causes data base events to be executed at the server rather than at the desktop. Scripts function much as a batch stream does on a mainframe and are used in the same ways. Exhibit 4-7-3 provides guidelines for estimating the time to develop triggers and scripts.

The estimates assume experienced data base personnel to develop the specifications and the code and to complete the testing.

ESTIMATING THE LIFE CYCLE

Once the construction phase estimates have been developed, it is possible to work forward to develop the estimates for the project as a whole.

The construction phase of a project usually runs in the range of 40% to 45% of the entire effort. The design, prototype, and data model phase is in the range of 20% to 25% of the total effort. Integration and acceptance testing runs in the range of 20% to 25%, assuming the use of such automated testing tools as SQA Teamtest. Implementation takes the remainder.

MANAGING THE PROJECT ESTIMATE

Delivering the project within or close to the estimate is always an issue, no matter what technique is used for estimation. The process of managing the project and meeting expectations continues to become more difficult and complex. This is especially true in the rapid application development (RAD) approaches that are being used for client/server projects. In a RAD environment, the commitment to the client/user to thoroughly document each portion of the work is vital, to prevent runaway situations. In addition, the approval of deliverables must be tightly constrained to avoid constant reapproval of work products, with the attendant increases in time and cost. Therefore, it is recommended that the client/user have one approval of events, designs, test results, and test plans. The client or user feedback is then incorporated into the revised deliverable, and that deliverable is approved without further review.

SUMMARY

Developing project estimates has been and will continue to be a combination of methodology, savvy, experience, and witches' brew. This chapter has provided a practical estimating methodology which can be blended with experience and "brewed" to produce a reliable estimate of a project's time and cost.

Section 5

The Reengineering Approach to Systems Integration

The integration of human beings, processes, and technologies into more agile and responsive organizations is perhaps the single most important objective of the strategically thinking executive. Significant advantages are available to those who integrate these critical elements into a seamless and harmonious process, but with the scope of integration efforts encompassing far more than just technology, the need for new and far more comprehensive management practices is becoming increasingly more important.

The authors in this section present a broad view of managing and reducing the risk of the integration process, including such major topics as assessing integration's impact on traditional information technology (IT) functions, organizing the IT function as a high-performance IT group, and reducing the process's risk. The material provided is both pragmatic and stimulating and will surely be valuable to every project manager and practitioner.

5-1

The Systems
Integration Life Cycle:
A Reengineering Approach

MICHAEL A. MISCHE

With process reengineering and organizational transformation becoming mainstays for many organizations, the major performance issue for information management and technology (IM&T) executives and professionals is speed—that is, speed defined as the rapid delivery of enabling solutions and responsive and cost-effective IM&T services to an increasingly sophisticated and demanding user community. Providing cost-effective computing solutions has always been a major goal for the IM&T professional; however, with reengineering and systems integration projects taking on increasing importance and megadollar proportions, the issues of rapid development and deployment of integrated solutions and enabling technologies is elevated to a new level of prominence and visibility in the enterprise.

The realities of systems integration, as discussed by several authors in this book, are that these projects are large, complex, and often mission critical to any operational reengineering effort. The advances in technologies combined with the needs created by a less-hierarchical organization create a new set of urgencies and priorities for the delivery of systems that bear little resemblance to historical precedents.

Several key factors drive the need for the accelerated delivery of enabling solutions. At one time, technology created competitive advantage; today, however, systems are easily replicated. Therefore, competitive advantage through technology is short-lived—and lessening. Organizations cannot wait three, four, and five years for traditional systems development and integration methods to deliver enabling solutions. By the time solutions are fully implemented and available for users, their needs, management sponsors, and technologies have all changed.

Another factor is technology is advanced and adaptable. At one time, large mainframes, special rooms and environments, and many people were required to support complex programs and interfaces; today, the computing environment has evolved and contains smaller, easier-to-support, and far friendlier technologies that are rapidly deployable.

Also driving the need for accelerated delivery is improvements in the quality of technology. Historically, some of the more fundamental measures of IM&T success included such criteria as technically correct programs, rapid response times, and 99% systems availability. Today, processing accuracy, technically correct applications, and system availability are fundamental "givens": they are prerequisites and no longer the predominant measures of an IM&T organization's effectiveness. Therefore, users and executives have a different set of expectations for IM&T performance that are beyond the traditional technical performance measures.

In addition, organizational transformation and operational reengineering are redefining entirely new applications and data needs. In the process, they are creating new opportunities for using technology and leveraging human resources.

End users are also simply smarter and more comfortable with technology. They understand technology and what is and is not possible. As a consequence, users are far more demanding of the systems provided and discriminating of the services, resources, and costs of the IM&T organization. They are holding the IM&T organization and the quality of systems and services delivered to much higher standards.

As a consequence of these changes, there is a need for a more flexible and accelerated approach to systems integration and the deployment of technology-enabling solutions. This chapter explores some of the issues associated with integrating systems in the context of reengineering and offers an overview of a new life-cycle methodology for the systems integration and development processes. This model is called the Process-Driven Integration Model (PDIM).

THE PROCESS-DRIVEN INTEGRATION MODEL

For the most part, the historical systems development life cycles (SDLCs) were restricted to use by the IM&T department for the development of new systems. In a sense, systems integration projects have relied on the traditional SDLC methodologies—or some derivative of the traditional SDLC—as a guide for developing systems integration work plans and managing the integration process. The classic SDLC provided a formal structure for developing mainframe and midrange systems, a structure that included a number of phases and work steps. Some methodologies had as few as four phases, some had 12 or more.

Over the course of time, variants to the traditional models were introduced and evolved to include computer-aided systems engineering (CASE) tools and joint user designs to support the systems integration process. Without question, the traditional SDLC methods were relatively effective in aiding the development of classic mainframe-based systems and facilitating the integration of function-specific systems. However, systems integration projects in support of reengineering are proving to be an entirely different matter. Whereas the classical methods benefited from a limited number of mainframe and proprietary technologies and methods, systems integration in support of operational reengineering and organization transformation involves the complete spectrum of technology, applications, data structures, and significant organizational and operational changes to the IM&T organization. Whereas the development process and integration of systems were once restricted to a finite set of applications and related data, integration in the reengineered enterprise requires cross-functionality and portability of technology and applications in support of leaner and more agile operations and organizational structures.

For most organizations attempting to reengineer, the traditional methods for systems development and the development project life cycle are no longer effective models for the management of the system integration processes that are required to support new cross-functional and leaner organizations. They are too slow, rigid, mechanical, and bureaucratic. The needs of reengineering and organizational transformation mandate a more accelerated approach to systems integration and providing enabling technologies to support the systemic changes related to reengineering. Hence, in systems integration and reengineering there is a need to bridge the gap between traditional SDLCs and the contemporary demands for the rapid delivery of integrated solutions.

Readdressing the concepts and principles of the traditional SDLC model in the context of reengineering is fundamental to effective project management and successful integration. A new model must provide for a fast-track integration process that successfully leverages technology to realize and support process reengineering and organizational transformation. To be effective in supporting reengineering, the new Systems Integration Methodology (SIM) must provide for the successful melding of reengineering and systems development. It must be process driven, linked to and consistent with an overall information technology (IT) architecture that supports the five tenets of reengineering, as discussed in Chapter III-1.

The following characteristics help to define the SIM for reengineering. Building off Chapter III-1's concepts for reengineering, this chapter presents a new template for managing and performing the systems integration project in the context of reengineering. The model is composed of a five-phase life cycle. The manager must:

1. Establish the integration environment.
2. Assess integration requirements and applications design.
3. Assess data requirements and design integrated data warehouses.
4. Develop and prototype integrated systems.
5. Implement integrated systems.

Exhibit 5-1-1 provides an illustration of the PDIM's five phases. This approach provides for the management of activities, practices, and progress reporting in a process-driven methodology. The model can be applied to the most complex as well as the simplest of integration efforts. Each of the five phases has a specific purpose, objective, work plan, work products, and level of complexity and risk associated with it. The main objective is to create speed and integration to support reengineering. The model comprises over 125 primary work steps and 35 major work products.

The five-phase model for systems integration provides a number of advantages for the organization performing integration and reengineering. First, it is a convenient method for framing the process and scope. Second, the model can be used to organize the activities and resources assigned. Third, the model provides for easy tracking and reporting of activities and progress. It is designed to be performed concurrently with reengineering and supports a concept called continuous integration.

Integration process acceleration is accomplished through the combination of several techniques that are implicit to the process-driven methodology. The process-driven approach to systems integration achieves the rapid deployment and delivery of integrated systems by using many of the reengineering concepts that are applied to business processes and the principles of concurrent engineering. The major principle is to view the systems integration project as a continuous and harmonious process rather than distinct or separate pieces that somehow must fit together. In this regard, the PDIM has the following key attributes:

- Use of concurrent engineering practices and techniques.
- Adoption and adaptation of business process reengineering principles.
- Front-loading of critical human resources into the design and assessment stages of the integration process.
- Use of quality function deployment (QFD) techniques to ensure that design changes are recognized early in the integration process and are minimized later in the effort to help ensure efficiency.

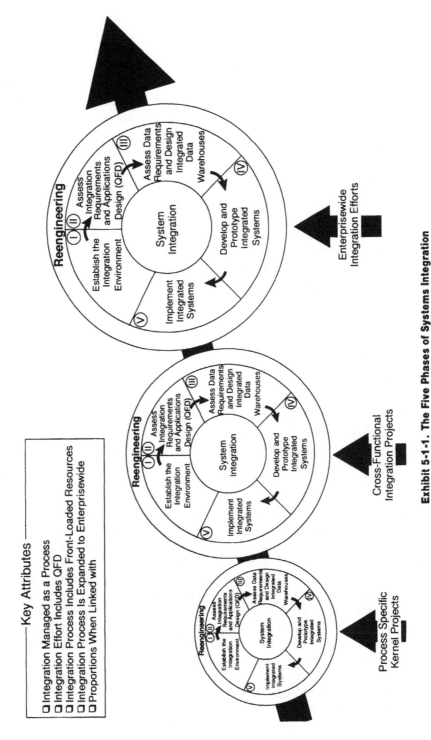

Exhibit 5-1-1. The Five Phases of Systems Integration

The need for front-end loading of critical customers and resources is important to the PDIM. In the traditional model, resources are gradually increased or ramped-up: resources are at a peak near project completion. In contrast, the PDIM assigns and applies the necessary user, external, and internal resources early in the process. This allows a number of activities to be performed concurrently and facilitates the exchange of knowledge and information that supports the design process. Acceleration in the integration process is supported in a number of ways. First, greater emphasis is placed, during the first two phases, on developing a knowledge of the technical aspects of the integrated environment and in developing a robust understanding of the analytical and functional needs of the integration process. As discussed throughout Section 4, a fluent knowledge of the target technical environment and the new to-be processes (i.e., to be supported by the integrated environment) is absolutely essential to the success and efficiency of the integration time line. These phases can be performed concurrently by two different components of the project team with the results converging, as appropriate.

Second, Phase 3 can be performed simultaneously. During this phase, data bases and data requirements are commonly identified, defined, and assigned to warehouses and residences in the organization's overall information architecture.

Third, user training, system testing, and performance tuning are all done concurrently in Phase 4. The processes are performed in the context of prototyping, which allows the convergence of users and developers for testing and fine-tuning integrated systems and business processes. Exhibit 5-1-2 compares the PDIM and the traditional life-cycle methodology in light of the resources and time to completion each requires. In general, the PDIM should help accelerate the integration project by as much as 40% when used with effective project-management practices and the proper CASE tools. The resources under PDIM are also applied much earlier and in greater numbers than in the traditional methods.

The other major characteristic of this approach is the use of QFD, a technique that gained prominence in the later part of the 1980s and has gained significant momentum in the 1990s, especially in the automotive and airline industries. QFD is integral to any efficient product and process design and can be adapted to the systems integration process. In their forthcoming book, *Customer Integration: The Quality Function Deployment Leader's Guide for Decision Making,* authors Doug Daetz, Bill Barnard, and Rick Norman define QFD as "... A systematic planning process that was developed to help a project team bring together and manage all elements needed to define, design, and produce a product that would meet or exceed customer need."

Thus, QFD is primarily a process designed to integrate the "voice of

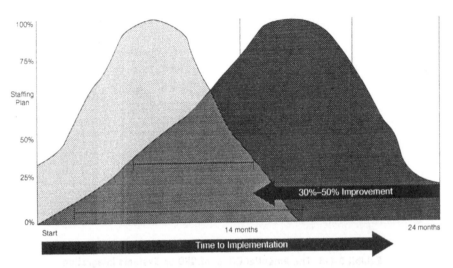

Exhibit 5-1-2. The Potential Effects of Front-End Loaded Integration Projects

the customer" into the design process. This is critical to accelerating the implementation of systems integration efforts because it can identify changes to the base design of the target environment early in the process. The earlier changes are identified, the better, as early detection limits leapfrogging and scope creep, which both can be fatal to the systems integration process. Exhibit 5-1-3 provides an example of QFD's potential impact on the changes and change-management process of the systems integration as compared to typical methods.

Essential to the QFD process is the creation of cross-functional teams and the front loading of the appropriate resources to the project. The creation of cross- functional teams in QFD, as in business process reengineering and organizational transformation, is fundamental to ensuring that all salient issues and target attributes are visible to the design team. The process ensures that data requirements have been interpreted properly and that relative data is factored into the design. The process also provides for the incorporation of definable measurements and results into the integration process early in the effort. These are important attributes in managing the integration project and ascertaining the project's progress and quality. The combination of these attributes make QFD an integral component to the accelerated systems integration effort. Some of the other benefits of the QFD process in systems integration include:

- Accelerated design and development times.
- Lower project-development costs.
- Greater stability of product design.

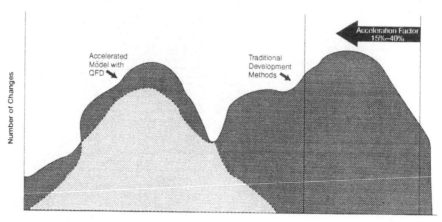

Exhibit 5-1-3. The Potential Effect of QFD on Systems Integration

- Improved use of resources and higher contribution for those resources.
- More complete, visible, and defined target environment and results.

The accelerated approach to systems integration requires the melding of the above techniques with the five-phase methodology. In deploying resources and budgeting time, emphasis is placed on techniques designed to ensure that changes are minimized and that long-lead-time activities are identified as early as possible. These activities help ensure that the staffing is adequately applied and that outsourcing long-lead-time activities is given consideration. As a result of using the accelerated approach, the actual implementation of the integrated system becomes a routine activity rather than an event. Implementation and deployment become natural and logical extensions to the overall process and are de-emphasized, in contrast to historical approaches. The following distribution of effort is suggested for the five phases of systems integration. The manager must:

1. Establish the integration environment. Percent of effort: 10% to 15%..
2. Assess integration requirements and applications design. Percent of effort: 25% to 30%.
3. Assess data requirements and design integrated data warehouses. Percent of effort: 25%.
4. Develop and prototype integrated systems. Percent of effort: 25 % to 30%.

5. Implement integrated systems. Percent of effort: 5% to 10%.

The total distribution of the manager's time should equal 100%.

PHASE 1: ESTABLISHING THE INTEGRATION ENVIRONMENT

The purpose of this phase is to create the technical environment for systems integration and establish that environment's support and operating structure. The technical environment includes processors, servers, routers, operating systems, personal computers, workstations, memory, controllers and peripheral equipment, data base environments, and networks. The technical environment typically involves a three-tiered computing architecture and distributed computing environment, which include mainframes or midrange consolidating processors, LANs and WANs, and workstations in the form of personal computers (PCs).

Key Processes

The following key processes are performed in this phase. The staff:

- Conducts initial technical environment orientation and training.
- Develops the physical technical-environment plan.
- Creates the physical technical environment.
- Develops the technical environment's connectivity and networking plan.
- Acquires the technical environment.
- Installs the CPU technical environment.
- Installs all system software related to the technical environment.
- Installs the peripherals.
- Conducts the technical environment tests.
- Conducts the technical environment connectivity tests.
- Conducts the technical environment tests.
- Conducts all the peripheral tests.
- Validates the tests.
- Establishes all working areas and partitions as appropriate.
- Refines the instruction sets as necessary.
- Establishes final configuration of all peripherals.
- Readies the production environment.
- Certifies the production environment for application readiness.

Key Work Products/Results

This phase, upon completion, should produce or demonstrate the following:

- An integration-ready production environment.
- A certified operational environment.
- Trained operations and support personnel.
- Trained applications personnel.
- An enabled applications environment.

PHASE 2: ASSESSING REQUIREMENTS AND APPLICATION DESIGN AND CAPABILITIES

This phase is designed to develop formal functional requirements and technical specifications for the integration of systems and data. It usually involves the acquisition, design, and installation of software to support reengineered business processes and an integrated systems and data environment. During this phase, a comprehensive assessment is performed to formally document user requirements that enable reengineering. These are linked to immediate needs and new to-be operations and processes.

Once the requirements are documented and new to-be processes defined, a search is performed for enabling software that supports the new processes and end-user requirements. The process usually includes an assessment of both internal software that currently supports a process and external sources that are available from third-party vendors. In performing the assessment, the gaps between the functionality of existing software and the needs of reengineered processes are identified and evaluated in the context of technological feasibility, need, cost-performance, and return on investment. The functional requirements for process reengineering and systems integration are compared and mapped to the specific capabilities of the existing systems that are targets subject to the integration effort. Once mapped, the gaps are assessed and ranked according to priority for severity, and plans are developed to address them.

In performing this phase, the integrator must develop estimates for several key areas, including the number of concurrent users who will be performing functions on the integrated systems, estimates of transaction volumes to be supported by the integrated environment, and data volumes. These are used to direct where, within the overall IT architecture, the applications and data should reside. This sizing is especially important when considering the use of a client/server integration strategy.

Key Processes

The following key processes are performed in this phase. The staff:

- Develops functional test cases and flows based on to-be process and integration needs.
- Develops transaction tests.
- Develops flow tests.
- Identifies functional linkages.
- Develops detailed assessment procedures.
- Develops assessment criteria and expected results.
- Develops documentation procedures.
- Identifies and acquires application software systems and products.
- Confirms application software systems and products.
- Loads application software products, as needed.
- Develops final application software production assessment procedures.
- Establishes an application software production testing schedule.
- Performs an application software assessment.
- Documents performance characteristics.
- Assesses the performance to standard benchmarks and reengineering targets.
- Documents performance results.
- Performs comparative assessment:
  - By function.
  - By module.
  - By existing system.
- Documents assessment results.
- Determines interface requirements:
  - To and from flows.
  - Timing and frequency.
  - Files and data.
- Determines enhancement requirements.
- Identifies key changes to existing systems and acquired packages.
- Evaluates the impact of discrepancies and changes on the project and budget.
- Performs retests, as necessary.
- Readies the application software environment for integration.

Key Work Products/Results

This phase, upon completion, should produce or demonstrate the following:

- Application software products reconciled to contract.
- A readied application software system environment.
- An initial application software testing and validation.
- An integrated application software testing and training environment.
- An integrated application software production performance profile.
- A comprehensive report on comparative assessment.
- Integration and interface requirements.
- An application- and data base-enhancement list.
- Data base requirements.
- An economic assessment of changes.

PHASE 3: ASSESSING DATA REQUIREMENTS AND DESIGNING INTEGRATED DATA WAREHOUSE ENVIRONMENT

Phase 3 of the systems integration methodology is designed to identify, define, redefine, and rationalize data elements necessary for the realization of systems integration. This is perhaps one of the more rigorous processes. It directly involves end users in creating consensus definitions through the comprehensive mapping of existing and to-be data elements to the required and to-be logical and physical structures. The process also requires a normalization of all data definitions and a rationalization of all data elements.

Key Processes

The following key processes are performed in this phase. The staff:

- Establishes data dictionary standards, including:
  - Developing dictionary and definitions.
  - Determining logical structures.
  - Determining physical structures.
  - Identifying logical and physical relationships and dependencies.
  - Documenting any unique or unusual data structures, dependencies, uses and sources.
- Identifies all pertinent existing data elements, including:
  - Populating data dictionary with definitions.

- — Determining logical data structures.
- — Determining physical data structures.
- — Identifying logical and physical relationships and dependencies.
- — Documenting any unique or unusual data structures, dependencies, uses, variants, and sources.
- Performs data mapping, including:

  - — Identifying common data and equivalents.
  - — Identifying data discrepancies.
  - — Determining rationalization data requirements.
  - — Determining normalization data requirements.
  - — Developing discrepancy resolution plan.
  - — Developing rationalization and normalization plan.
- Performs rationalization and normalization, including:

  - — Determining data normalization rules.
  - — Determining data rationalization rules.
  - — Developing data rationalizations and normalizations.
- Develops an integrated data dictionary.
- Develops an initial integration data conversion and population plan.

Key Work Products/Results

This phase, upon completion, should produce or demonstrate the following:

- A data map.
- Rationalized data.
- Normalized data.
- A data dictionary.
- A data base conversion and population plan.

PHASE 4: DEVELOPING PROTOTYPE AND INTEGRATED SYSTEMS

Phase 4 of the systems integration methodology is designed to develop, test, install, and document integrated systems that support reengineering and new business processes. Systems are developed for the integrated environment and are tested through a prototyping process that links them with new business practices. The prototyping process tests overall systems performance by simulating transactions volumes and the environment that the systems will support.

During Phase 4, a series of comprehensive unit and full-system-integration tests are performed on the communications network, applications, and data bases.

In developing new applications, the integration process is facilitated by the use of CASE tools, such as PowerBuilder by Power Soft Corp. These tools have capabilities that support systems development through the use of templates, formal rules, and standard presentation techniques. The integration effort also benefits from the use of project management software and data base definition and development tools.

Key Processes

The following key processes are performed in this phase. The staff:

- Develops overall integration, programming, and interface specifications.
- Develops integration work plans, including:
  - A conceptual overview.
  - A program structure.
  - Input/output shells.
  - Files accessed/generated.
  - Frequency/timing.
  - Control programs.
  - Data structures.
- Develops programs and performs integration, and develops:
  - Presentation standards.
  - Logic standards.
  - Programming specifications.
  - Programs and applications.
- Documents programs for:
  - System/program.
  - User/development.
- Performs tests, including:
  - Developing test procedures.
  - Developing test criteria.
  - Performing tests.
  - Documenting results.
- Resolves all discrepancies.
- Validates all programs for production use.

- Places validated interfaces in pre-production library.
- Finalizes all documentation.

Key Work Products/Results

This phase, upon completion, should produce or demonstrate the following:

- Integration and programming specifications.
- Final programs.
- Production programs and interfaces.
- A secure preproduction library.
- Formal test plans.
- Established criteria.
- A formal acceptance.
- Substantiated documentation.
- Audit trails.
- Documented procedures.
- A preproduction fine-tuning list.

PHASE 5: IMPLEMENTING INTEGRATED SYSTEMS

The objective of this phase of the project is to prepare, convert, and migrate to the integrated application and data base environment. The integrated systems environment represents the culmination of the systems integration process. During this phase, enabling solutions are delivered, and users are trained to exploit fully the capabilities of integrated technology and applications.

Key Processes

The following key processes are performed in this phase. The staff:

- Develops formal testing procedures and scenarios, including:
  - Functionality tests.
  - Transactions tests.
  - Compliance tests.
  - Substantive tests.
  - Data relationships and validity.
  - Access control.
  - Interfaces, linkages, and feeds.

— Special tests.
- Develops specific transaction loads (i.e., volumes) for performance testing.
- Develops testing plan.
- Develops documentation procedures, including:
  — Error recording.
  — Error resolution.
- Establishes performance documentation.
- Develops sign-off procedures from the:
  — Testing team.
  — Project managers.
  — Project steering committee.
  — End users.
  — Quality assurance.
- Develops the preproduction fine-tuning list.
- Establishes the data-conversion schedule.
- Develops the data-conversion standards and documentation.
- Develops a detailed data-conversion plan.
- Develops the data-conversion and testing plan.
- Performs data conversion and populates the data-base structures.
- Performs the following tests and validates test results:
  — Transactions tests.
  — Validation tests.
  — Accuracy tests.
  — Reads/translation tests.
  — Population tests.
- Documents test results.
- Isolates validated data bases for preproduction.
- Documents data bases.
- Develops data administration and management standards.

Key Work Products/Results

This phase, upon completion, should produce or demonstrate the following:

- Populated data bases.
- Validated data.
- A documentation trail.

- An updated data dictionary.
- Populated data repositories.
- Completed programs and integrated systems.
- A completed multitiered computing environment.

SUMMARY

The systems integration process is essential to the reengineering efforts of any process and organization. The approach described above represents an outline for accelerating the systems integration process. The major objective of any integration is the effective melding of technology, applications, and data into a framework that supports cross-functional business processes, accessibility to data, and leaner organizational structures.

However, as in reengineering, speed of delivery is of the essence. The rapid delivery of systems is necessary to support reengineering and to fully exploit the investment in technology. In this regard, the methodology presented must be supported through the use of rapid development techniques and tools and CASE products.

5-2
Managing the Systems Integration Project: A Reengineering Approach to Project Management

MICHAEL A. MISCHE

E ffective technical and project management is neither trivial nor simple. Systems integration implicates far more than legacy systems and new "go to" technologies and affects more than the traditional information management and technology (IM&T) organization. Reengineering and systems integration efforts extend into the very culture and structure of the enterprise and must be managed formally.

The complexities of systems integration and reengineering projects require an effective project management methodology that addresses not only the technical aspects of the project but also organizational and cultural changes. Unfortunately, many integration and reengineering efforts do not meet their targets, and the methods used to manage those projects are in part a contributing factor to their failure. This chapter discusses the strengths and weaknesses of traditional project management methods and provides a new framework for managing the systems integration project.

THE NEED FOR A NEW APPROACH TO PROJECT MANAGEMENT

To avoid the pitfalls that plague many integration efforts, IM&T organizations are sending legions of analysts and programmers to project management and quality improvement training courses. Although these steps are appropriate and, in many cases necessary, the typical two- or three-day project management seminar is inadequate in providing any meaningful assurance that the integration project will be completed in an efficient and cost-effective manner. Recognizing the benefits of external resources, IM&T organizations are turning to third-party systems integrators and management consultants. As discussed in Section 7, engaging a systems integrator to facilitate the integration process helps avoid many of the common problems associated with systems integration; however, external consultants and systems integrators are hired to do a specific task for a limited time and are quickly gone from the scene after the project is over. They do not have to live with the consequences of the process. Internally, many IM&T organizations lack the global perspective, political empowerment, and broad-based business knowledge necessary to manage the multidisciplinary approach to systems integration. Thus, it is common to see some combination of resources dedicated to the systems integration project.

The experiences of many organizations in the systems integration and reengineering arena emphasize the need for a more effective method for managing these highly complex and inherently risky projects. As discussed in Section III, mounting evidence indicates that 70% to 80% of systems integration and reengineering efforts end in failure, at least in the sense that projects fall short of expected goals and results.

The traditional project management methods that were so effective in managing classic systems development and operational improvement efforts are simply not up to the rigors of cross-platform and cross-functional systems integration. The project management needs of a systems integration and reengineering project are very different today, and the classic project management methods were not designed to be used in projects that involve cross-functional processes, rapid migration paths to new technologies, and the inevitable flattening of the organization. Organizations can no longer rely solely on Gantt charts and dozens of MIS personnel huddled in cubicles around terminals to manage multimillion-dollar efforts.

With systems integration projects spanning several years and running into the tens and hundreds of millions of dollars, senior leadership is well justified to ask, "Why integrate?" In many cases, the wholesale replacement of systems and outsourcing of the entire IM&T function could be less expensive and far more effective in supporting the organization.

Responsive project management considers the integration process systematically and anticipates the project's effects on the entire organization rather than on just the individual applications, IM&T department, hardware, and software. Where traditional project management processes have focused on individual elements of project management, such as work plans, PERT charts, and project team-reporting relationships, managing systems integration endeavors requires the management of systemic change. The pace at which most systems integration projects move mandates that resources, new "go to" technologies, project problems, and issues be identified in an accelerated, preemptive manner, as opposed to a reactionary manner. Project management for systems integration must be both opportunistic and preventive. Managing these megadollar efforts requires a framework that melds the best of the tried-and-true, traditional project management practices with new measurements and a process that provides for the constant examination of the functional relationships and technical dependencies among key project variables and processes. The project management practices for systems integration must address not only the typical IM&T technical issues but also the rate of organizational change that is related to systems integration. Thus, there is a preemptive dimension to the systems integration management process that must be provided.

A MODEL FRAMEWORK FOR MANAGING THE SYSTEMS INTEGRATION PROJECT

A starting point for managing systems integration projects requires a definition for project management. Defined as a process, project management is a coordinated set of processes that provides for:

- The effective management of project resources.
- Optimized application of project resources.
- Correct, consistent, and timely measurement and communication of the project's status.

The model framework for managing the systems integration project has five components:

1. A Comprehensive Project Work Plan (CWP).
2. A Project Management and Tracking System (PMT).
3. Project Team Organization (PTO).
4. Quality Assurance Process (QAP).
5. Employee Training and Process Simulation (ETS).

Collectively, the five components are important because they provide:

- Empowered work groups.
- A uniform discipline and structure to follow and manage resources.
- A structure that emphasizes quality in the project management process.
- A balance between traditional practices, which concentrate on control and reporting, and the modern need to be anticipatory in the project management process.

- A process to measure and manage the rate of project progress and technological and organizational change effectively.

Project Team Organization

An effective and functional project team is paramount to the success of any integration effort. It makes little difference how encompassing and refined the project work plan is if the tasks are assigned to a team with mediocre abilities or insufficient time.

Conversely, it is tremendously demoralizing to a potentially effective team to be harnessed by an ineffective work plan or inept project leadership. As discussed in Chapter 5-3, the project teams should be staffed by cross-functional and multidisciplinary individuals who are technically competent and proficient in their knowledge of the business.

Effective project management requires that a multidisciplinary and cross-functional team structure be used to integrate key technology requirements, organizational responsibilities, and business processes. Process integration requires an empowered team structure for the governance and management of the project. The organization structure for integration is composed of specialty teams, each dedicated to a specific purpose and need. The specialty teams include key leadership, management and knowledge workers. Sponsorship, accountabilities, and responsibilities for the results of the project and its process are shared across disciplines. Not all the project teams work simultaneously throughout the process of integration. The project plan, team specializations, project phases, and individual work schedules drive the timing and sequencing of the teams. Certain teams, such as those with a highly specialized focus and finite purpose, disband once their work has been completed. Other teams have a constant role through project completion. The structure also lends itself to the inclusion of customers, trading partners, suppliers, and business alliance members to function in a team member capacity.

Because each integration and reengineering effort is unique, there are no set rules for either the number and types of teams required to support and manage the integration process effectively or for the number of individuals assigned to a team. In general, it is best to limit the number of individuals assigned to any one team to no more than 10. Ideally, teams should be composed of three to seven people with one core team responsible for the entire effort. The teams are integrated into the overall working structure through the management practices and tools incorporated in the CWP and PMT, including direct reporting by individual team leaders, oversight by the project steering committee, structured weekly, bi-weekly, monthly, and quarterly briefing sessions and a consistent training program for all project team members in the practices and processes of project management.

The organization structure for the integration and reengineering effort should be diamond-shaped rather than hierarchical. This structure is flatter and more conducive to the simultaneous coordination of many specialty teams and helps to create a more seamless communication flow among teams and individual team members. It also supports the integration of skill sets, responsibilities, and knowledge and helps ensure the proper balance of these factors as well as organizational levels throughout the systems integration project.

THE COMPREHENSIVE PROJECT WORK PLAN (CWP)

The nucleus of systems integration project management is the CWP. As discussed in Chapter 5-1, the CWP contains all of the tasks, staffing assignments, deliverables, dependencies, prerequisites, milestones, and estimated effort necessary to complete the integration process. The CWP is organized into six components.

Project Overview

The initial component of the CWP provides for a narrative description of the project. Providing a broad conceptual overview is important because it establishes the formal rationale, expectations, and parameters in which the project is performed. It also provides for a common basis from which to assess the project and serves as an effective reference point when communicating about the project. The major items included in the overview include:

- Objectives, scope, and intended results of the integration effort.
- Organization of the project's phases and life cycle.
- Estimated level of effort and types of resources required.
- Interdependencies among tasks and work products.
- Description of project teams and responsibilities.
- Expectations and anticipated benefits.
- Timing and calendar for the effort.

Project Phases

The second level of the CWP provides a high-level description and representation of the integration project's life cycle. Organizing the effort into discrete life cycle phases such as those described in Chapter 5-1 creates an effective and convenient working structure. For each phase of the integration project, narrative and supporting representations are developed describing:

- Objectives, purpose, and scope of the integration effort, with linkages to business objectives and organizational benefits.
- Specific activities to be accomplished.
- Timing and relationships among phases, activities, and tasks.
- Estimated level of professional effort for each phase and the distribution of that effort.
- Major deliverables and work products to be developed.
- Staffing assignments at the project team level.
- Measurements and standards for evaluating the effectiveness of the phase and quality of work performed.

Developing a clear representation of the phases and work plan for the integration project is important to understanding the structure of the project and measuring results and progress.

Detailed Project Work Plan

The third element of the CWP provides a detailed description of how the systems integration work is to be performed. This involves the development of detailed work plans, activity lists, tasks, work steps, and staffing assignments for the integration project. The detailed project work plan includes:

- Descriptions of the detailed activities, tasks, and work steps to be performed.
- Detailed descriptions of the individual work products and project deliverables to be developed.
- Timing requirements and key assumptions about the tasks and the development and presentation of deliverables and work products.
- Identification of dependencies and critical path issues related to the work plan.
- Identification of individual responsibilities, task assignments, budgets, and labor estimates.
- Assessment of critical path items and the inter-dependencies between tasks, deliverables, and project team assignments.

Project Team Organization

The fourth part of the CWP provides the identification and description of the project teams assigned to perform the integration project. The description includes:

- A definition of the project team and its role, purpose, and charter in the integration effort.

- The activities, tasks, and work steps assigned to each team and their respective dependencies and timing.
- The assignment of the work products to be produced by each team.
- An estimate of the number of days of commitment for each team to the project and the dependencies among the teams.

There are a number of different models for organizing resources for the integration project. Chapter 5-3 provides an example structure.

Individual Team and Task Assignments

The fifth component of the CWP provides for the assignment of individual team members to a particular project team, tasks, and project deliverables. Specific topics addressed in the individual team and task assignments include:

- Individual tasks and work product assignments.
- Days or estimated hours of commitment to complete each task and work product.
- Timing and schedule for completing tasks and work products assigned and their dependencies to other products and activities.

Daily To-Dos

The sixth and final level of the CWP is the daily to-do list, which summarizes all the activities that are assigned to team members to be worked at or completed on a daily basis. It is, in a sense, a production schedule for the integration process. Because of the fluidity of the systems integration process, the daily to-do lists are subject to changes. As such, they not only consider major changes in the work plan and immediate needs and opportunities, but also anticipate the availability of personnel and factor them into the individual schedules of each member of the project team.

An important but almost always overlooked measure is the rate at which the project team works compared to the planned activities. Developing this measure requires the capturing of data on the time spent by the individual project team member directly on the assignment as compared to time spent on other activities and total available time. This measure is called the effectivity rate.

THE PROJECT MANAGEMENT AND TRACKING SYSTEM (PMT)

The second component of the systems integration project management structure is the PMT, which facilitates the tracking and management of the CWP and individual work plan tasks. A project management process composed of uniform procedures and forms designed to capture information about the individual activities and results of the project, PMT can be automated and used with any third-party project management software. The PMT requires training and discipline if it is to be used effectively to manage and measure the progress, status and quality of the integration effort. The PMT helps ensure the continuous calibration of actual results achieved to the original project plan through the use of consistent progress tracking metrics.

Managing the systems integration process requires a combination of individual status reporting and detailed work plan updating procedures. Systems integration projects require the constant assessment of the following items:

1. *Consistent application.* The practices, procedures, and methods used to enable project management must be consistently applied, with no exceptions or deviations from schedules.

2. *Procedures and practices.* Project management involves the constant application of procedures and recurring practices that are designed to capture, normalize, and assess data about the project and individual project team member performance.

3. *Multiple inputs.* Systems integration projects have a multitude of activities and processes under way at any one point. As such, project management must provide for multiple sources of information and evaluation points.

4. *Measurements.* An effective project management process must not only report on progress but measure and assess it against established baselines. The measurements must be valid, meaningful, and most importantly, consistently applied.

5. *Constant calibration.* Once measured, the project must be constantly evaluated and calibrated to its intended goals, expectations, objectives, and original and revised schedules for completion and financial requirements.

Communicating the progress and disposition of the systems integration process is key to managing the activities of those assigned to the project and the expectations of the organization. Systems integration projects necessitate that communications be meaningful, frequent and in a standard format that is easy to comprehend, use, and reference. Exhibit 5-2-1 provides an illustration of the PMT process.

The major elements of the PMT process include:

- PMT practices and procedures.
- PMT project management structure and reporting framework.
- PMT management reports.

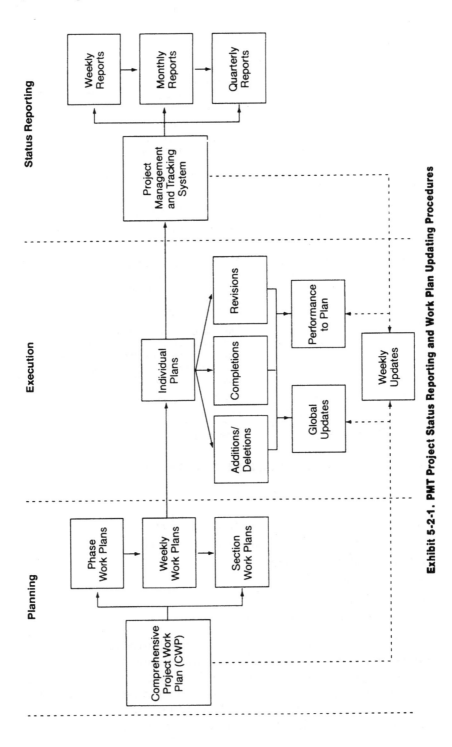

Exhibit 5-2-1. PMT Project Status Reporting and Work Plan Updating Procedures

Number	Form	Description
1.0	Comprehensive Project Work Plan (CWP)	The CWP provides the detailed work plan, task subtasks, work products, start and stop dates, assignments, deliverables, and estimated level effort for all aspects of the project. The CWP is updated weekly and is PC based.
2.0	Team Leader Reporting Procedures	Team leader procedures are formatted procedu for reporting on individual project team activitie accomplishments, and changes to the CWP. Th procedures are updated weekly and are used t coordinate individual project team member acti
3.0	Team Member Reporting Procedures	Team member reporting procedures are format forms and procedures used to communicate, tr and monitor the individual activities of personn assigned to the project. It is completed weekly individual project team members.
4.0	CWP Task/Work Product Additions	The form is used to describe minor additions a deletions of individual work plan tasks on the o project. The task, estimated effort, staffing implications, and start and stop dates are provi The form is completed weekly.
5.0	Change Management Control	The change management form is used to desc and assess the impact of a significant change i project's scope, work plan, or staffing plan. It is as necessary and requires the approval of the project steering committee
6.0	Project Status Reporting Structure/Formats (PSR)	The PSR is a standard set of procedures and f that summarizes the project's status and communicates progress to management and th project team.

Exhibit 5-2-2. PMT Forms Description

- Team leader reporting procedures and forms.
- Team member reporting procedures and forms.
- CWP updating procedures.
- CWP tacking procedures.
- Change management processes.
- Standard metrics and measures.

Exhibit 5-2-2 shows the various forms and procedures that are used in the management of the project and their frequency of use.

The project management and tracking system (PMTS) is a coordinated process of procedures, forms and practices for managing the systems integration project. The PMTS can be used to support a variety of projects and is especially effective in those which are large and complex

PMT Component	Form	Frequency of Use				
		Weekly	Biweekly	Monthly	Quarterly	As Needed
Project Management Tracking System (PMTS)	---	•	•	•	•	
Comprehensive Project Work Plan (CWP)	1.0	•	•	•	•	•
Team Leader Reporting	2.0	•				
Team Member Reporting	3.0	•				
CWP Updating Procedures	4.0	•				
CWP Additions/Deletions Procedures	4.0	•				•
Change Management Procedures	5.0					•
Project Status Reporting	6.0	•		•	•	
Presentations to Project Steering Committee (PSC)	---				•	

Exhibit 5-2-3. PMT Forms and Frequency of Use

and have a number of resources assigned to them. The PMT is composed of six forms. (See Exhibit 5-2-3.) The forms are used throughout the life of the project to collect and organize a variety of data related to the actual performance of the systems integration. The major requirement for use of this project management process is discipline. That is, the project team must constantly use the forms and procedures to collect, organize, and present data about the integration process to accurately reflect the progress and quality of the effort. Exhibit 5-2-3 provides a summary of frequency of using the various forms. The process of updating the PMT involves the completion of time logs and task tracking forms on a periodic and recurring basis. The updating process is illustrated in Exhibit 5-2-4.

Form 1 of the PMT is the Comprehensive Work Plan (CWP). Because the CWP is different for each systems integration effort, a sample is not provided here. However, Chapter 5-1 provides a discussion of the systems integration life cycle, which can be used as a model for developing the comprehensive work plan for systems integration.

Form 2 of the PMT is used for team leader reporting, to aggregate and summarize the individual activities and progress of a specific project team (see Exhibit 5-2-5). The form is completed by individuals assigned to a project team, submitted to the team leader and aggregated to provide a

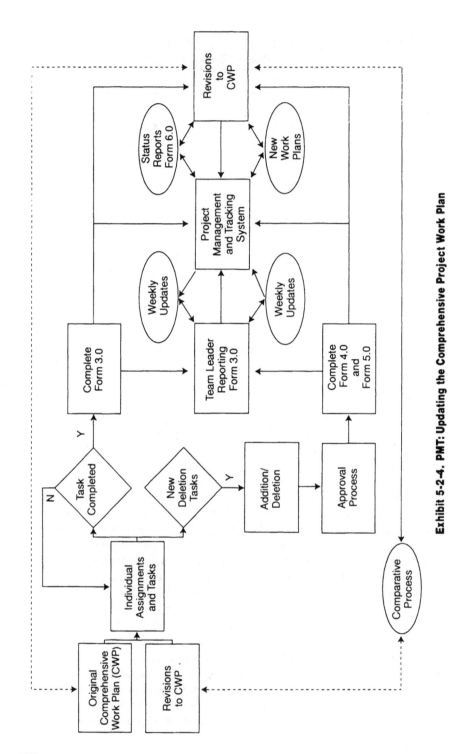

Exhibit 5-2-4. PMT: Updating the Comprehensive Project Work Plan

representation of work performed and completed. The form reconciles work accomplished to the work planned and assigned for a specified period of time. Form 3 is used to collect detailed information and updates for individual tasks and work steps that are assigned to specific individuals. (See Exhibit 5-2-6.) The form also collects information about how the individual team member is performing to the preplanned budget, the status of individual productivity and additions and deletions to the individual's workloads. One of the most important data elements captured by this form is the summarization of unplanned activities and interruptions which can adversely affect individual productivity and the status of tasks and work steps.

Form 4 is used to collect information that will update the CWP. The form summarizes additions, deletions, and extensions to the work plan including estimated effort, new start and completion dates, and project team assignments. (See Exhibit 5-2-7.)

Form 5 is used to facilitate change management. As discussed in Chapter 4-6, change management is integral to any successful integration effort. Form 5 is designed to collect, describe, and assess the impact of changes to the work plan. The form summarizes important information such as economic impact and the effect of the change on the critical path of the effort. (See Exhibit 5-2-8.)

Form 6 provides a standard structure for summarizing the key criteria and information about the systems integration effort. The standard format ensures consistency and continuity are achieved in assessing the status of the project and in communicating the results of the project. (See Exhibit 5-2-9.)

Examples and instructions for each of the forms that comprise the PMT are provided below.

Using the PMT to Update the Plan

Implicit to using the PMT successfully is the updating of the CWP and the production of status reports. Specific PMT procedures include:

1. *Team leader reporting procedures.* Team leader procedures are formatted procedures for reporting on individual project team activities, accomplishments, and changes to the CWP. These procedures are updated weekly and are used to coordinate individual project team member activities.

2. *Team member reporting procedures.* Team member reporting procedures are formatted forms and procedures used to track and monitor the individual activities of personnel assigned to the project. They are completed weekly by individual project team members.

3. *CWP task/work product additions.* This weekly form is used to de-

Project Name:		Form 2.0 Team Leader Reporting

Reporting Period: From   (1)   To   (End)

Your Name: (2)   Self-Explanatory   **Extension:**

Project Team Leader: (3)   Self-Explanatory   **Extension:**

Project Status Report

Activities Accomplished This Period			Activities Accomplished to Date		
	Tasks	Hours		Tasks	Hours
Activities Completed	(10)	_____	Original Plan	(4)	_____
Additions	(11)	_____	Completed to Date	(5)	_____
Deletions	(12)	_____	Additions to Date	(6)	_____
			Deletions to Date	(7)	_____
			Net Tasks Not Completed	(8)	_____
			Tasks in Process	(9)	_____

Exhibit 5-2-5. Team Leader Reporting

280

Forms Description

Form Title: **Team Leader Reporting**

Form Number: **2.0**

Form Description

Purpose
To Summarize Individual Activity and Progress by Project Team and Leader

Used By	Reference Contacts
Project Team Leaders Project Management Group	

Format Description

Reference Number	Description
Reference Number	**Description**
1. For Week Ending	Time Reporting Period
2. Your Name	Your Name
3. Project Leader	Project Team Leader's Name
Project-to-Date Totals	
4. Original Plan	Number of Tasks and Hours in the Original Plan
5. Completed to Date	Number of Tasks Completed Project to Date
6. Additions to Date	Number of Tasks and Hours Added Project to Date
7. Deletions to Date	Number of Tasks and Hours Deleted Project to Date
8. Net Tasks Not Completed	Net Number of Tasks to Be Completed Project to Date
9. Tasks in Process	Number of Tasks Currently in Process
Current Period Totals	
10. Completed	Tasks Completed During the Current Reporting Period
11. Additions	Tasks and Hours Added to the Plan During the Current Reporting Period
12. Deletions	Tasks and Hours Deleted from the Plan During the Current Reporting Period

Exhibit 5-2-5. (*Continued*)

Effective Date:_____

Procedure

Procedure for

Project Leader Reporting and Individual Time Sheet Collection
Facilitates Individual Project Team Leaders in the Completion of Weekly Project
Team Leader Reports and in Collecting Weekly Reports from Team Members

Procedure Supports	Frequency
Forms Used: 3.0, 4.0 Forms Completed or Submitted by This Procedure: 2.0.	Weekly

Responsibility	Reference Contacts
Individual Project Team Leaders	

Procedure Steps

Example Procedures for Collecting Team Member Reports:

1. Collect all time sheets from project team members (Forms 3.0 and 4.0) by 2:00 pm every week.

2. Verify that all team member time sheets have been collected.

3. Review all time sheets for hours worked, estimates to complete, revised start or end dates, status of activities, unplanned activities, and comments.

4. If necessary, identify any new tasks or projects (see Form 4.0) that develop. The new tasks or projects should be documented using the procedures provided on Form 4.0.

5. Resolve any questions or issues that arise from entries on the time sheets with respective team members before submitting to project controller.

6. Approve all time sheets for use by initialing the upper right-hand corner.

7. Submit to the project controller no later than Monday at 10:00 am.

Example Procedures for Completing the Summary Sheet:

1. Enter the time reporting period in Field Number 1 for week ending. Reporting periods begin at 7:00 am on Monday and end at 5:00 pm the following Sunday.

2. Enter the total number of tasks that have been completed for this project in Field Number 5, Completed to Date. This number can be obtained from the Activity Status, Time, and Usage. Report by counting the number of tasks that have a status of "Complete."

3. Enter the total number of tasks and total hours that have been added to this project in Field Number 6, Additions to Date. This number must be accumulated each week, for the duration of the project.

Exhibit 5-2-5. (*Continued*)

4. Enter the total number of tasks and total hours that have been deleted from this project in Field Number 7, Deletions to Date. This number must be accumulated each week, for the duration of the project.

5. Calculate the net tasks not completed, Field Number 8, For This Project. This total is derived by adding additions to date, Field Number 6, To Original Plan, Field Number 4, and then subtracting from that sum completed to date, Field Number 5, and Deletions to Date, Field Number 7.

6. Enter the number of tasks that are currently in process in Field Number 9, Tasks in Process. This number can be obtained from the Activity Status, Time, and Usage report, by counting the number of tasks that have either a status of start or a percent complete number in the status column.

7. In Field Number 10, Completed, enter the number of tasks that were completed during the current reporting period. This number must be obtained by comparing the total number of completed tasks from the Activity Status, Time, and Usage reports from the current and previous period.

8. Addition for the Current Period, Field Number 11, must be derived from the Task Maintenance forms received from the team members for the current reporting period.

9. Deletions for the Current Period, Field Number 12, must also be derived from the Task Maintenance Form 4.0, received from the team members for the current reporting period.

Exhibit 5-2-5. (*Continued*)

scribe minor additions and deletions of individual work plan tasks on the overall project. The task, estimated effort, staffing implications, and start and stop dates are provided.

4. *Change management control.* The change management form is used to describe and assess the impact of a significant change in the project's scope, work plan, or staffing plan.

5. *Project status reporting structure and formats (PSR).* The PSR is a standard set of procedures and metrics that summarize the project's status and communication progress to management and the project team.

Translating updates into an effective project is almost always a problem because of the high level of discipline that is required; however, updating the plan can become routine, indeed mechanical, if the correct practices, procedures, and discipline are in place.

Using the PMT to Measure Progress and Status

Using a consistent and valid measurement system is important because it supports trend analysis, isolates problems, indicates progress and helps

Project Name:	Form 3.0
	Team Member Reporting

Reporting Period:	**From**	(1)		**To**	(End)
Your Name:		(2)	Self-Explanatory	**Telephone:**	
Project Phase or Section:		(3)			

Using the Individual Work Plan from the Comprehensive Work Plan, Complete the Following:

							This Period				
Status as of: _____							**New Accounts/Estimates**				
(4) Task #	(5) Plan	(6) Account	(7) Estimate to Completion	(8) Start Date	(9) End Date	(10) Status	(11) Account Use	(12) Estimate to Completion	(13) Start	(14) End	(15) Status
3.2.1.10 Database (Example)	40	32	12	1/1/95	1/10/95	A	8	4	1/10	1/10	A
(16) Unplanned Activities -Mtgs.				—	—		32				
(17) Total Actual Usage This Period: 40											
(18) Comments or Outstanding/Anticipated Problems											

Exhibit 5-2-6. Team Member Reporting

| Form Title: | Project Team Member Reporting |
| Form Number: | 3.0 |

Form Description

Purpose
Used Weekly by Each Member of the Project Team, IM&T, and Functional, to Report on Tasks, Effort, Activities and Accomplishments.

Used By	Reference Contacts
Individual Project Team Members	

Format Description

Reference Number	Description
1.	Time Reporting Period: Monday, 7:00 am, Through Sunday, 5:00 pm
3.	Project Description
4.	Task Number and Task Description
Project-to-Date Totals	
5. Planned	Number of Hours Planned to Work on the Task
6. Actual	Actual Hours Incurred on the Task Project-to-Date
7. Estimate to Complete	The Latest Estimate, in Hours, to Complete the Task, as of 5:00 pm of the Previous Week
8. Start Plan/Revision	Planned Start Date of the Task
9. End Plan/Revision	Planned End Date of the Task
10. Status	Status of the Task (Blank = Not Started, Compl = Complete, Strt = Started)
11. Actual Usage	Actual Hours Incurred on the Task for the Reporting Period
12. Estimate to Complete	New Estimate to Complete in Hours
13. New Start	New Start Date for the Task
14. New End	New End Date for the Task
15. Status	Status for the Task
16.	Unplanned Activities Description Section
17.	Total Hours Worked During the Reporting Period
18.	Comments or Anticipated Problem Description Section

Exhibit 5-2-6. (*Continued*)

Effective Date:_____

Procedure

Procedure For
Project Team Member Reporting

Procedure Supports	Frequency
Forms Used: 3.0 Forms Generated: 2.0	Weekly

Responsibility	Reference Contacts
All Project Team Members	

Procedure Steps

Procedures for Completing the Time Sheet:

1. For each task worked during the reporting period, enter the actual hours in Field Number 11, Actual Usage. Enter whole hours only.
2. In Field Number 12, Estimate to Complete, enter the number of hours remaining to complete the tasks. Enter whole numbers only.
3. If the actual start date of the task will be different from the planned start date, enter the new start date in Field Number 12, New Start.
4. If the end date of the task will be different from the planned end date, enter the new end date in Field Number 14, New End.
5. If the task is completed during the reporting period, enter a C or Comp in Field Number 15, Status.
6. Describe any unplanned activities in Section 16, unplanned activities, not related to the project that were worked on during the reporting period. These unplanned activities categories are vacation, sick days, holiday, training not related to the project, or tasks not related to the project. Include hours worked on each unplanned activity and a brief description, if necessary.

   The time worked on new tasks that has been approved to be added to the plan should not be entered in this section, but entered on the time sheet provided for this purpose (see form 4.0).
7. Total the hours incurred on the project-related tasks listed on the time sheet and the hours from unplanned activities and enter in field 17, Total Usage for Period. This total should equal the total work hours for the week.
8. Enter any comments or problems in section 18, Comments/Outstanding or Anticipated Problems.
9. Each project team member receives a time sheet for each week that includes tasks that you have been assigned to work on. The time sheets are to be completed and turned in to the appropriate team leader for each project by 2:00 pm Friday.

   Time sheets are distributed on Tuesday morning of the current reporting period. Reporting periods begin on Monday at 7:00 pm and end on the following Sunday at 5:00 pm.

Exhibit 5-2-6. (*Continued*)

Project Name:		Form 4.0 CWP Maintenance New/Deleted Tasks								

Reporting Period: From   (1)   **To**   (End)

Your Name: (2) Self-Explanatory   **Extension:**

Project Team Leader: (3) Self-Explanatory   **Extension:**

Project Status Report

Activity Number (If Known)	Task Description	Original Budget				Revised				
		Actual	Est. to Comp	Start	End	Status	Est. to Comp	New Start	New End	Assign To
(4)	(5)	(6)	(7)	(8)	(9)	(10)	(11)	(12)	(13)	(14)

Comments: For additions to the project plan, indicate predecessor and successor activities.

(15)

Exhibit 5-2-7. CWP Maintenance

Form Title: CWP Maintenance **Form Number:** 4.0	

Form Description

Purpose To update the CWP

Used By All Project Team Members	**Reference Contacts**

Format Description

Reference Number	**Description**
1. Reporting Period	Dates Covered by Form
2. Your Name	Person Completing Form
3. Team Leader	Team Leader's Name
4. Activity Number	Task Number
5. Task Description	Title of Task
6. Actual	Actual Hours Expended to Date
7. Estimate to Complete	Estimated Hours to Complete Task
8. Start	Original Start Date
9. End	Original End Date
10. Status	C = Complete, D = Delete, N = New Task, R = Revision to Existing Task
11. Estimate to Complete	Estimate to Complete Task, Revised
12. New Start	New Start Date
13. New End	New End Date
14. Assigned To	Project Team Member's Initials
15. Comments	

Exhibit 5-2-7. (*Continued*)

Effective Date:

Procedure
Procedure for

To update existing tasks in the CWP, revise existing estimates, designate tasks as complete, designate precedessor/succesor activities, provide additions and deletions, revise work schedules, change work step/tasks responsibilities and assignments, and revise start and stop dates.

Procedure Supports Forms used: N/A Forms generated: 4.0	**Frequency** Weekly or as Needed
Responsibility All Project Team Members	**Reference Contacts**

Procedure Steps

Procedures

1. Item 1 – Complete reporting date

2. Item 2 – Complete your name

3. Item 3 – Complete project team leader's name

4. Enter 4 – (activity number) and 5 (project description) for the project to perform task maintenance.

5. Item 10   Enter the code for the type of change/project maintenance to be performed:

   – Enter an N to add a task to the project.
   – Enter a C to show the completion of a task.
   – Enter an R to change or revise an existing project task.
   – Enter a D to delete an existing project task.

6. Enter 5 (task description) as indicated on the form. The task description is required for all task maintenance.

7. Indicate 7 (start date) and 8 (end date) for the task. This information is required only for adding new tasks or changing the dates for revised tasks.

8. If a new task, enter the new task number.

9. Provide information about the resources assigned to the task in section 14 (resource assignments). Resource information is required only for adding new tasks or for changing the resource assignments for revised tasks.
   – Indicate the number of 11 (hours) planned for each individual assigned to the task.

Exhibit 5-2-7. (*Continued*)

Procedure Steps (Continued)

Procedures

10. Use Section 15, comment, for any explanatory information about the changes being performed. For added tasks, indicate the existing predecessor task that the added task must follow.

11. Submit the completed form to the team leader.

12. The team leader reviews the task maintenance form and must approve it before it can be entered into the project plan.

13. Task maintenance is entered into the CWP before the entry of time sheet information.

Exhibit 5-2-7. (*Continued*)

---

to manage expectations. There are sixteen critical measurements of progress and status for the systems integration project.

1. Number of tasks actually accomplished this reporting period and to-date as compared with those planned in the original and revised CWP.

2. Hours or days actually expended this reporting period and to-date as compared with those planned in the original and revised CWP.

3. Percentage of actual completion of tasks, effort, and work products as compared with those planned in the original and revised CWP.

4. Number of work products actually completed as compared with those planned in the original and revised CWP.

5. Project milestones and phases actually completed as compared with those planned in the original and revised CWP.

6. Tasks added this reporting period and to-date to the CWP as compared with those planned in the original and revised CWP.

7. Tasks deleted this reporting period and to-date from the CWP as compared with those planned in the original and revised CWP.

8. Hours/days of effort added this reporting period and to-date to the CWP as compared with those planned in the original and revised CWP.

9. Hours/days of effort deleted this reporting period and to-date from the CWP as compared with those planned in the original and revised CWP.

Number	Form	Description
1.0	Comprehensive Project Work Plan (CWP)	The CWP provides the detailed work plan, tasks, subtasks, work products, start and stop dates, assignments, deliverables, and estimated level of effort for all aspects of the project. The CWP is updated weekly and is PC based.
2.0	Team Leader Reporting Procedures	Team leader procedures are formatted procedures for reporting on individual project team activities, accomplishments, and changes to the CWP. These procedures are updated weekly and are used to coordinate individual project team member activities.
3.0	Team Member Reporting Procedures	Team member reporting procedures are formatted forms and procedures used to communicate, track, and monitor the individual activities of personnel assigned to the project. It is completed weekly by individual project team members.
4.0	CWP Task/Work Product Additions	The form is used to describe minor additions and deletions of individual work plan tasks on the overall project. The task, estimated effort, staffing implications, and start and stop dates are provided. The form is completed weekly.
5.0	Change Management Control	The change management form is used to describe and assess the impact of a significant change in the project's scope, work plan, or staffing plan. It is used as necessary and requires the approval of the project steering committee.
6.0	Project Status Reporting Structure/Formats (PSR)	The PSR is a standard set of procedures and forms that summarizes the project's status and communicates progress to management and the project team.

Exhibit 5-2-7. (*Continued*)

---

10. Work products added this reporting period and to-date to the CWP as compared with those planned in the original and revised CWP.

11. Work products deleted this reporting period and to-date from the CWP as compared with those planned in the original and revised CWP.

12. Tasks actually initiated and in-process this reporting period as compared with those planned in the original and revised CWP.

13. Financial reconciliation of capital and expense outlays and com-

	Date: (1)
Project Name:	**Form 5.0** **Change Management**

Reporting Period: From:_____(2)_____
                                      To:_____(2.1)_____

Your Name:_____(3)_____
Project Leader:_____(3.1)_____

Description and Sponsorship:_____(4)_____

Rationale and Benefits:_____(5)_____

Economic Impact:

Estimated Hours:[_____(6)_____]   Estimated Earliest Start:____(13)____

Estimated Resources:$____(7)____   Estimated Latest Start:____(14)____

Estimated Costs:_____(8)_____   Estimated Earliest End:____(15)____

New H/W Costs:_____(9)_____   Estimated Latest End:____(16)____

New S/W Costs:_____(10)_____

Other Costs: [_____(11)_____]

Total: $_____(12)_____

Sensitivity Analysis:

Overall Impact on Project:____(17)_____

Overall Economic Impact:____(18)_____

Business Risk:_____(19)_____

Approvals: (20)

_____   _____   _____

_____   _____   _____

_____   _____   _____

Exhibit 5-2-8. Change Management

Effective Date:	
Procedure	
Procedure for	
Documenting changes to the project, software, scope, and team. It should be completed when any material change to scope or requirements is contemplated.	
Procedure Supports Forms Used: Forms Generated: 5.0	**Frequency** As Needed
Responsibility Project Management Team	**Reference Contacts**
Procedure Steps	

Procedures

1. Complete items 1 through 5, as specified.

2. Complete items 6 through 16, stressing economic impact and impact on human resources.

3. Complete items 17 through 19, with particular emphasis on business risk.

4. Complete item 20. All changes to the project's scope, timing, budget, and resources must be formally approved.

Note: All changes to the project scope, requirements, design, timing, and budget must be reviewed and approved before performing any changes or establishing commitments and new dates.

Exhibit 5-2-8. (*Continued*)

Forms Description	
Form Title: Change Management	**Approved:** _____
Form Number: 5.0	**Not Approved:** _____

Form Description

Purpose
To document the potential impact and decisions regarding changes to the project's scope, timing, and resources.

Used By	**Reference Contacts**
Project Management Team Forms Generated: 5.0	

Format Description

Reference Number	Description
1. Date	Date Completed
2. Reporting Period	Period of Reporting
3. Your Name	Your Name
4. Project Leader	Project Leader's Name
5. Rationale Benefits	Description of Benefits and Rationale for Change
6-16	The impact on the project's budget and timing are described here.
17-19. Sensitivity	Description of the Business Risk and Risk to the Project
20. Approvals	

Exhibit 5-2-8. (*Continued*)

Project Name:	Form 6.0 Status Report Format

Project Name: From: _____ (1) _____ To: _____

Prepared by: _____ (2) _____ **Submitted to:** _____ (3) _____

Summary of Status: (4)

Accomplishments This Reporting Period			Comments
	Number	**Hours**	
Tasks Completed	(5)	(5)	
Products Completed	(6)	(6)	
Tasks Added	(7)	(7)	(11)
Tasks Deleted	(8)	(8)	
Tasks Active	(9)	(9)	
Tasks Started	(10)	(10)	

Accomplishments To Date			Planned Activities
	Number	**Hours**	1. 2. 3. 4.
Total Tasks Assigned	(12)	(12)	5. (17)
Total Tasks Completed	(13)	(13)	6.
Total Tasks Added	(14)	(14)	7.
Total Tasks Deleted	(15)	(15)	8.
Total Tasks Active	(16)	(16)	9.
Tasks Started	(17)	(17)	10. 11.

Issues Requiring Management Attention:

(18)

Signature:_____

Exhibit 5-2-9. Status Report Format

| Form Title: | Status Report Format | Form Number: | 6.0 |

Form Description

Purpose
This form is used to provide a high-level activity summary and status report on weekly progress. The form also serves as a template for monthly and quarterly reporting.

Used By	Reference Contacts
Project Management Team Project Leaders (Selectively)	

Format Description

Reference Number	Description
1. Reporting Period	Reporting Date (Usually Weekly)
2. Prepared by	Your Name
3. Submitted to	Reporting Relationship
4. Summary of Tasks	Narrative Description of the Status of the Project
5. Tasks Completed	Number of Tasks Completed This Reporting Period
6. Product Completed	Number of Work Products or Milestones Completed This Reporting Period
7. Tasks or Hours Added	Number of New Tasks and Hours Added to the CWP This Period
8. Tasks/Hours Deleted	Number of Tasks and Hours Deleted from the CWP This Period
9. Tasks/Hours Active	Number of Tasks and Hours That Are Underway, in Process and Active During This Reporting Period
10. Tasks/Hours Started	Number of Tasks Initiated This Period
11. Comments	Any Pertinent Comments Related to This Reporting Period
12. Total Tasks/Hours Assigned	The Number of Total Tasks Assigned for the Entire Project
13. Total Tasks/Hours Completed	The Number of Total Tasks Completed, Inclusive of This Reporting Period, for the Entire Project
14. Total Tasks Added	The Total Number of New Tasks Added to the CWP from Project Inception Through This Reporting Period
15. Total Tasks Deleted	The Total Number of Tasks and Hours Deleted for the Project
16. Total Tasks Active	The Total Number of Tasks Active for the Entire Project
17. Planned Activities	Activities Planned
18. Issues	Any Issues
19. Signature	—

Exhibit 5-2-9. (*Continued*)

mitments as compared with those planned in the original budget and project description.

14. Revised CWP as of the update date based on items 6, 7, 8, 9, 10, and 11.
15. Distribution of labor by resource type (i.e., programmer, analyst, manager, or user) and use as compared with those planned in the original and revised CWP.
16. Forecasted labor needs as compared with those planned in the original and revised CWP.

Using the PMT for Reporting and Communication

The final component of the PMT is status reporting and communication. To be useful in tracking progress and meaningful in communicating that progress, status reports must be produced in a periodic and timely manner and must be consistent in style, format and content. The status reports for the systems integration project must address six topics:

1. Project summary.
2. Accomplishments this reporting period.
3. Planned activities next period.
4. Issues requiring management attention.
5. Financial summary.
6. Other items.

Developing the status report in this format helps to consolidate the project into a meaningful framework and can isolate the issues and major accomplishments. The report should be supported by exhibits and an analysis of the sixteen measurements of progress.

QUALITY ASSURANCE PROCESS (QAP)

The QAP component is the fourth component of project management for systems integration. Creating a quality assurance standard and measurement process for the integration project is difficult. There is, of course, an extensive list of statistical measures available to determine the quality of software design and performance. However, measuring quality and ensuring quality are two very different activities. Measuring the quality of the process in the integration effort generally involves the evaluation of ten major project attributes:

1. Quality of communications.
2. Quality of process management and project tracking.

3. Quality of training and end-user ability to use the system fully.

4. Quality of individual performance, attitudes, and relationships.

5. Quality of management reporting.

6. Consistency with or compliance to internal, American Institute of CPAs (AICPA), and other management or regulatory agency standards, requirements, and guidelines.

7. Consistency with the organization's TQM initiatives.

8. Continuous refinement and improvement of the work plan, staffing plan, and milestones.

9. Quality of team performance.

10. Quality and usefulness of metrics.

As a component of project management, quality assurance must be embedded into daily project management practices. Quality is the responsibility of the project steering committee, EDP auditing and independent advisors, project manager and project team leaders, and individual project team members. Assessing the quality of the integration process can be supported partially through the use of the terminally ill integration project (TIIP) score described in Chapter 4-1, which is designed to assess the key components and project management practices of the integration effort to provide both a prospective and perspective view of the process.

If an external systems integrator is used, the quality assurance process may be outsourced to the integrator by adopting the quality standards and practices of the integrator. Expanding the role of the EDP auditor in the systems integration project can also help improve quality and the overall effectiveness of the project management process. Some of the major activities that the EDP auditor can perform include:

- Providing perspective on AICPA, regulatory, and control issues as related to the implementation effort.

- Ensuring that project management standards are maintained as appropriate.

- Providing independent perspective on the quality of the process used to manage resources and implement the project.

- Serving as a consultant to the project team for the interpretation of regulatory requirements, assessing alternatives, providing perspective on enterprisewide issues, and evaluating the applicability of professional standards as related to the design, execution, and implementation of the project.

- Conducting specific tests of validation and documentation for the work that was performed and the benefits derived.

- Performing periodic reviews of the project management practices, standards of work produced, documentation generated, and compliance to work plans.
- Confirming project status for the project steering committee.
- Participating with IM&T and end-users in "sign-offs" of work process results, work products, deliverables, and acceptance tests.
- Participating in comprehensive unit and integrated testing of the software, data bases, and conversion processes.
- Participating in comprehensive, integrated load testing of software in a full-stress transactions testing environment.
- Performing special audit tests for data integrity, conversion accuracy, and system security, including logical and physical access controls.
- Contributing to work plan development, monitoring, and weekly status reporting.

EMPLOYEE TRAINING AND PROCESS SIMULATION (ETS)

The fifth and final element of the project management model involves employee training together with system testing and process simulation. A major deficiency in the management of many integration projects is the transition from the state of working on the integration to one of actual integration, implementation, and sustained production. Often deficiencies and many postimplementation problems can be traced to the training and education of the work force in the use of the new systems and changes in business processes. Despite the millions of dollars spent on employee development, most organizations remain ill-equipped to train employees in large and complex integration projects. In systems integration projects, the reeducation and redirection of employees are important to the success of the project and use of integrated systems.

To help accomplish the transition to an integrated systems environment, employees should participate in education and training programs specifically tailored for the integration effort and the introduction and use of new business processes and technologies. This education must extend beyond the usual user class about new software and hardware and into areas related to process change. Training for the implementation and use of integrated technologies and reengineered organizations must be accomplished at three levels.

1. General project orientation.
2. Project team training.
3. Specialized team training.

General Project Orientation

The general orientation level should provide an overall perspective on the background, history, approach, resources, expected benefits, milestones, processes, etc., associated with the integration effort. Emphasis is directed to presenting the metrics used to measure both quality and progress.

This initial level is designed for a general audience of employees, including executives. Fundamental to the overall success of the project and project management process are the application and use of the project management standards and practices as well as the effective use of the CWP and PMT project management methods.

Project Team Training

The second level of training is directed toward employees and individuals assigned to the systems integration team who are responsible for performing according to the work plan. The second level of the education program is related to the practices of project management and the effective application of those practices to the integration and reengineering project. PMT education concentrates on providing the essential information and knowledge on how to use PMT to its fullest advantage.

Specialized Team Training

Finally, comprehensive training occurs in the third level. This training is highly specialized and is intended for those employees who are actually going to use the integrated systems and new processes. As such, it is concentrated and focused on individual topics, such as use of the new software, data base design and accessibility, implementation and use of new processes, and computer and network operations.

SUMMARY

Managing systems integration is a full-time and specialized endeavor. These projects are very large, extremely complex and demand a high level of management attention. The process described in this chapter takes a very fundamental approach to project management. It requires a structured work plan, formal staffing assignments and the constant discipline to capture information and update and report on the effort. The key performance metrics described in this chapter will help ensure that any systems integration project is effectively measured and that its progress is accurately assessed.

5-3

Considerations and Costs of the Client/Server Integration Strategy

MICHAEL A. MISCHE

C lient/server computing—and the need for distributed end-user computing and high-powered workstations that support systems integration and reengineering—have become major management and operational issues for many information management and technology (IM&T) executives. An ongoing and passionate debate centers on the merits, disadvantages, and rationale for migrating into the client/server environment and providing integrated, enterprisewide computing solutions that are client/server based. A number of factors and considerations fuel the debate and help form the strategies:

- Technology and a more computer-fluent and technology-dependent work force are making enormous strides in leveraging technology to enable more agile business processes. As a consequence, users want technology at the point of job-step execution in a real-time mode rather than some distant, batch-processed system.

- The advances in technology have placed significant computing power and self-sufficiency at the fingertips of knowledgeable users. As a result, users want to manipulate and adapt computing power to their immediate needs and are constantly seeking creative ways to use technology to enable reengineering opportunities.

- IM&T organizations are under increasingly greater pressures to deliver results—now, not five years from now. Traditional methods are too slow and too bureaucratic to support accelerated user needs and rapid deployment to reengineered processes.

- Executives and leaders of organizations are looking for higher returns on technology investments and greater flexibility for using technology. Historical returns on technology are dismally low.

Client/servers are perceived to be less expensive to acquire and maintain.

This chapter provides a perspective and overview of some of the issues, drivers, and costs of integrating, migrating, and developing systems in the client/server environment.

THE CLIENT/SERVER INTEGRATION STRATEGY

Using client/server technologies to support systems integration and provide for an integrated computing environment has generated quite a bit of press and research. Some of the research and certainly much of the opinion point to an almost herd mentality in migrating away from mainframes and onto client/servers. Advocates purport creating enterprisewide solutions through the implementation of third-party software and client/server technologies. Proponents cite the "openness," portability, and programming facilities of client/servers, which appeal to technical personnel, and the computing power, systems delivery speed, and ease of use, which appeal most to functional users.

A Definition of Client/Server Environment

At one time, the combination of these characteristics were significant enough to allow client/servers to be thought of as the strategic computing technologies of the future, replacing the traditional information technology (IT) computing solutions and architectures. However, the gap created by these capabilities and their benefits has closed significantly, and the trend of migrating off of mainframes has abated in recent years while multitiered computing architectures have emerged.

Indeed, worldwide mainframe sales increased in 1994. Other studies concentrate on the advantages of maintaining a multitiered computing environment. The issues notwithstanding, client/server systems have a distinct and important role in the computing environment of today and into the early years of the twenty-first century. However, the mainframe, which was once thought to be a relic and replaceable by client/servers, is far from dead. Mainframes, in some form or another, have had and will continue to have an important role in enterprisewide computing—and they will be around for quite a long time.

Before this chapter explores the potential and implications of the client/server environment and its role in the systems integration process, a useful definition places the client/server environment in context. The definition used in this chapter incorporates the conceptual, technical, and application aspects. It is implicit that the definition includes the concepts of openness and distributed processing. Specifically, the client/

server environment is an end-user computing environment that is deployed in a manner that is interoperable, networked, open, and distributed, is supported by programmable desktop computing technologies that request services and resources from other devices on its network, is equipped with graphical user interfaces (GUIs), and is designed to serve data and applications that are rationalized and homogeneous to the end-user community in a transparent and on-demand manner.

The Three-Tiered Technology

A number of elements must be present to support the client/server environment and end-user. Technology is paramount. In the context of this definition, the use of a three-tier architecture, as described in Chapter 3-3, is implied. This architecture includes:

1. A consolidating processor, usually in the form of a mainframe or midrange computer.
2. Communications networks, in the form of local and wide area networks (LANs and WANs) linking the tiers together.
3. Workstations, in the form of personal computers and workstations.

Component connectivity and reliability are perhaps the most essential elements of the client/server computing environment.

Accessible, Rationalized, and Distributed Data

Second, the data that resides in the client/server environment must have three qualities. It must be accessible, rationalized, and distributed. Typically, in the three-tiered technology architecture, there are three classes of data: enterprisewide data, which is reposited in warehouses; process-enabling data, which is resident in general warehouses accessed by a specific community of user-owners, and personal or desktop data, which is specific to an individual or small group of users.

Ingredients of Client/Server Applications

Client/server applications must also have a number of elements. Inherent in the client/server application is the use of consistent screen presentations, navigation tools and paths, databases, structures, and application logic. It is also important to provide for security and processing/data recovery in the client/server environment. Thus, the client/server environment is defined in a way that includes the interoperability of technology, applications, data and end-users. The definition exhibits five key elements:

1. Distributed applications and data.
2. Server control and management over services and data.
3. Server control and management over conflicts and contention that may result from user requests for service.
4. Communications between users, clients, and servers.
5. Cooperative interactive sessions.

As the technology deployment experience is still evolving, so is the definition for client/server computing changing. This state of flux will continue until standards are uniformly adopted.

The Importance of Open Systems

The concepts and principles of openness are important to the client/server integration strategy and the definition discussed previously. This definition has adapted the Institute of Electrical and Electronic Engineers' description of the open environment, which says that the open computing environment is a "comprehensive and consistent set of international information technology standards and functional standards that specify interfaces, services, and supporting formats to accomplish interoperability and portability of applications, data, and people."

For purposes of this chapter, openness is defined as having three critical qualities: interoperability, scalability, and portability. Essentially openness has two components:

1. The distributed computing environment (DCE).
2. The open network environment (ONE).

Both of these are explored in more detail in Chapter 6-2. The DCE aspect has also been adopted by the European Commission for unification standards and, therefore, has the potential to be the de facto standard.

The UNIX Operating System

From an operating system perspective, UNIX is one of the mainstay operating systems of the client/server arena and, as such, demands some specific consideration. First developed by AT&T in 1969, UNIX is coupled with the concepts of client/server computing and open architecture. However, UNIX alone does not mean open, nor does it guarantee openness. Historically, UNIX has been predominantly a niche operating system supporting engineering, education, scientific operations, and other vertical specialties; however, since the mid-1980s, third-party vendors and developers have been making a concerted effort to provide more commercially oriented applications. Nonetheless, the vast majority of business products, application system vendors, and users are still non-UNIX, but that is changing. Indeed, a proliferation of vendors are offering products in this market over the past few years.

The openness of UNIX has yielded to subtle differences among vendors. Today, there are reportedly some 200 different varieties of UNIX and several standard-setting bodies, including the Open Systems Foundation (OSF) and UNIX International (UI), both of which are reportedly working toward common codifications and definitions.

The Many Varieties of UNIX. Some of the more popular varieties of UNIX include UNIX by AT&T, AIX by International Business Machines (IBM) Corp., ULTRIX by Digital Equipment Corporation (DEC), and XENIX by Microsoft. The majority of the UNIX variants are derivatives of the University of California at Berkeley in the form of Berkeley Software Distribution, much of which was developed—and is available—under public domain.

As a result of so many variants, however, UNIX operating systems are, in reality, very much like the proprietary closed systems against which they compete. No two varieties are exactly alike, and porting applications and data between two UNIX systems requires that the processors and systems software, including hardware, database management systems, and security systems, be exactly congruent. Is portability more direct and simpler in the UNIX world of client/server systems, as compared with the traditional alternatives? Most definitely, but differences do create issues, extra effort, and investment. The major advantage of the UNIX environment is that it takes advantage of RISC architecture.

THE RATIONALE FOR CLIENT/SERVER INTEGRATION

The working definition has been established; this section discusses the rationale for adopting the client/server computing environment and systems integration strategy.

Client/Server Computing Empowers the End User

Perhaps the most commonly cited reason for using client/servers to integrate systems is that end users like it and want it. Without question, applications are more accessible, data is more accessible, and personal computing power is provided; end users feel and indeed have the control over the data and applications that they require to support their function. From this perspective, end users believe they are empowered by this deployment strategy. However, these desirable qualities are usually insufficient by themselves to justify the effort and investment of adopting this strategy.

Client/Server Computing Accelerates the Systems Development Process

Perhaps the other two most commonly cited reasons for adopting a client/server strategy center on delivery of applications and services and, more subtly, the performance of the traditional IM&T organization. Proponents of the client/server integration strategy suggest that this environment can, but does not always, accelerate the delivery of enabling solutions to users. Also, the client/server strategy can, but does not always, help distance the end user from the bureaucracy of the traditional IM&T organization.

This latter point may, in fact, be one of the most compelling and viable reasons for explaining much of the user enthusiasm surrounding the potential for client/server integration and migration. Users are far more knowledgeable and less dependent on the IM&T organization for support today than in the past. They are also less tolerant of the slow development cycles, backlogs, and functional rigidity of traditional IM&T organizations and management practices. In some situations, end users have the perception that the traditional IM&T structure and technologies are inhibitors to change, and the more removed and decoupled the user and systems are from centralized IM&T control, the more advantageous it is for the user.

Client/Server Computing Makes Financial Sense

Finally, there is a financial rationale that client/servers are less expensive to acquire, maintain and develop applications for, as they are more open and less dependent on traditional IM&T systems development methodologies and proprietary operating systems. (See Chapter 5-1 for a discussion on the systems integration life cycle methodology.) Another rationale is that the distribution of computing resources across a network leads to cost reductions.

Some of the other more commonly cited advantages or expectations for migrating to the client/server environment include:

- Increased end-user productivity levels.
- Improved decision making.
- Improved process integration.
- Support of reengineering.
- Improved employee morale.

- Improved strategic alignment of IM&T to the business's needs and direction.
- Freedom from proprietary operating systems and wider selection of products.

Client/Server Computing Is the Corporate Philosophy

For many organizations, the rationale for using a client/server integration strategy appears to be largely one of philosophy and not necessarily entirely based on technical or financial criteria. Some have adopted the multitiered strategy. Others have concluded that mainframes and midranges are "dead" strategies and have disbanded the use of mainframe and midrange computers altogether.

For example, one company, which expressly stated that the mainframe and midrange computing environments were "Jurassic," is reportedly spending close to $70 million to implement a third-party client/server integration strategy. The project is to span seven years, and one is well justified to ask how any economic payback can possibly be achieved. By the time the system is complete and fully implemented, the company will have to migrate to newer solutions. Given the investment and effort associated with client/server integration efforts, this creates somewhat of a dilemma, as the majority of nontechnical objectives can be satisfied by a number of alternatives. This is certainly the case for much of the rationale cited above. In this sense, the real issues with respect to the philosophical advantages and benefits of client/servers appear to be centered on:

- Longevity of the technology and operating systems (i.e. useful life).
- Determining what applications, data, and computing power should be distributed to the client/server environment (i.e., tactical deployment).
- The ability of the client/server integration strategy to adapt (i.e., flexibility) to the rapidly changing needs of the organization (i.e., rapid development and deployment of system solutions).
- Costs of computing as measured by both direct and intrinsic methods.

THE RATIONALE AGAINST CLIENT/SERVER INTEGRATION

For most organizations, it would appear that the rationale for adopting the client/server environment is relatively situational. That is, valuing the economic benefits of better decision making and improved morale are difficult to do. Isolating initial acquisition costs are simple, but estimating true ongoing development, integration, and maintenance is difficult.

Too, not all applications, databases, and users are candidates for client/server integration. In some instances, such as the processing of vast volumes of homogenous transactions, integrating systems using client/server technologies may not be desirable. Performance is a factor.

Similarly, providing client/server integration to support nonessential overhead functions may be an inappropriate application of the technology and one that yields no economic benefit. However, in certain situations, there may be some technical reasons driving the client/server integration strategy, especially when the current technology has far outlived its estimated life span, and it simply does not make any sense to continue with it or replace it with like in kind. In these instances, it is not unusual to find the technology and applications in place to be antiquated, and wholesale replacement is clearly in order.

EVALUATING CLIENT/SERVER PERFORMANCE

Another issue with respect to using client/server solutions is complexity and performance of the client/server environment. To make the client/server environment work requires a number of technologies and products as well as the melding of multiple vendors and technologies into a coherent and highly reliable network. As discussed in Chapter 1-1, client/server technology is systems integration in its purest sense in that it requires making different pieces of equipment work in concert together. This combination of technology is expensive, highly complex, and difficult to support and demands specialized skill sets.

Because of the complexity of the networks, the composite nature of equipment, and shifts in managing data and computing assets, reliability and the client/server environment's ability to support mission-critical applications are issues. Because distributed applications and databases are inherently more difficult to manage, they also present challenges in providing version control, maintenance, and individual server performance.

Another issue is that although client/server processors are fast and have tremendous capabilities, much of their activity in a transactions-intensive application can be dedicated to managing I/O. Thus, the client/server environment presents many of the same capacity management and planning issues as does the traditional computing arena. In developing a systems integration strategy and deployment tactics, four key performance issues must be considered:

1. The quantity of data to be processed.
2. The number of active users that will be requesting services in a concurrent manner.

3. The estimated number of interactions between clients and servers.
4. The number of transactions to be processed.

The combination of these factors could easily result in an overall increase to the costs of computing when all factors are considered.

THE COSTS OF CLIENT/SERVER COMPUTING

The debate surrounding the costs and benefits of client/server computing is equally as torrid as that which revolves around the superiority of client/server computing over traditional mainframe and midrange solutions. Developing valid and credible cost comparisons is difficult, and calculating any form of return on investment (ROI) is enormously challenging, if not impossible. Proponents of the client/server integration strategy cite the costs of implementation and ongoing maintenance as economic incentives for adopting the technology. Usually, processing power and acquisition cost comparisons are made between mainframes and client/server technologies. Comparisons are also made between software development and acquisitions costs.

However, much of the cost comparisons, which are based on millions of instructions per second (MIPS), initial acquisition costs, and software development costs can be misleading. Typically, the majority of studies are confined to the obvious, or "hard," IT costs alone. They do not consider the overall implications of client/server integration on the user community, growing transactions volumes, and distribution of data on an enterprisewide basis. The omission of these factors can be a very significant error in reasoning. This is like comparing the costs of buying a home to the costs of home ownership—very different situations that yield very different results. In evaluating using client/server systems as a means of supporting systems integration, consideration must be given to the total costs of the client/server environment, including hardware, software, networks, and users.

The financial implications appear to be potentially significant when client/server systems are involved, and a number of studies are confirming that the overall long-term operating costs for the distributed client/server strategy are significantly higher than the traditional environments. For example, one study found that the average monthly cost for a UNIX-based workstation was 34% greater than that for an AS/400.

The issue in comparing the costs of the client/server integration strategy involves the identification and rationalization of valid cost components and items, not just elementary hardware and software costs. There are many hidden costs, the largest of which appears to be labor. Labor is defined by direct IM&T skill set development, external resources, and most importantly, the end user. This component of cost is often overlooked or rationalized in the justification because it is assumed to be there anyway. That is true, but the opportunity costs and effort required to realize integration using this strategy are enormous and simply must be factored into the overall costing equation. The successful deployment of a client/server integration strategy is almost always accompanied by major changes in the IM&T organization and user responsibilities. These changes involve significant costs.

Cost models developed and studies performed by the Gartner Group also support the conclusion that the client/server integration strategy is potentially far more expensive than the traditional environments. For example, the Gartner Group developed a model that indicates that the costs for supporting a multiuser client/server environment ranges between $48,000 to $66,350, per client, over a five-year period.

Another third-party study indicates that the average cost of supporting a user in the client/server strategy runs approximately $5,800 per user, per year, versus $2,800 in an alternative midrange environment. These comparisons are significant. In almost all cases, studies indicate that the total costs of IM&T appear to increase as a result of adopting the client/server integration strategy. These conclusions are further supported by a recent IBM study, which indicated that client/server computing was not less expensive than the alternative environments. Based on these conclusions and the author's direct experience, it appears that client/server computing costs are not offset by any significant economic benefits or increased ROI that can be attributed to technology changes alone. Therefore, financial justification based on hardware and software assumptions alone do not appear to justify the investment in client/server technology. Furthermore, the intangible benefits of improved productivity, greater morale, and better decision making are hard pressed to surpass the costs, especially when the application is restricted to overhead functions alone. Therefore, combining operational and organizational reengineering with systems integration in a client/server environment is essential if financial justification and technology leverage are important.

How Development Time Affects the Cost/Benefit Analysis

Another important issue in the assessment of client/server integration strategy is the speed at which solutions are developed and delivered. As discussed in Chapter 5-1, speed is the most essential element in the delivery of enabling technology solutions to end users. An often-cited rationale for using client/server systems is that the speed of development and overall development cycle is accelerated. Studies and experience tend to indicate that development cycle may be faster, but only for certain types of applications. Issues are how much faster and whether the speed of delivery creates any tangible shareholder advantage or economic value to the organization.

For example, IBM indicates in a study on client/servers that the average development time for their study group was 1.6 years, for relatively noncomplex applications. Furthermore, over two-thirds of the projects surveyed in the IBM study failed to be delivered on time, and only half were performed within budget. Many of the same issues and problems attributed to traditional systems development and systems integration projects also plague the client/server environment. For example, development speed and product quality are contingent on a number of factors. The quality of personnel assigned to the task, work plan, tools involved, project management, complexity of application, and a number of other factors all influence the results of integration in the client/server, midrange, and mainframe worlds alike. However, it very well could be that the experience—or lack thereof—is one of the most influential factors on costs.

The complexities and duration of client/server development and implementation projects provides another dilemma: speed. As indicated, speed of development and solution delivery are incentives for the move into a client/server environment. However, there are situations and cases in which client/server projects are reaching 36, 60, and even 72 months. In these situations, the time differential is certainly not a factor for a comparative basis, and the benefit and rationale for the strategy become more difficult to evaluate. A general rule is if development and reengineering efforts span more than 20 months, they become susceptible to many external influences. If they exceed 36 months, the propensity for failure and vulnerability to management changes and technology shifts is extremely high. Therefore, one has to constantly challenge the development cycle, especially in the realm of client/server systems.

KEY FACTORS TO CONSIDER

The implementation of a client/server integration strategy is, as many organizations have learned, fraught with many difficulties and hidden expenses. The cost of hardware and software is only the tip of the iceberg when it comes to the total cost of the client/server integration strategy, developing systems, or deploying the technology. Typically, when the organization makes the decision to use a client/server integration strategy, five outcomes are virtually guaranteed:

1. *The total cost for information technology rises.* That is, spending on IT increases as a result of adopting a client/server computing and integration strategy.

2. *The implementation of a client/server integration strategy redistributes greater processing power to the end user.* Do end users need all that power, and are they able to manage the computing resources on their desktops? Processing power in end user computing is like horsepower in an automobile: after a point there are diminishing returns. The rider only needs to go so fast or pull so much!

3. *Both end user and IT labor requirements and costs, as measured by training, learning, and ongoing maintenance represent the majority of initial and long-term costs.* In a client/server world, end users are not just users of software and hardware, they are co-owners, co-managers, co-administrators, and co-maintainers. They must learn how to administer their systems, back up their files, navigate through GUIs, and manage desktop IT resources. For all but a few organizations, the transition to a client/server integrated environment is one not only of technology but also of direct accountability for managing the computing resource. Personnel who once were adverse to computing or dependent on their internal IT organizations find themselves in the direct line of technology and the management of that technology. To manage this redistribution and power, users must be trained in IM&T management and administrative practices. The IM&T organization itself must be realigned and retrained to support this redeployment of resources and power. The issues can be complex, and the costs can be high.

4. *Client/server systems cannot be fully exploited if deployed in a patchwork of systems without a clear integration strategy that links applications, data, and users into a multitiered environment.* Deployment without an overall architecture and an understanding of how the client/server integration strategy supports the strategic and reengineering needs of the organization leads to dilution of effort and ROI. When this occurs, investments in technology and human support systems become redundant.

5. *Deployment without an enterprisewide maintenance policy for security, application version control, and data leads to enormous complications and divergence.* A secondary, but essential, consideration is the management and control of applications and data in the client/server strategy. The deployment of applications and data requires significant cost and investment in version control, data maintenance, security and auditability.

Thus, as with any computing strategy and environment, there are advantages and disadvantages with moving to a client/server integration strategy. Quite simply, not all organizations, applications, and IM&T organizations are ready and capable of supporting the client/server integration strategy. It is situational to the organization, functional needs of the business, and applications involved. Moving to a client/server integration strategy is also not the solution to a poorly managed or performing IM&T organization. Nor is it the answer for low-performing IM&T financial return and end-user disenchantment. The problems of managing the client/server integration strategy are greatly compounded, because the technology is dispersed, data and applications are distributed, and the technology is complex. From an organizational perspective, the client/server integration strategy demands a more sophisticated IM&T organization and set of management practices than do the traditional computing environments and integration methods. As such, it is clear that the client/server computing environment requires formal infrastructures and disciplines similar to those of the traditional systems integration and computing methods and environments. However, the way in which the technologies are managed and the IM&T organizational elements and disciplines are structured are very different. Therein is the major challenge for many organizations attempting to adopt this strategy.

Successfully integrating systems in the client/server environment requires formal technology management procedures, professional staffing and a responsive organizational structure. Within this structure are a number of key operational and management elements, which must be present to optimize the investment and capabilities of this technology. First, the client/server integration strategy must be managed as a logically consolidated asset and physically distributed resource. (See Chapter 3-3 for definitions.) Secondly, the strategy must have a clearly delineated architecture that depicts how the technology is to be used and deployed in the enterprise. Finally, it must have certain infrastructure elements, which are fundamental to the management of the client/server integration strategy. They include:

- Fail-safe methods and practices for the management and control of versions and releases of software.
- Highly refined policies and procedures that provide for capacity planning, monitoring, and management.
- Change-management procedures for software and hardware.
- A formal procedure and structure for reporting, managing, and resolving problems.
- A formal chargeback or market-based fee-for-service/use accounting system.
- A comprehensive system of audit checks and controls that enforce security and data safeguards and provide virus protection and disaster recovery.

- Attributes of the high-performance IM&T organization.

SUMMARY

The evaluation of and decision to use client/server technology to achieve integration should be driven by business needs, not by technology orientation or vendor preferences. If the business needs mandate and support client/server systems, then managers must proceed with a carefully scripted plan and well-thought-out architecture. The bottom line is that organizations considering the implementation of a client/server integration strategy must be objective and realistic about the use and selection of technology. The key issues and questions that should be addressed include:

- Will the organization be at the leading edge or the bleeding edge, and why?
- What applications will be integrated using the technology?
- Do sufficient resources exist to adequately support the process?
- Have the essential infrastructure elements been implemented to support the integration effort and postintegrated environment?
- Does the organization have the culture to support client/server strategies? Can it create the culture?
- Does the organization have a full financial understanding of client/server strategies?
- Finally, what is the real financial return on the investment?

5-4
The Implications of Systems Integration on the IS Organization

JAGDISH R. DALAL

D emands of today's business require that the information systems (IS) organization develop a long-term strategy for addressing the information needs for the 1990s, which provides an impetus to the chief information officers to develop a longer-term strategy built on a vision of where to take the IS organization rather than just reacting to the tactical requirements of the moment. In fact, this longer-term view of the IS function, coupled with the boldness of the strategy, can either strengthen or weaken the IS position and role—and therefore affect that of the chief information officer (CIO).

THE IMPACT OF SYSTEMS INTEGRATION ON THE IS FUNCTION

As systems integration strategies are identified and evolve, most IS organizations must critically evaluate their capabilities for delivering the IS requirements of the business. Several factors combine to drive this analysis:

- Ever-increasing demands of the business change along with the profitability-driven cost containment.
- Fast-changing technology requiring both the introductory as well as end-of-life strategies for various technologies in use.
- Requirements of significant investment in introducing and managing the new technology.
- Ever-increasing demands of changing technology and business requirements, which put pressure on the IS staff to maintain current skills while dedicating the necessary time—and therefore expense—to learn new skills.

- Above all, a reduced cycle time of delivering business-support solutions in a cost-effective manner.

These factors encourage the CIO to consider, or in many cases, reconsider, systems integration as a strategy in delivering business solutions. However, before the systems integration can be labeled a "bold strategy for change," the CIO must consider the strategy's impact on the organization as a whole.

Implementation of a systems integration strategy without a concomitant change in the organizational aspects of IS is a recipe for disaster. Literature is full of examples of systems integration projects going awry through the lack of a holistic view of the future organization. This chapter discusses the implications a systems integration project has for an organization, examines the management of change and transition in the organization, and offers various models of organizations in which systems integration is a part of the systems-delivery strategy.

DEFINING SYSTEMS INTEGRATION

In the context of this chapter, systems integration is defined as an IS environment in which internal or external self-managed teams deliver the business solution by assembling IS components from various sources. The emphasis in this definition is that the solution assembly either can be done by an independent team within the IS organization or can be purchased as a skill from the outside, which is the more common solution and a basis for significant business for such organizations as Andersen Consulting and EDS. However, one must not waive the establishment of an internal systems integration function from the strategy.

For the purposes of this chapter, the systems integration strategy is assumed to be not all-encompassing for the entire IS effort but pertains only to a portion of the IS requirements. In this respect, a total systems integration implementation should be viewed more as an outsourcing activity than as one of the IS strategies of future.

Depending on the business needs and the long-term IS strategy, there are several dimensions of the systems integration strategies that can be implemented. These strategies depend on the organization's opportunity for integration as well as the degree of integration required in the IS environment. The opportunity for integration is where the IS environment would allow the CIO to consider any aspect of systems integration. The degree of integration is a factor of the existing architecture and environment and the vision of the future. These alternatives are shown in Exhibit 5-4-1.

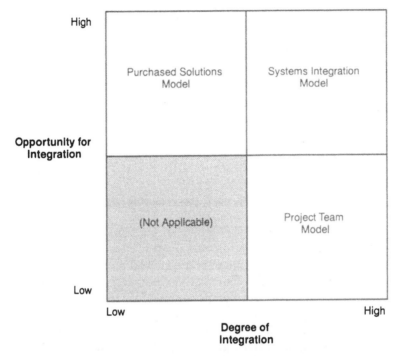

Exhibit 5-4-1. Systems Integration Opportunity Model

SYSTEMS INTEGRATION AND CORE COMPETENCIES

One of the opportunities for systems integration requires the organization to have a thorough understanding of their IS core competencies, which are skill sets considered to be the "key differential" for the organization. In other words, these can be neither matched nor exceeded when being acquired from outside the company for the same value. IS organizations must develop their visions of the core competencies that they would retain or cultivate as part of their long-range IS strategy. This clear understanding of the core competencies determines the type and level of systems integration that the organization requires.

Again, the presumption is that the IS organization has developed a long-range IS strategy and has established core competencies based on strategy and realism rather than wish lists.

Organizational Model

An organization is more than the structure published as part of some organization chart. Exhibit 5-4-2 shows the dimensions of the organization that must be considered when designing an organization model.

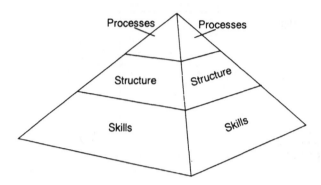

Exhibit 5-4-2. The Dimensions of an Organization

Skills. These represent the current as well as the visioned state of an organization's set of skills. The skill definition for the future includes the identification of the core competencies. Project management, data architecture and management, and networking are examples of an organization's skills. These skills do not necessarily have to be possessed by everyone in the organization; generally, they are not. Organizational structure and processes help define who would possess these skills and how they would be utilized.

Structure. The structure of an organization would define both the formal and informal relationships necessary to accomplish the tasks. These structures can be finitely defined, as in an organization chart, or can be evolved through an informal definition, as it would in a team environment.

Processes. If the structure of the organization is the bones, processes are the nerves, helping the organization to function effectively, efficiently, and most importantly, predictably. Processes include procedures, methodologies, and policies. One cannot imagine the functioning of an organization without processes such as:

- Rewards and recognition.
- Planning.
- Project management.
- Customer relations.

Exhibit 5-4-3 reviews the IS organization from the above perspectives when considering the implications of systems integration as a strategy.

Organizational Dimension	Without Systems Integration Strategy	With Systems Integration Strategy
Skills	• All required skills are represented in the organization; only in certain circumstances are they supplemented by an external source (i.e., use of contractors). • Skill separation and expertise maintenance are key factors in designing an organization's structure. • Skills development and tracking are major organizational issues.	• Work is contracted based on the business requirements rather than skills retained. • Skill identification and maintenance are not issues, unless they are part of the envisioned core competency. • Skill assessment and systems integration vendor effectiveness are material only for selection. Delivery of the product is the determining factor.
Structure	• Structure is defined for the optimum delivery of product, based on efficiency and effectiveness. • Structure is generally hierarchical, even if the "team" concept is used to manage the work effort. • Structure requires the organization to enable both horizontal and vertical integration of effort.	• Structure is defined for the maximum value in delivering the "total" solution. • Structure needs to be "fluid" to accommodate various approaches—and projects of—systems integration. • Structure can be hierarchical and yet accommodate systems integration activities.
Processes	• Processes define the vertical control and horizontal integration of services and products. • Processes promote "standardization" on a vertical basis (i.e., from project to project) to create efficiency.	• Processes provide the horizontal integration of systems integration projects within the IS organization. • Processes promote "ease of integration" of systems integration projects to each other as well as to the existing environment.

Exhibit 5-4-3. The IS Organization

SKILLS AND SYSTEMS INTEGRATION

The skills required to manage systems integration are generally different from those used in managing traditional effort. Some skills necessary to manage systems integration effort with success include:

- *Planning.* A critical skill in managing the systems integration effort to ensure that the effort is part of the longer-range vision, planning is essential to develop a multiyear systems plan that clearly defines the deliverables and the endpoints. Knowledge in use of various project planning and management tools and techniques is also required.

- *Architecture and standards.* To manage the systems integration effort, the IS organization must have a well-defined architecture for the endpoint and a set of standards that are compliant with the architecture. Without these skills in the organization, the integration of the delivered project with other systems becomes a challenge and quite expensive. Established architecture and standards also ensure that the output from a systems integrator is compliant with the endpoint vision.

- *Project management and integration.* Necessary to ensure the delivery of systems as planned, this skill allows people in the IS organization to "plan their work and work their plan." An organization wishing to manage systems integration projects must ensure the overall integration of all efforts. In this respect, this is "meta" systems integration skill. This includes management of convergence of efforts and managing the transition of systems effectively. There can be circumstances in which an organization can require the systems integrator to perform this task in addition to delivering a solution. However, this would be judged risky, as it may create the scenario of "the fox watching the henhouse."

- *Work scope management.* This allows the organization to manage the deliverables of the systems integration team. The skill set includes contract development and management as well as vendor relations. Required skills also include creating a thorough understanding of business needs, business environments, technology enablers, and the environment as proposed by the systems integrator. Effective communication skill is one of the necessary "soft" skills required to manage the relationship with the vendors and customers.

These comprise the set of core competencies for an IS organization setting out to implement a systems integration strategy. Either these competencies exist in the organization and are leveraged, or they are built as part of setting up the new IS organization. Consideration of these factors helps define and drive the design of the new organizational model.

STRUCTURE AND SYSTEMS INTEGRATION

As companies implement acquisition through systems integration, the organization's structure needs to be modified. Experience has shown that the organization structure needs to be fluid to ensure that as systems integration activities change, the structure changes along with it.

Traditional information management organization structures are established to ensure vertical and horizontal integration of work output and that the core skills are not replicated in the organization, providing maximum leverage for the investment in those skills. Successful assimilation of systems integration activities require a new way of managing the output as well as the skills of an organization.

The Core-Competency-Centric Model

Exhibit 5-4-4 contains examples of three types of models of organization that can successfully implement systems integration activities. In this model, the organizational competencies are retained with the capability to deliver other systems efforts. This type of organizational model is appropriate when:

- The information management organization has critical, retained, core competencies or has a requirement to build a strong set of core competencies. For example, an organization migrating to new business processes or new technology platforms may choose to retain some of the core skills.
- Systems integration deliverables are one of the sources of output to satisfy customer requirements.
- The systems integration effort is more closely managed by the information management organization.
- The customer set is diversified or customers have a strong influence on the deliverables while the requirements are integrated in their output. In such an instance, the customer account management function provides a critical linkage between the customer requirements and delivery organization.

The Atomic Organization Model

In Exhibit 5-4-5, which shows the atomic organization, various systems delivery projects are established on an equal basis. This type of organizational model is appropriate when:

- The information management organization has critical, retained, core competencies or has a requirement to build a strong set of core competencies. For example, an organization migrating to new business processes or new technology platforms may choose to retain some of the core skills.
- Systems integration deliverables are the primary source of output to satisfy customer requirements.

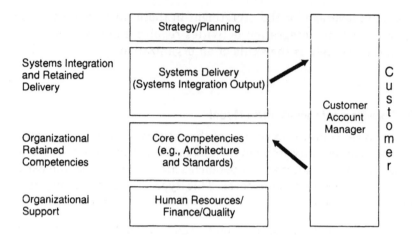

Exhibit 5-4-4. The Core-Competency-Centric Organization Model

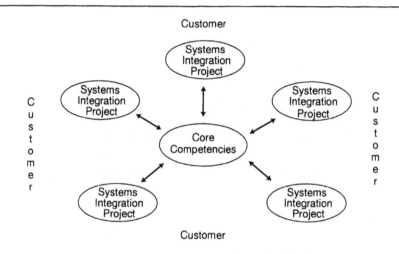

Exhibit 5-4-5. The Atomic Organization Model

- The systems integration effort is more closely managed by the customers.
- The customer set is diversified or customers have a strong influence on the deliverables but are not interdependent for their output.
- The systems integration strategy is driven by customers or a decentralized IS organization. In these cases, the core-competency organization is generally called corporate IS.

The Vertically Integrated Organizational Model

The organizational competencies are not critical to the organization. This type of organizational model is appropriate when:

- The information management organization does not have a need for core competencies. For example, an organization that has a well-defined set of standards and architecture does not require a strong core competency to manage the system integrator.
- Systems integration deliverables are the only sources of output to satisfy customer requirements.
- The systems integration effort is more closely managed by the customers.
- The customer set is diversified or customers have a strong influence on the deliverables but are not interdependent for their output.

PROCESSES AND SYSTEMS INTEGRATION

As described above, the systems integration effort requires that the organizational processes are modified and implemented. Organizational processes—both business and management—that need to be modified include:

- *Human resource-related processes.* These include recruiting, position placements, such as promotion and job placements, performance management, reward, and recognition. Modifications to these processes need to consider the organizational model being implemented and its impact on the resident staff. Careful analysis of existing human resource profiles in light of the vision of the future ensures that the introduction of systems integration does not create employee dissatisfaction.
- *Planning and project management processes.* Organizations must modify and enhance their planning and project management skills and should modify the processes for the systems integration activities. For example, a more formal project reporting process must be established. Audit/inspection processes must be developed and implemented to validate the in-process status of the systems integration project.
- *Systems acceptance process.* To ensure that the systems integrator's output has met the requirements, the systems acceptance processes have to be modified, or at least made more formal. Processes include testing, compliance checking to standards and architecture, and production ability to support.
- *Systems maintenance and support processes.* Based on the system

maintenance and support strategy adopted by the organization, these processes have to be modified to ensure that there is a "clean" interface with the systems integrator. In some instances, these processes are adopted directly from the way the systems integrator has agreed to provide them.

ORGANIZATIONAL TRANSITION CHALLENGES

The biggest challenge in managing the transition results from defining the new set of skills and understanding its implication on the current set of employees. In some instances, this requires developing and implementing a skills management program. Key aspects are:

- Identification of new skills and number of employees required for each skill category.
- Assessment of existing skill set of the employees.
- Development of a skill transformation program, such as recruiting and training.
- In extreme cases, reduction in force of employees possessing skills of the past and inadequately prepared for retraining and reeducation.

Even after the new organizational model is established, the challenge still remains as to how to manage the transformation to the new model successfully. Change in the organization, like any change, creates stress and discomfort. Also, as in managing any other change, most of the tools and techniques can be applied to ensuring success. Some of the challenges of managing such an organizational transition are:

- *Communication.* Organizational communication is vital in managing the change. As the systems integration strategy is revealed, employees have a lot of questions about the strategy, their role in the organization, and work processes affecting their own work. Organizational change specialists recommend that communication should continue through the entire change, even if there are no answers that can be given at that moment.
- *Process and policy changes.* One of the key dimensions of defining an organization is development and definition of policies and processes for integrating the systems integration effort. Ideally, these processes and policies are developed before the organizational change is effected. Even when there are no newly defined processes in existence, employees should be told of the types of processes and policies to be developed and implemented.

SUMMARY

The demands of today's business environment and the associated pressures on information management to transform itself is creating opportunities to use the principles of systems integration as a strategy. A well-planned and executed approach to integrating the systems integration strategy in an existing organization provides long-term benefits to the corporation. There are several organizational models available for consideration, based on the profile of the information systems organization, that would ensure that the introduction of systems integration activities do not create havoc in the organization. All aspects of an organization—skills, structure, and processes—must be considered when developing a view of the future organization. Several industry benchmarks exist that can be studied before designing and implementing the new organization. Questions to be asked to ensure success of the new IT organization include:

- Is there a long-term strategy supporting the implementation of systems integration activities? What is the case for action for systems integration?
- Is there a clear understanding of how the systems integration services will be used?
- What are the strengths and weaknesses of the current organization? What are the core competencies of the current organization? What are the core competencies of the envisioned organization?
- What industry benchmarks have been studied in developing the implementation of these strategies? What industry benchmarks have provided learning for managing the transition?
- What metrics of success have been identified to know the organizational effectiveness post systems integration? What metrics of success have been identified to manage the transition?

Section 6
Integrating Technology, Data, and Applications

The integration of communications, data, applications, and technology provides the infrastructure for the melding of human and organizational systems. Despite the advances in technology and the increasing ease of end-user computing, the integration of technology still remains a intricate, high-risk, and expensive process. Much is at stake when integrating human and organizational systems: one cannot be fully effective without the other. This section explores not only the technology issues and implications of integration but also the impact of integration on the end-user.

Because of recent technological advances, the authors have emphasized client/server integration issues, although the mainframe and midrange arenas are clearly still relevant topics. A number of different points of view, including that of the end-user, project manager, and systems integrator, are provided. As indicated by one of the authors, the integration process is not just a project; it is a journey. The material covered in this section is bound to be interesting as well as informative.

6-1
Improving User Productivity Through Systems Integration

RUI J. FIGUEIREDO

The search for a definitive, quantitative measure of computer-user productivity has to date been unsuccessful. The perceived growth in white-collar worker productivity has generally matched the evolution of electronic data processing and particularly paralleled the development of the personal computer (PC) from its simple beginnings as an individual's single-purpose workstation within the traditional company hierarchy to its role today as a multifunctional node in the central nervous system of a global, electronically networked organization. This change in the role of information technology (IT) has begged for a completely new definition for how an organization measures productivity and gauges the degree of success resulting from the deployment of new technologies and application systems.

In the past, the very nature of an organization's hierarchical structure contributed to its inward focus, when efficiency within functions (e.g., departments) was perceived to be of greater importance or had a higher visibility than the organization's overall effectiveness. Measuring productivity using this myopic view no longer makes sense, however, in a networked (i.e., flattened) organization, with its more complex and more directly cross-functional dependencies. White-collar productivity, especially when it relates to the impact of IT, is often subjective and consequently hard to measure. As a matter of fact, this subjectivity has complicated the accurate identification of any aggregate productivity gains from IT investments over the last two decades. National productivity surveys conducted as recently as 1992 show that massive corporate investments in office technology frequently fail to achieve commensurate increases in white-collar productivity. For every success story, numerous failures abound, including examples of significant IT investments that were abandoned because they did not yield expected results or did not deliver anticipated productivity gains.

Although poor measurement techniques make it difficult to gauge the degree of success or failure of some of these ventures, many of the failures are actually caused by resource constraints (i.e., not committing the proper personnel and financial resources) or poor management (i.e., the lack of management commitment, inadequate project management, or misdirected resources). For example, field office managers in one of the nation's largest insurance companies were still spending an inordinate amount of time manually entering routine information into the computer system—after the system had been redesigned to optimize worker efficiency. Apparently, the redesign had been initiated to help optimize the back office, but the developers, in their hurry to complete the project, had failed to consider the primary users of the system (i.e., the front-office field managers).

Users are relatively quick to acknowledge whether an application system has made their jobs easier, because they can see how well an application helps them use information to perform better or even differently. Several factors intuitively contribute to these perceptions:

- *Ease of use.* Access to information through user-friendly interface standards [e.g., graphical user interfaces (GUIs)] and the ability to use a variety of tools through a single interface, such as the Microsoft Windows desktop, can make learning the use of multiple applications and functions not only simpler but fun, as well.

- *Response time.* When requests for information can be satisfied quickly through robust systems and applications, users are quick to perceive this as timely access to information. Time spent waiting for the system to unlock is frustrating to users, as is the time spent leaving one application and starting up another to initiate a needed function that is not available in the original application.

- *Information accuracy or reliability.* Information accuracy and consistency can determine how much a user will trust the reliability of an application and consequently how much use an application will get.

- *Information availability.* Information that can be accessed when and where it is needed can also contribute to a user's sense of being productive. Availability implies sharing knowledge across an organization and can give employees the opportunity to perform tasks that were otherwise unattainable.

Collectively, these factors are the most important when measuring productivity; they are also used to determine a significant portion of the value that an organization is looking for to justify technology investment costs. This portion of the value equation involves more than reduction in administrative and product manufacturing expenses; it also focuses on leveraging personnel skills and minimizing the effects of employee turn-over by providing more challenging working conditions for both professionals and factory workers. This in turn can instill in employees a greater sense of responsibility and ownership for the tasks they perform in their day-to-day jobs.

APPLICATION INTEGRATION

In today's networked organizations, integration of functions requires information to be easily accessible by various departments and business units in many ways. The "information is power" perception that may have once caused individuals to withhold information is no longer tolerated, and companies are continually attempting to extend their information systems to a broader user base, be they employees or business partners. The ability to access this information easily leads to empowering people, because it broadens the knowledge base within which users operate and upon which they make their decisions. Barriers inherent in the old hierarchies tend to break down naturally, allowing communications to move at a faster rate across organizations, with less information filtering and therefore less potential for misinformation or misinterpretation.

The old "stove-pipe" organization dictated that information flow was vertical as well as horizontal; decisions had to flow up through supervisors and managers before being handed off across functions. This significantly reduced the speed at which communication occurred between someone who had the information and someone who needed it. Today, sophisticated data communications techniques are empowering people to interface directly with their information sources and have minimized the need for information to flow vertically. This has, in turn, resulted in faster communications and less time spent waiting for information delivery. Individuals use this extra time to be more productive and contribute more effectively to an organization's drive for success.

The need to network data across functions is more easily satisfied when information is delivered through integrated application systems. Information is entered once, where appropriate, and then accessed by many users, wherever it is needed. Data and the methods used to access the data (i.e., naming and system standards) are defined uniformly across the organization. The information architecture required to support this level of integration and the tools used to deploy the strategy are keys to how successfully applications can be built to satisfy the data and functional needs of various system users. Systems developed without considering the architecture and deployment tools tend to be developed in a vacuum, and the productivity gains anticipated by users are often offset by the introduction of methods or tools that force processes to operate unnaturally, that is, processes are unnecessarily iterative or outside the scope of users' natural work routines. This in turn, limits the anticipated gains. Two factors, information architecture and the related application system deployment strategy, play distinct roles in application integration.

Information Architecture

Applications that have been designed around a reliable data resource are inherently accurate and consistent. A reliable information architecture is the result of constant evaluation and definition of the business data integrity and process flow. Without this evaluation and refinement, automation may cause only slight gains in speed at the individual task level and none at the overall process level. For example, one large company in the Midwest automated all aspects of its manufacturing process from order entry to shipping. When the system went live, order entry clerks could key in an order quickly, but the overall order processing still took the same amount of time to complete, because the tasks associated with having the discount level manually verified against the customer data base had not changed. This turned out to have been an architectural decision—enabling a single data record to be shared across multiple applications—that could have been implemented up front at a very low cost.

A formal evaluation of business data and process flow can help identify ways to reduce the time and effort required to perform routine, business-related activities. Organizations can optimize one or more aspects of their business operations and consequently develop the ability to shorten product life cycles (i.e., time to market). Such an evaluation operates on three basic principles:

- The identification of opportunities for simplifying unnecessarily complex tasks.
- The elimination of unnecessary or redundant tasks.
- The identification of the remaining opportunities for automation.

The process definition that the first two items yield should result in a fairly accurate model for how information is used within an organization and where it is needed. This knowledge can, in turn, provide accurate guidelines for improving the efficiency and effectiveness of business processes through sound data-placement decisions and task automation, the third item on the list.

Application Deployment Strategy

A deployment strategy consists of determining which tools, techniques, and technologies are required to build and maintain applications as well as to manage how these applications are implemented, supported, and used. Central to this strategy are the concepts that include object sharing (i.e., the sharing of code objects) and code reusability, an extension of object sharing.

Objects can be pictures (e.g., images), documents (e.g., text, word-processing files, and spreadsheet data) or even common business rules (e.g., logic blocks or lines of code that define how specific business processes operate within the organization). Object sharing espouses the one-time creation and update of objects by the functions that have direct ownership and responsibility for the objects' integrity and accuracy. These objects can be retrieved and used across the entire organization, thereby minimizing unplanned redundancy and the potential for data inconsistency. Code reusability, as an extension of object sharing, can contribute to:

- *A reduced development life cycle.* The code is developed once and reused in multiple places within as well as across applications.
- *Logical, or business rule, consistency across applications.* The exact same logic can be applied within different areas of the application and organization.

It is easier to maintain the accuracy and consistency of information if there are fewer areas of the data that need to be synchronized or fewer steps to accomplish a task. Minimizing the number of steps in a process automatically minimizes the potential for errors.

For example, in older, more traditional systems, cross-department sharing meant using files to transfer information from one system to another. The receiving system's accuracy and integrity could only match the originating system's accuracy and the processes used to transfer the information. The many functions within an application taken for granted in mainstream technologies, such as automatic data base recovery and two-phase commit support for remote transaction processing, had to be designed, developed, implemented, and supported by in-house programmers. The complexity of these routine functions needing to be applied at all interface points, coupled with the number of "hand-offs"—often custom developed—that had to occur from the originating system to the final destination, left the overall process exposed in multiple areas. Consequently, the information available at the end of a multiple hand-off process was more susceptible to errors and inaccuracies than the original information.

Deploying applications in today's distributed environment requires up-front planning on how the available system management functions can be used. Although there is a technical solution (i.e., the ability to code and connect any functions) and many routine (and complex) functions are vendor-supplied, the final solution's manageability determines its longevity and value. Therefore, before an application is deployed, the organization must know how it will be managed and supported to ensure its long-term viability and integrity. Planning must include general administration (i.e., allowing/denying access to information resources) and ongoing, day-to-day support (i.e., monitoring backup integrity and having recover processes in place for minimum disruption to the business). However, the technology to support other types of application management requirements in a distributed environment, such as automated change impact assessment, software distribution, and inventory tracking, are not mature enough today. Ultimately, managers need to understand that until these technologies have matured further, organizations that deploy distributed applications should rely on a strong mix of manual in-house processes and some level of automated support tools and techniques. These in turn, should be complemented by a common-sense, best-practices approach to ensure the fundamental stability and reliability of the application over time.

APPLICATION FUNCTIONALITY

Increasing application functionality by itself does not necessarily imply end-user productivity gains. Experience has shown that even intuitive front ends and GUI standards require that staff receive a minimum level of training, so they can fully exploit the available functionality. For example, many users experience a definite drop in productivity when Windows is first introduced to their desktops. Much of this initial loss of productivity can be attributed to human nature—aversion to change, learning something new, what is appropriate for one person does not necessarily benefit someone else, and personal preferences. In the long run, however, experience has also shown that once users are familiar with one Windows application, other Windows applications are more easily learned, as the basic functionality is the same. The ease with which this functionality can be accessed comes with training and use. This is important in determining user familiarity with the application and, consequently, to what extent the application will be used.

PC access techniques using GUI standards are designed to provide intuitive access to a wide range of functionality, an ease of access that must also be carried through at the custom-application design level. Other characteristics further define a set of minimum design standards within the GUI environment:

- Using color or pictures to depict certain types of functions (i.e., making "drill down" fields a certain color so the user knows there is more data available for display by selecting—drilling down on—the field).

- Using color or pictures to depict certain types of information or trigger a certain response in users (i.e., displaying negative numbers in red and positive numbers in green to focus attention more on one value).

- Placing radio buttons or icons that perform certain types of functions in the same specific location on all forms (i.e. placing the Exit icon always in the top right hand corner of all screens).

- Defining certain symbols or icons to represent the same function across all applications, for example, using a printer icon to depict the print function.

Additionally, standardizing the development and operating system platforms within organizations can significantly increase the productivity of:

- Systems personnel, when fewer, better trained, more focused support resources can be leveraged over a wider user base.

- End users, who can leverage the skills learned on one application using a standard front-end presentation technique (i.e., Windows based) to many other applications with the same type of front-end presentation technique (i.e., other Windows-based applications). This implies that one-time training costs and ongoing work experience can be leveraged across functions. The result is a reduction in the organization's cost for training, a shortening of the learning curve, by allowing users to catch up faster, and providing users with the ability to focus on business solutions rather than the technology being used.

Standards being defined and adopted at all levels of the software industry facilitates vendor delivery of user-centric, functionality-rich IT solutions. The insulation of users from the more routine, repetitive tasks that computers typically perform is made possible by layering the architecture into discrete components. Standard protocols at various levels of the architecture play key roles in tying consecutive layers together. One example of this type of layering is occurring in the Open Data Base Connectivity (ODBC) market, in which Microsoft is attempting to make ODBC the access standard for tying desktop applications to heterogeneous, back-end data, that is, information stored in a variety of data base management systems (DBMSs), on a variety of hardware platforms. This allows client software developers to focus on the functional richness of their products and build applications independent of the back-end data base management system, whereas data base management system vendors focus on the robustness and high-performance requirements of their shared data base systems (i.e., transaction throughput, backup/recovery, concurrent processing controls, and two-phase commit integrity).

Additionally, such technologies as Windows, coupled with Object Linking and Embedding (OLE), are examples of other minimum standards that have been defined at the higher levels of the architecture to provide users with seamless access to multiple functionality through a single application or desktop screen. OLE provides the ability to embed objects (i.e., documents, images, and text) within objects or link them to other externally stored objects from within an object, regardless of the application that created the embedded or linked objects. The applications that operate on these objects can be automatically launched by selecting the objects.

One example of the value of this type of capability is its use in displaying compound documents, such as image, text, and data on one screen. Compound documents are particularly useful in engineering change notification applications, in which pictures of products often need to be displayed with a bill of materials and other relevant textual information (e.g., the assembly process).

Various other types of cross-application interface standards, such as MAPI (Microsoft Application Program Interface) and VIM (Lotus's Vendor Independent Messaging), have allowed the emergence of many niche players that cater to every type of functional requirement imaginable. These interfaces provide the connection points for allowing users to weave their way through seamless multiple applications and various types of functionality with relative ease. As more plug-and-play functionality is introduced at the various layers of the IT architecture—what the industry has started to call objectware—the productivity of the custom application developers in delivering application solutions and the productivity of end users in using these applications must also increase proportionally.

SUMMARY

Finally, it is the combination of application functionality and integration that can increase user productivity dramatically. This is delivered through client/server architecture, in which distributed PC functionality can be married with the central data management and information-sharing capabilities of robust data base servers.

DBMSs today have the ability to manage the integrity and accuracy of structured (i.e., data) and unstructured (i.e., free-form textual) objects, as well as the various types of relationships and dependencies that exist between them. Techniques that support data replication (i.e., server to server) and remote data access (i.e., server to server and workstation to server), also enable a variety of geographically dispersed nodes, such as servers and workstations, to use this information. Distributing functionality and a significant portion of the required processing resources to a user's desktop can only enhance a user's creativity and the ability to contribute to the business effectively.

Even though the systems management area of client/server (i.e. change-impact analysis, software distribution, and licensing) still needs to be adequately addressed, the client/server paradigm is here to stay. It supports the view of a technology that complements distributed functionality with a well-designed, robust information architecture and makes users even more productive. Such an environment encourages an openness and versatility in the way people interact with each other, because it reflects the way that people naturally work in networked, geographically dispersed organizations.

6-2
Integrating Technology, Data, and Applications in the Open Systems Environment

SUNDAR SUBRAMANIAM

The term *client/server* has become the buzzword of the 1990s. Just about every software and hardware company in the world is claiming to be client/server oriented, selling client/server products, or producing reports on what client/server is. Having been involved in what is now referred to as client/server technology for a number of years, the author of this chapter has seen many successful and unsuccessful implementations. What separates successful efforts from failures is clear: having the proper application architecture to support development and integration are key to a successful client/server strategy.

THE CLIENT/SERVER ARCHITECTURE

Though client/server architecture can be very complex, organizations have essentially two kinds of client/server infrastructures from which to choose: two-tiered and three-tiered. Each has distinct characteristics and qualities. Briefly, the two-tiered architecture contains two computers, a client and a server; areas of logic are combined on the client, and data is stored on the server. A three-tiered computing environment may have only two computers but always has three separate layers of components: the graphical user interface (GUI), all application logic, and the data. Two of the most commonly asked questions about this architecture are, "Why must the functionality be separated from the GUI?" and "Why is functionality not built within Powersoft or SQL Windows?" Unfortunately, the most powerful reason client/server applications should not be built for two-tiered environments is that so many organizations have tried and found that the two-tiered approach simply did not provide enough processing power on the desktop.

Two-Tiered Client/Server

The reasons for the existence of this architecture are evolutionary. Initially, the file system portions of the data layer were separated into a server (e.g., Novell). Users then wanted to centralize more functions on the server and put the data tier (i.e., the data base) on a server (e.g., an SQL server). The remaining portion of the application was left on the client. The advantage of this architecture was that it fit the hardware [personal computer (PC) and local area network (LAN)] architectures. The problems and issues that arose due to this architecture are:

- *A lack of scalability.* The language in the data base is usually difficult to code and maintain, and applications tend to be written primarily on the client. This means that if the logic is complicated, the PC becomes bogged down with processing, and the applications do not scale very well. The PC is doing the presentation as well as the logic and coordination. Large memories and processors are required to support the environment and avoid contention.

- *A low level of manageability.* Because two-tiered architectures require a connection from every client to every server on the network, scalability and manageability are low. When performance problems occur, they are difficult to isolate and remedy. (See Exhibit 6-2-1 .)

- *An inability to leverage client/server cost performance.* Because the cost performance of the client/server environment, as measured by cost per millions of instructions (MIPS) is decreasing as a result of parallel processing, the business logic is best left on the server. However, the limitations of the two-tiered architecture prevents effective use of the server and thereby compromises the architecture's cost performance.

- *A low ability to manage change.* Change occurs at breakneck speed on both the presentation (e.g., 3270, VT100, DOS, Windows, NEWTON, and voice) and data (e.g., flat files, hierarchy, networked, relational, object) layers. A two-tiered architecture locks the presentation and data layers and does not allow changes for presentation or type of data base.

- *Reuse.* Most application programmers would like to reuse business solutions written in previous applications, a difficulty in two-tiered architectures, because the logic is embedded in either the presentation or data base.

- *An inflexibility to incorporate best of breed.* Fourth-generation lan-

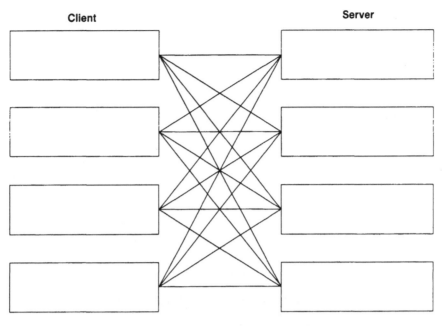

Exhibit 6-2-1. The Two-Tiered Architecture

guages (4GLs) used in the presentation tier are general-purpose languages. Programmers must be able to use the appropriate hardware and software tools and languages to optimize applications. Any expert system should be written in a language optimized for expert systems.

Finally, within a two-tiered system data access is typically developed using a series of application-programming interfaces (APIs) specific to the GUI, and the language used to develop business logic functions is also quite proprietary. For example, Oracle Card requires that Hyper Script be used; Powerbuilder requires PowerScript.

The Three-Tiered Architecture

In a three-tiered architecture, the components are divided into three layers: a presentation layer (i.e., a GUI that displays information, receives user input, and makes requests), an application logic layer (i.e., a set of functions that perform the business logic of an application and manages its technical workings), and the data access layer (i.e., that contains the application data and access to it). Each of these layers must

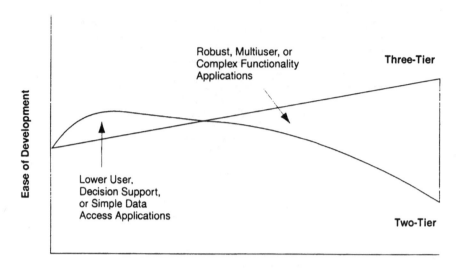

Complexity of Application

Exhibit 6-2-2. Development in Two-Tier Versus Three-Tier

---

be logically separate; the code that comprises the presentation layer does not have business logic functionality embedded within it. (See Exhibit 6-2-2.)

For example, a PC's GUI might be developed in Visual Basic, Powerbuilder, or another GUI builder for Windows and executed from the desktop. The functionality for performing calculations, heterogeneous data base joins, and sorting, however, resides on a separate UNIX or OS/2 server, with access to a data base or mainframe. Another reason for the separation of application logic onto a different series of servers is to use the industry standard client/server interface, Open Software Foundation's Distributed Computing Environment (OSF/DCE). An OSF/DCE remote procedural call (RPC) could be issued by a Windows DCE product directly from any Windows application to software servers within UNIX, VMS, or a range of other environments.

Three-tiered architectures are the accepted standard for enterprise-wide deployment. The logical architecture is partitioned into three logical layers: presentation, functionality, and data. Physically, this can be mapped into one or more hardware machines. Three-tiered architectures overcome the major problems of a two-tiered architecture:

- *Scalability.* The application can be deployed in any physical configuration and is flexible enough to be modified easily, so any part that requires more power or disk space can be easily accommodated.

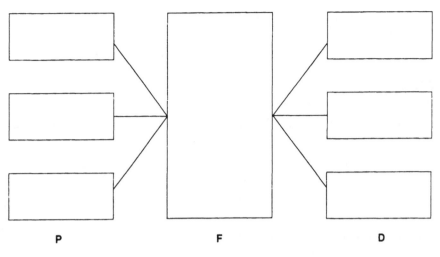

P   F   D

Exhibit 6-2-3. The Three-Tiered Architecture

- *Manageability.* The functionality tier is separate from the presentation and data, so it can act much like a telephone switch, arbitrating individual calls. This reduces point-to-point connections and eases manageability (see Exhibit 6-2-3).
- *An ability to leverage any advance in hardware performance.* The application can physically be redeployed in any fashion, so any advance in technology can be incorporated into a system with little modification (e.g., if a massively parallel machine allows functionality to be run more efficiently, that portion of the application can be redeployed on the new machine).
- *An ability to manage change.* A change occurring in the presentation or data layer can be immediately incorporated into the application.
- *Reuse.* The functionality is separate from the presentation and data, so it can be reused in many applications.
- *An ability to incorporate best-of-breed technology.* The ability to use different languages and tools allows for incorporation of best-of-breed technology.

Additionally, a project team can be organized by their special skills. People who understand the business and the user concentrate only on the presentation. Traditional data processing programs work on the functionality, and the data base analysts work on the data tier. Because these functions are all logically separated, the users can work in parallel with special skills and tools.

The architectural framework can be used to address practical considerations, such as standards and the application to real-world problems.

STANDARDS

One of the biggest benefits of a three-tiered architecture is the ability to swap components. For this to work practically, an industry standard has to be adopted for three main features:

- *Naming.* How does any one piece find the other piece of the application; how does a client find a particular server?
- *Security.* How is the security between the different pieces maintained?
- *Communications.* How do the different components communicate with one another?

The industry standard also has to be available on all the platforms (e.g., MVS, VMS, UNIX). The only current standard for these three base services is OSF-DCE, which specifies the security, naming, and communication mechanisms for a heterogeneous environment. The only alternative is to use a proprietary mechanism, such as a messaging architecture. In recent months, the communication mechanism has come under considerable scrutiny, with issues arising as to whether a messaging or RPC-based communication mechanism is more appropriate.

Remote Procedure Call (RPC)

A remote procedure call is the analog of a subroutine call that happens over two programs. The client calls a subroutine that actually runs in a different program.

Messaging

Messaging architectures usually give the user a set of calls that allow one program to communicate with another program:

- send message
- get message
- print name
- send message (i.e., get name, name)
- get message (i.e., name)

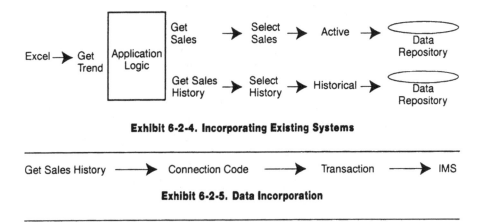

Exhibit 6-2-4. Incorporating Existing Systems

Get Sales History ⟶ Connection Code ⟶ Transaction ⟶ IMS

Exhibit 6-2-5. Data Incorporation

- print name

Because messaging has one function to send messages and another to receive, multiple messages can be sent to various servers and all the results can be retrieved later, allowing for concurrent execution of multiple servers. RPCs themselves are synchronous; the client has to wait until its server is done before continuing. While this itself could be a deficiency, such standards as OSF-DCE specify threading (i.e., parallel execution), allowing a client to call multiple RPCs simultaneously. Most commercial implementations, such as NFS and OLE, use RPCs for their implementation of communications. Three-tiered architectures can be implemented with standard RPCs or messaging.

Incorporating Existing Systems

The three-tiered architecture abstracts the three tiers, making it possible to incorporate existing systems as blank boxes (see Exhibit 6-2-4).

If the historical information exists on a legacy system, the data component can be changed without affecting the architecture (see Exhibit 6-2-5). This allows a connectivity component to be incorporated without affecting the rest of the application architecture.

The three ways to connect to existing systems are:

- *Connect to the application.* Also known as screen scraping, this method allows the connectivity server to emulate a user of an existing system using a terminal, such as a 3270 or VT100.
- *Connect to the transactions.* Also known as peer to peer, this method allows the connectivity server to call existing transactions (e.g., ICS transactions) without having to go through the screens.

345

- *Connect to the data base either directly or through a gateway.* This allows a server to connect directly to a data base, such as DB/2 or MVS, using protocols, such as DRDA. If the remote data is nonrelational, translators, such as EDA/SQL, allow an SQL transaction to be translated to nonrelational formats, such as IMS.

However, each of the three methods of connecting legacy systems has advantages and disadvantages. The data base connection allows the most flexibility but requires the existing transactions to be abandoned. Peer to peer allows reuse of transactions but is more difficult to build than screen scraping, which is the easiest to build but the hardest to maintain.

The following case studies, adapted from case studies prepared by the Open Environment Corp., provide additional insights into the client/server development and implementation process.

CASE STUDY: COLUMBIA NATURAL RESOURCES

Whereas other companies only talk about replacing their mainframe systems with client/server architectures, Columbia Natural Resources has done it—smoothly, efficiently, and in a matter of months. At a point when many businesses are still struggling to implement distributed processing, the Charleston, W. Va., oil and gas firm has successfully rolled out three key applications and is aggressively planning a complete migration from the mainframe.

CNR faced all the usual migration challenges, sifting through the many platforms, network protocols, operating systems, and data bases. As with most companies, CNR's staff of mainframe specialists needed reschooling in everything from UNIX to front-end application development. What made the difference? CNR added one more important decision to its list: once the company settled on a relational data base and front-end application development tool, it added a third, middle tier to its distributed architecture. CNR used Open Environment Corp.'s OEC Toolkit to place all of its code in a middle layer, effectively freeing the company from dependence on data base or application-tool vendors and dramatically reducing programming time and making for a swift implementation cycle.

When the one of the nation's largest natural gas utilities, Columbia Gas System, began reevaluating the mainframe's role in early 1993, CGS gave its subsidiaries the go-ahead to create their own computing environments. They could continue to tap into the company's mainframes or build new architectures, when justified. For CNR and the other subsidiaries, the prospect was exciting—and daunting. Since its inception in 1985, CNR had run a traditional IBM mainframe shop, relying on its parent company's large-system computers to store the information for managing more than 6,000 oil and gas wells located on 30,000 lease parcels in 10 states. Nearly two dozen applications ran remotely from the IBM ES 9000 and 3090 at Columbia Gas's service facility near Columbus, Ohio, and the information systems staff had years of experience with IBM's DB2 relational data base.

CNR was not immediately sold on switching to distributed computing. Maintaining the mainframe environment held plenty of appeal from both programming and financial perspectives. Cost projections, however, began to mount. Monthly expenditures to support a mainframe at the West Virginia facility amounted to 95,000, including the hiring of five systems programmers. Perhaps most important, the money would be invested in older technology that offered little flexibility in the way of programming and end-user contribution.

CNR bet its future on client/server architecture. By dispersing its processing power, it could leverage the benefits of low-cost PC hardware and GUIs (GUIs) on the desks of 200 employees. More important, breaking out of the glass house offered the company a chance to poise itself for growth and compete more nimbly in the competitive oil and gas industry. After years as a Big Blue site, CNR was enthusiastic about creating an open system; however, it envisioned taking the openness one step further. In addition to being free of proprietary hardware and operating systems, CNR's programmers wanted to avoid designing applications to the specifics of a particular front end or relational data base.

To help select the most cost-effective client/server solution, the company solicited suggestions from several consulting services, and Hewlett-Packard Co. (HP) devised the most attractive recommendation. The plan called for CNR to concentrate on the development of applications, while HP focused on the architecture, supervising the move from OS/2 to Microsoft Windows clients and designing both a frame-relay information network to connect CNR's 16 field offices and a network print environment. Sponsored by HP, CNR sent executives and members of the IS department to classes in open systems-oriented programming at Open Environment Corp. in Cambridge, Mass. There, CNR learned the advantages of a three-tier client/server architecture. Conventional client/server schemes require the system's front and back ends to communicate using DMBS-specific protocols. The OEC Toolkit does away with those limitations, supplying communication among the presentation, application, and data base programming logic over generic RPCs. Because the logic resides in the isolated middle level, companies are free to change the elements of the system without the expensive and time-consuming steps of recreating the application.

Maintaining applications investments also made important business sense. By not locking into a specific vendor's way of doing things, CNR would be building a system designed for change. The use of OEC tools would allow CNR's IS department to develop the new applications more speedily. As a result, the company could quickly begin to reap the benefits of more efficient, cost-effective computing.

In the summer of 1993, CNR began laying out the design for its new system. It purchased two Hewlett-Packard HP 9000 UNIX-based mini-computers. The G40 model serves as the IS department's development machine. The I70 model, the processing equivalent of a low-end main-frame with 30G bytes of storage space, acts as the applications server and connecting point for CNR's 16 field offices. For its corporate data base, CNR chose Sybase SQL, which uses SQL, the de facto industry standard. To create desktop applications that run over the data base, it chose Pow-ersoft Corp.'s PowerBuilder. The desktop environment was beefed up with the addition of more than 100 486 PCs. Printing capabilities were dramatically streamlined with a Hewlett-Packard F100 laser printer. The 100-pages-per-minute machine prints directly from the mainframe over the HP 9000, from the HP-UX servers, and from the Token-Ring LAN, on which employees use packaged and custom applications.

In November, after a brief three months of development time, which also included time spent mastering SQL and UNIX, the IS development team unveiled its first application. Replacing a hard-to-use terminal-based application, the Windows program tracks data on the company's 300 pick-up, drilling, and heavy-rigging trucks. The application shows off the advantages of the three-tiered architecture. When field supervisors, who are responsible for vehicle costs, click the mouse on a menu item, the action triggers an RPC that queries the data base for the requested information and displays the information on the screen. SQL statements are written just once as RPCs. If the statement or calculation is needed for another application, the programmer can reuse it. Unlike program-ming done in COBOL, there is no need to rewrite the code each time it is used.

Two months later, the IS department took the wraps off another critical application, a gas-pricing tool used by the planning department. Perhaps even more exciting is the third and most strategic application, which has captured the true spirit of distributed computing. Originally written to help the company manage the thousands of wells covering 2.3 million acres of land, it generated such an enthusiastic flood of user suggestions that it has been expanded into a complete facilities-management tool. Now, in addition to monitoring production cycles, sort-ing data by state, county, and well tender, and tallying the cost and profit of each producing well, the application has become an all-around operations and maintenance program, storing facts and figures that previously had been informally tucked in the desk drawers.

The migration task is far from over as 20 applications remain to be converted, but as other businesses continue to talk about the competitive advantages of open architecture, CNR is making them a reality.

CASE STUDY: JOHN DEERE HEALTH CARE AND UNITED BEHAVIORAL SYSTEMS

As the architects of President Clinton's health care reform plan know all too well, administrative paperwork and insurance overhead deserve much of the blame for spiraling health care costs. As the reform debate rages on in Washington, however, two Midwestern health care providers are taking the lead in showing how new technology can significantly cut administrative costs.

John Deere Health Care Inc. and United Behavioral Systems are pioneers in the brave new world of "electronic commerce." Simply put, the companies do business together through a new breed of client/server application that spans each company's enterprise network. The transenterprise application is part of UBS's Statistical Tracking and Evaluation of Patient Services (STEPS), the grand prize winner of *Datamation* magazine's 1994 client/server solution contest. Built with Open Environment Corporation's OEC Toolkit, the STEPS application is helping the two companies hold down health care costs by eliminating the endless paperwork and claims-processing bottlenecks that occur when verifying the insurance eligibility of new patients.

Moline, Il.-based John Deere Health Care Inc. is the wholly owned health care maintenance organization (HMO) subsidiary of the Deere & Co. farm equipment empire. With the rising popularity and competitiveness of HMOs, John Deere Health Care needed to add mental health care coverage to its expanding base of services, especially in the treatment of substance dependency. But instead of developing its own in-house program, the company found it more cost-effective to contract the services from UBS, a mental health care provider owned by the multibillion dollar United Healthcare Corp.

The only glitch in this partnership was in verifying that one of the more than 280,000 John Deere Health Care enrollees visiting a UBS clinic was indeed eligible for mental health coverage.

At first, UBS administrators had to hand-verify each claimant by searching through computer printouts of John Deere's mainframe data. This paper-based approach quickly proved its inefficiency and was replaced by a direct 3270 terminal connection to John Deere Health Care's IBM mainframe, which contained all of the enrollees' health care data.

The remote terminal-to-mainframe approach posed a new set of problems, however. The UBS staff had to learn an entirely new application to retrieve only a fraction of the information. With multiple users logging into the mainframe remotely from several locations, telecommunications costs skyrocketed, and security became an issue, as well.

Ironically, the solution was already in place, just waiting to be exploited. UBS had developed STEPS with the OEC toolkit as a way of registering new patients and scheduling meetings with doctors. It was a proven production application, serving 80 users at its Minneapolis headquarters and in several UBS clinics. The John Deere Health Care staff indicated they wanted to build a similar application.

In less than six weeks, the John Deere Health Care IS staff was trained in the OEC Toolkit environment and jointly developed its first application with UBS. Because STEPS is based on OEC's three-tier architecture, which separates the GUI front end and data base back end from the business logic, the companies' development teams were able to work independently of each other.

OEC's three-tier approach is based on the Open Software Foundation's Distributed Computing Environment (DCE), a vendor-neutral model for distributed, multiplatform computing. Following this model, UBS built a Windows-based client application using PowerSoft's PowerBuilder GUI development tool. It gives UBS users a single, easy-to-use graphical interface to every information system they may deal with, including John Deere Health Care, United Healthcare and several other providers who have contracted their mental health care services with UBS.

On the back end, enrollee information is delivered from a Sybase relational data base running on an IBM RS/6000 UNIX system in John Deere's IS data center. Programmers used the OEC Toolkit to generate the communications code and to build RPCs to the Sybase server from the application and presentation logic. Whereas the application's graphical front-end at UBS was built with PowerBuilder, the John Deere Health Care staff built their own GUI interface with Microsoft's VisualBasic to test the application. To users, the connectivity and communications is transparent. John Deere's RS/6000 is essentially another server in UBS's vast client/server network. When a John Deere enrollee requests treatment, a UBS administrator can now immediately verify his or her eligibility by simply clicking on John Deere Health Care from a pull-down menu. Behind the scenes, the eligibility request is automatically carried to an HP UNIX server running the OEC Toolkit at John Deere's data center in Moline over an OEC Broker. The eligibility status is gleaned from a Sybase data base on the server, and the Broker returns its findings to the STEPS application running on UBS' UNISYS server. The user can instantly verify insurance coverage, dramatically reducing the time, communication costs, and frustration of navigating through an archaic, remote 3270 application.

Both companies are already working on extending the underlying application logic to other areas of their business. John Deere Health Care, which has 17 sites in four states, wants to provide instant eligibility verification to all its clinics and other third-party partners. UBS wants to take the client/server application beyond eligibility checking to include claims status, referral information, and benefits authorization, among other tasks.

CASE STUDY: FREDDIE MAC

Every four seconds, the Federal Home Loan Mortgage Corp. (Freddie Mac) makes the American Dream of home ownership possible, purchasing nearly one out of every six residential mortgages in the US each year. This $450 billion company converts a steady diet of home mortgages into securities, which it markets to Wall Street. By linking Wall Street with Main Street, Freddie Mac sustains a stable mortgage credit system and reduces the mortgage rates for home buyers. It is also one of the industry's best examples of how to turn a high-volume operation into an efficient, solid financial performer by deploying client/server technology.

With the help of a unique client/server application developed with the OEC Toolkit, Freddie Mac is saving nearly $500,000 in after-tax profits for every hour the system stays up and running. The company also recently set new records for financial performance, purchasing over $230 billion in mortgages.

Freddie Mac's market is huge and intense. It competes head-to-head with the Federal National Mortgage Association (Fannie Mae) for more than two-thirds of all residential mortgages in the US. In this competitive context, Freddie Mac's IS department is a high-stakes environment in which literally millions of dollars ride on the uninterrupted flow of information from the company's computer systems and outside data sources. As its secondary mortgage business boomed over the last few years, however, the company's aging mainframe buckled under the pressure. Regular nightly reports started taking longer to generate on the mainframe; reports that couldn't completely print overnight consumed daytime computing resources. The system started crashing more often and required more frequent maintenance.

Within five months, Freddie Mac rolled out a new client/server application called Gold Server with OEC Toolkit to replace the mortgage-purchasing application on the mainframe. OEC's three-tier architectural approach segments the GUI front end and the data base back end from the business logic. Freddie Mac built both the user interface and local application logic using Microsoft's Visual Basic. A Sybase relational data base running on a UNIX server, which acts as the point of convergence for data sources and runs the business logic, fuels the client application over a TCP/IP network. Gold Server enables users to gather up-to-the-minute interest rates, stock market activity, and other market data from a variety of sources, and quickly process the mortgage-to-securities transactions that generate the company's revenue.

The OEC infrastructure allows data to be passed between the PC, UNIX, and data base environments. OEC provides the communications layer to translate data among the different languages. To the user, the presentation layer makes the data appear local and from one source; however, behind the scenes, OEC-generated RPCs provide centralization of the application's core business logic so that information gleaned from the different data sources can be quickly delivered to the Visual Basic graphical interface. The OEC Toolkit, which is modeled after the Open Software Foundation Distributed Computing Environment (DCE), also provides distributed management, naming services, and security to the Gold Server application.

Segmenting the user interface and data base from the data and business logic also enabled Freddie Mac to achieve its number-one objective with Gold Server: uncompromising reliability and availability. The buying and converting of mortgages into securities is the company's most mission-critical function, with hundred of thousands of dollars at stake should the computer system shut down even for a short period. Requiring no extra work by application programmers, OEC's three-tiered approach enabled Freddie Mac to set up a client/server environment with full protective redundancy built in at several locations. A warm backup OEC server at another site is always ready to kick in automatically if the primary server should fail.

The three-tiered approach also freed Freddie Mac from becoming wedded to any one vendor's product. Because the GUI, application logic, and data bases are separable but interoperable components in the Gold Server schema, Freddie Mac can change any one of them at any time without changing the application. For example, one function of Gold Server enables the Freddie Mac customer service staff to give customers timely quotes on home mortgage interest rates. Because interest rates are always changing, they need to be able to change the application rules regularly to keep the quotes current and accurate.

By putting the business logic on the middle tier of the application, Freddie Mac is able to change the application very rapidly without having to rewrite the Visual Basic code or restructure the Sybase data base.

Alternately, they can easily replace the interface or data base in the future without affecting the application logic.

SUMMARY

Application management is the area in which the differences between two- and three-tiered architectures become most apparent. In a simple application, two tiers can be effective, but once the need to join and manipulate information from multiple data sources comes into play, performance rears its ugly head. For example, the corporate headquarters for a chain of stores wants access to sales and inventory information and integrate it so executives can see historical and projected sales and inventory, broken down by location, region, week, month, quarter, and year. Their historical sales information is stored on an Informix data base, current sales information is on an Oracle data base, and the inventory maintenance system is on an IBM 3070 mainframe. The information the user retrieves is to be displayed on their desktop PCs through a GUI, such as PowerBuilder or VisualBasic.

In a two-tiered client/server environment, the Informix data base would be accessed from the user's desktop. The information would be retrieved and stored, and the user would connect again, this time to the Oracle data base, where the same sequence—access/retrieve/store— would occur. Then a third request would be made to the mainframe. Only after all the requested information had been stored would it then be combined, sorted, and displayed. Obviously, in this scenario, response time and performance could be a real issue.

In contrast, the three-tiered client/server model would allow the user to issue a single request directly to the application functionality server, which would in turn generate three simultaneous queries for information from both data bases and the mainframe. After the requests were returned, information would then be sorted by the functionality server and results passed back to the GUI.

6-3

Integrating the Client/Server Environment: A Case Study

R. JACK MUNYAN

GORDON COUTURIER

Although a variety of reasons may prompt a company's decision to switch from mainframe computers to networks of PCs, important generic lessons can be learned from companies that have already bitten this bullet. The reasons most often cited for making a conversion are meeting executive pressures to do more with less and creating a system that is more flexible and adaptable to the economic changes occurring in the industry. Additional reasons for making this change include increasing corporate competitiveness, improving total information accessibility and availability, providing information in an easier-to-use format, working with the perception that this new concept will result in cheaper computing, and satisfying the allure to be on the cutting edge, while avoiding the bleeding edge, of change.

A CASE IN POINT

Many companies have traveled the road of client/server conversion, and valuable lessons have been learned in each case. Plan Services, Inc. (PSI), a subsidiary of Dun & Bradstreet and one of the largest third-party administrators for small business group life and health insurance, started to feel the effects of the economic recession on their business. With the advent of increased managed-care activity, tighter government regulations, mandated benefits, and other industry trends affecting customers, senior management realized that PSI was facing significant challenges in its marketplace.

355

Management believed they not only had to reengineer their business but also their information systems (IS) organization. As Robert Weissman, president of Dun & Bradstreet Corporation, stated, "We have to do more with less; we must make our processes cheaper but better." Cliff Bateman. president of PSI, reinforced this by saying, "Systems must enable our business to adapt to a rapidly changing market quicker and with less cost." PSI defined business reengineering as "fundamentally changing the way work is performed in order to achieve radical performance improvements in speed, cost, and quality." Concurrent with business redesign was the need for technological capabilities to support the goal of performance improvements.

Implementing the Business and Technological Infrastructure

To respond to this need, PSI completed a business strategic plan and formed a group called LINX, a name arbitrarily chosen, to work on the technological aspects. The LINX process was responsible for the implementation of the business and technological infrastructure to support the reengineering effort. The LINX mission was to build flexible, dynamic business systems driven by the goals of Dun & Bradstreet Plan Services, Inc. that respond to competitive business needs.

LINX was comprised of a dedicated team of four employees and a former outside consultant hired by PSI to be in charge. LINX followed an information engineering (IE) approach to complete its mission. IE was supported by tools, techniques, and procedures, enabling PSI to align information, business, and technology strategies. The first step was the development of the information strategic plan (ISP), which provided PSI with the necessary road map to ensure that the LINX mission was completed.

Linx reported to a steering committee comprised of nine senior executives who provided direction and maintained overall responsibility for the ISP development. The team also conferred with the Technology Advisory Board, comprised of technology management, for assistance with the technology strategy. By September 1991, the ISP was completed. The current vice-president of data processing stated that a guiding rule of the ISP was "Business side is the driver; no technology for technology's sake."

Establishing and Confirming Business Strategies

The first order of business in the ISP process was to conduct a facilitated session with senior PSI executives to confirm PSI's goals and strategies, to identify the critical success factors that supported PSI's goals. PSI identified its corporate vision, strategic goals, and critical success factors as follows:

- *Corporate vision.* The vision is to be recognized nationally by customers, prospects, and competitors as the leader in providing quality marketing and administration of employee benefits.
- *Strategic goals.* The strategic goals are to:
  - Increase annual revenue and operating income growth.
  - Improve sales and marketing efforts to achieve revenue targets and become the market leader in the base business (small group and individual).
  - Establish stronger working relationships with risk-bearing partners.
  - Develop an improved image reflecting high-quality employees and excellent customer service.
  - Evaluate and focus efforts to expand business in areas that are outside the base business, as appropriate.
- *Critical success factors.* The factors that critically affect meeting corporate goals are:
  - Reduce monthly lapse rates.
  - Institute more stable pricing practices.
  - Increase monthly new-business production.
  - Develop stronger and enduring relationships with high-quality producers.
  - Identify and target business with good persistency and claims experience.
  - Expand successful aspects of the strategic business pilots throughout the organization.
  - Create marketing plans for each risk-bearing partner.
  - Divide the sales organization into dedicated teams for each risk bearer and manage the new organization by partnering.
  - Align PSI strategies and goals with risk-bearing partners.
  - Improve management of claims costs to ensure risk-bearing partner profitability.
  - Establish partnerships with providers of managed-care services.

Following that very important first step, the ISP process continued with literature scans, interviews with other organizations, and sessions with the Technology Advisory Board to identify the technological trends that would enable PSI to take advantage of industry advances. A facilitated session was conducted with senior technology management to identify fundamental IS principles that would serve as PSI's basic technology philosophies to support PSI's goals. An assessment of the current systems environment, including technology (i.e., hardware, systems software, and network), IS infrastructure, and applications was completed to ensure that ISP took advantage of current investments and defined appropriate migration requirements to new environments. Facilitated sessions were conducted with key customers to identify information needs, opportunities for improvement, business functions, and data that directly support PSI's critical success factors. Potential business reengineering activities were identified and included in the analysis. The target technical architecture was developed to provide a technical strategy to support PSI's critical success factors. Technology trends, IS principles, and the current systems environment were considered in the development process. Projects that focused on potential business reengineering activities to support PSI's critical success factors were identified. The strategic potential of each project was assessed and potential benefits identified to rank according to priority the projects for implementation.

ISP'S OVERALL RECOMMENDATIONS

ISP made several overall recommendations to support the previously stated PSI goals and related critical success factors.

Current Systems Assessment

Alternative development platforms (i.e., PC based) to offload development from the mainframe were evaluated due to extremely high-processor usage levels. Centralized inventories of leased and purchased products were developed to improve software management capabilities. Decision-making and approval processes were revised to allow better alignment of authority, responsibility, and accountability.

Enterprise Operations Analysis

Business functions, processes, and high-level data items needed to support the business were identified in a series of Joint Requirements Planning sessions. Over 500 information needs and opportunities for improvement were identified during these sessions.

Information Plan

The primary objective of the technical strategy was to downsize to a lower-cost, more customer-focused client/server environment using intelligent workstations, GUIs, and productivity tools to decrease PSI's dependence on the current IBM mainframe.

Thirteen projects were defined and categorized as high or medium strategic potential or required to support the IE-implementation infrastructure. Projects were ranked based on their strategic potential, overall benefit, and implementation-sequencing requirements. A resource plan outlining the resource requirements to implement the projects was then completed. The first client/server development project, Prequote System, was recently completed with very satisfactory results. The manager of the Prequote System project did state, however, that the learning curve was tremendous in the first project.

The target IS organization chart focused on improving IS's alignment with the customer by organizing the department into two main functions, customer support and technical support. Customer support consists of applications, ad hoc reporting, and problem resolution. Technical support focuses on back-office technical operations and support. The IS training plan focused on providing education to applications development associates to enable them to better support an IE environment.

THE ISSUES INVOLVED

The ISP provided a framework to proceed with reengineering the business using an IE environment. It is intended to be a dynamic document to be updated on a yearly basis, or more often if necessary, to ensure consistent alignment with PSI's strategic direction. The issues involved in the actual reengineering, however, require further detailed attention.

Mainframe Versus Client/Server Issue

Under a five-year lease purchase arrangement, COMDISCO bought the existing mainframe system for PSI from IBM by COMDISCO. The mainframe, serviced by Bell Atlantic, is fast approaching full capacity, and future work/applications will require additional power. Improvement in the mainframe capacity and performance exhibit step-function increases in expenditures. An upgrade of a 34% increase, from a 400E model to a 600E model, would cost $200,000 and increase the maintenance fee by $30,000 a year. Switching to a new technology mainframe (ES9000) for the added power requirements would require a $3,000,000 mainframe purchase, put utility fees at $24,000 a year, and place maintenance—free the first year—at $48,000 a year thereafter. Continuing with the existing system strategy would have little impact on user understanding and would not require additional training.

Actual/budgeted expenditures, as a percentage of total CIS figures over the prior two years and estimated for the current year, were arrived at. For LINX, the increase would be 2% for 1991, 26% for 1992, and 26% for 1993. For ERISCO, the mainframe, the increase would be 36% for 1991, 34% for 1992, and 31% for 1993.

The applications running on the mainframe were eight years old, on average. To meet new customer requirements, the applications need to be rewritten to provide new, user-requested maintainability of the programs.

The assistant vice-president of operations stated that PSI "should not just eliminate the mainframe but utilize it as an efficient data repository. The cost will be dramatically reduced once the mainframe is paid for." A client/server system configuration required the purchase of PCs, local area networks (LANs) and the development of new application programs. This technology provided GUIs that are used to improve person-machine interfacing. However, due to the new technology, considerable effort is required to train the user in the features available. With additional training, the user is able to assist in the development of new applications.

In the mainframe arena, development tools are available and well understood. In the client/server system, tools are available but are not as well understood, because of a lack of prior experience in the workplace. As the vice-president of data processing stated, "The tough part [in client/server systems] is choosing the correct development tools to get [usable applications]." A 1993 *Wall Street Journal* article, "In Downsizing from Mainframes to PCs, Unexpected Glitches Often Defer Gains," reported that even though hundreds of programs have been written to help run critical functions on mainframe machines, very little is available to run these same applications on lower-powered machines. In addition, the non-mainframe software has not been fully tested, which could cost companies millions of dollars in lost time or disappearing data.

Another major problem involved in downsizing is political. Many employees view this change as job threatening. During the time that the LINX group was trying to sell their client/server solution, they did a presentation showing the mainframe as a dinosaur. The manager of tech support said that this was demoralizing to those who worked in the mainframe side of the shop. As a result, it took about six months to convince people that their skills would be needed in the future and that their help was essential for a successful migration to client/server.

A 1991 IBM and IBM PCM user survey conducted by *Datamation* indicated that a significant percent of current mainframe systems are to be moved to distributed systems during the next several years. In PSI's case, LINX has evaluated the benefits of the trend in its organization. Although a gradual migration to the downsized environment was required because of technical limitations of available PC hardware to perform intensive batch operations and easy access to large, integrated transaction data bases, PSI has committed completely to its implementation. Existing, business-as-usual mainframe applications will continue to be run in the mainframe until replaced by projects specified by the ISP.

One of the client/server computing benefits is the use of low-cost PCs. However, once the estimates were obtained for all the pieces necessary to implement a production client/server LAN environment for PSI's needs, the actual costs turned out to be higher than the original estimates. The total cost for PCs, data base server, all miscellaneous equipment, and software amounted to over $1,200,000.

Response Time Issue

PSI believes that a major goal in the development of GUI-based client/server systems is to improve total information accessibility, availability, and usability. PSI considered total information turnaround time, or getting the right information to the right person at the right time, to be more important than individual transaction response time.

The design of the first client/server project was based on this assumption, and trade-off decisions between fast response and ease of use and better information usually were made in favor of the ease of use. Additionally, the project development team chose to use a development architecture consisting of precoded libraries, objects, functions, and program shell generators provided by a consulting partner. This architecture was chosen as a method of shortening the development time, lessening the developer learning curve, and providing program standards.

Actual transaction response times in the final system vary from a subsecond for a simple, keyed, data base search to 20 to 30 seconds for a complex presentation of calculated information, in which data is derived from several data base sources, including the mainframe. The average transaction for most system users is about 5 to 10 seconds.

System users, accustomed to mostly subsecond transactions on the older, mainframe-based system, voiced concerns about the ability of the new system. The project manager said, "Our users are uncomfortable waiting more than a second for computers to process their work. As long as they are actively working, they seem more comfortable spending one minute doing something involving multiple transactions than they are waiting 10 seconds for the system to do the same thing in a single transaction."

User management was informed of the response-time issue before the implementation of the system and decided to proceed with the implementation as planned.

The following tests were conducted with users of the system:

- Test A measured the total time to obtain a customer profile using the old system and then again using the new system. The test concluded that even though individual transactions responded faster on the older system, far fewer transactions were required to get the same or better information from the new system.

- Test B measured users' accuracy at looking up and identifying problem customer records in the old versus the new system. This test showed a significantly improved user accuracy rate from the old to the new system.

The Industry Client/Server Bandwagon Issue

In the late 1980s and early 1990s, a lot of hype appeared in the trade journals about the emerging advantages of client/server systems over mainframe-based systems. A client/server system is typically a LAN on which a large microcomputer performs the role of a file/printer repository and server for attached user stations, called clients. Clients access the server for file data or printing requests. The client/server system may be enlarged to multiple LANs connected by wide area networks (WANs) in which there may be multiple servers and many clients.

"MIS Execs: Pressure's Up to Downsize," appearing in the February 22, 1993, issue of *Communications Week,* summarizes the prevalent executive pressures, stating that there are three relative considerations to downsizing to client/server systems:

- MIS directors want to meet management's and end users' desires to distribute computing.
- Many new products and development tools are still in development, with vendors touting future availability.
- There is still a large installed base of mainframe systems to consider.
- The pressure to change is overwhelming.

PSI feels that the key benefits for users from client/server computing are increased productivity, reduced training costs, and improved information access. The benefits for the MIS/DP department of client/server computing are decreased computing costs, scaleable investment, multivendor hardware integration, and improved speed in delivering new applications.

However, many companies have run into unexpected problems. In a recent study, KPMG Peat Marwick reports that 70% of its clients reported many such problems and that a doubling of anticipated costs was not at all unusual.

The Training Issue

Shifting from a mainframe system to a workstation environment resulted in significant training and educational issues. PSI recognized the importance of providing employee/user training and documentation to both the business end users as well as the IS associates.

LINX assigned a full-time trainer to introduce PCs and PC concepts to the business areas several months before the implementation of the first production client/server application. Additionally, business users were provided training on productivity tools (e.g., spreadsheets and work processors) to support the LINX objective of empowering PSI associates to become knowledge workers.

A training plan was created to address the high-level training needs within application development and outlined courses and recommended timing and techniques necessary to ensure a smooth transition to the new environment.

As LINX progressed further along in the project plan, budget and timing became bigger issues, because of an economic downturn PSI was experiencing. Like most companies, the area of training is considered a good candidate for budget cutting. Considerable effort was expended to reduce excess staff but to maintain training for those remaining. However, the training plan was greatly affected, and most courses were postponed or canceled. To remain on schedule, LINX was faced with the question, "Is it better to hire new employees already proficient in the client/server area and are comfortable as trainers/change agents or to train the employees assigned to the current project?"

PSI took a two-pronged approach for their initial project; they moved some of their experienced designers into the client/server project and then contracted with an outside firm for some experienced client/server designers. After the initial project, the experienced designers were rotated back into their old department, the outside contractors were eliminated and other members of the department were trained for implementation of the next project. This was very effective. Although there are no right answers for these types of questions, they must be faced, studied, and resolved by striking a balance between meeting the professional development needs of the technology staff while operating within budget constraints.

PSI decided to train the project staff by hiring consultants familiar with the new environment. A member of the LINX group stated, "In doing so, we hoped to maintain the project delivery data and deliver a high quality system while transferring critical client/server knowledge from the consultants to our developers." Another member stated that this was a large expenditure, but it was necessary.

Like many other companies, PSI encountered a learning curve that was much longer than originally anticipated for the developers assigned to implement the initial client/server project. IS leaders must make other difficult decisions considering the substantial amount of time, money, and learning costs already invested in associates. Should experienced client/server developers be returned to their traditional development arena (e.g., mainframe and COBOL) so that inexperienced developers can have the same educational and project development opportunities? Should the experienced developers continue to be assigned to client/server projects, assuming that the experience gained from previous project participation would result in shortened development cycles and significant cost savings due to productivity gains? According to a member of the LINX group, "Striking a balance between meeting the professional development needs of the technology staff while operating within budget constraints continues to be a challenge for PSI's management team. We plan to give everyone an opportunity to use the new tools and techniques used for developing client/server applications, but we must do so while meeting corporate operating expenses objectives."

THE LESSONS TO BE LEARNED

Although PSI has accomplished their original objectives, many lessons learned could prove invaluable to those companies planning or in the process of a change from mainframe to client/server. These include:

- *Making sure that senior management is actively involved in the entire process.* PSI is an appropriate illustration of the positive effects of an executive order mandating that all levels of management participate from the very beginning. In fact, an executive level steering committee was set up and continuously involved in the decision making during the planning and subsequent stages of the project. This commitment facilitated the funding and movement of personnel to meet the needs of the project.

- *Developing an ISP that ties in with the overall business strategic plan.* The decision should not be solely based on technological reasons, and the business side should be the driver.

- *Doing the necessary investigation and analysis to ensure that a client/ server conversion is the right decision for the company.* The bandwagon issue should not cause pressure.

- *Keeping the learning curve in mind.* This is a tremendous change for many people who have spent years doing things the old way. This is especially true for the first project. Therefore, the first project should be small, so it can be completed on time and successfully. This improves users' confidence and morale and reduces the critics' ammunition.

- *Checking the experiences of those who have already ventured into client/ server.* Although development tools in the mainframe arena are available and well understood, tools in the client/server arena are not as readily available or understood because of a lack of prior experience in the workplace. Many of these have not been fully tested. Therefore, care should be taken in this area.

- *Involving users of the systems in design and testing.*
- *Anticipating a possible response-time issue.* PSI considered total information turnaround time, or getting the right information to the right person at the right time, more important than individual transaction response time. Design of the first client/server project was based on this assumption and trade-off decisions between fast response versus ease of use and better information usually went in favor of the latter.
- *Realizing that a major problem involved in this type of downsizing is political.* Many employees will view this change as job-threatening. Effective communication is essential.
- *Recognizing the significant training and educational issues.* Economic pressures could force companies to reduce or eliminate these vital and necessary expenses. This type of training is essential to successfully complete the conversion and should not be reduced.
- *Being prepared to answer such questions as:*
  — Is it better to hire new employees already proficient in the client/server area that are comfortable as trainers/change agents or to train the employees assigned to the current project?
  — Should experienced client/server developers be returned to their traditional development arena (e.g., mainframe and COBOL) so inexperienced developers can have the same educational and project development opportunities?
  — Should the experienced developers continue to be assigned to client/server projects, assuming that the experience gained from previous project participation would result in shortened development cycles and significant cost savings due to productivity gains?
- *Anticipating that the initial cost savings are not as great as first anticipated.* One of PSI's major reasons for proceeding with this project was to develop a system that was more flexible and adaptable to the changing environment. It was perceived by some as a cost savings move. With the cost of new hardware, new software, new development tools, and the training to use them, PSI could hardly have realized a cost savings over the old system. It did, however, achieve a system that had a cost advantage in terms of lower incremental cost in growing and adapting.
- *Being patient!* This transition takes much longer than originally

anticipated. This is a radical culture change and it takes time for any organization to adapt. It must be done, however, to remain competitive in this fast-changing environment.

SUMMARY

There is no one solution to the issue of downsizing, rightsizing, or any of the others. When management factors in the nature of the business, the organization of the business, the people issues, the corporate traditions/culture, and last but not least, the economics, finding a solution becomes an extremely complex issue.

The conversion from a mainframe to a client/server environment is complex. Although this chapter attempts to identify some lessons learned in the conversion accomplished successfully by just one firm, senior executives, striving for effective and strategic management of information systems, should benefit from these lessons to assure themselves that these areas have been covered. The challenges are political, environmental, technical, social, and economical. They must all be anticipated, recognized, and managed to successfully implement the conversion.

6-4

Integrating the Open Systems Environment: Reflections of a Project Manager

LEORA FROCHT

W hen dealing with applications development of open systems, the issues are numerous and varied. New systems are being developed from scratch based on new concepts; ideas that once could never be translated into a system now can be. Open systems tools make possible the development of applications that could never be done on mainframes or pre-Windows PCs. Therefore, design and implementation of brand new systems present their own set of issues just by virtue that they are new.

Members of the development team generally do not have as deep an experience base for open systems development. The technology has not been available for that long, and the tools are not very mature, mature tools having been in heavy use commercially for at least 10 years. Developers with many years of experience generally have their experience and tools in mainframes or pre-Windows PCs. Big changes in thinking patterns need to occur for experienced developers to adapt to open system and the client/server approach.

This chapter provides a personal look at the integration process and describes the experiences of one project manager in the integration of legacy systems into a new open systems environment.

OPEN SYSTEMS INTEGRATION

Integrated system development involves issues such as performance, communication, and compatibility. In the integration effort described in this chapter, the issues centered on seven major factors:

- User specifications.

- System performance and response time.
- Communications.
- Data accuracy.
- Resource contention.
- Ergonomics.
- Legacy systems.

The major technical and project components affected by these issues are graphical user interfaces (GUIs), data base, the programming language used, and the processes they spawn. Systems analysts and developers need tools to build, test, and measure these components. For example, when adding a communication protocol to an open system, the communication code needs to reside in the application code. The application code must properly set up the information so the communication protocol can handle it. Upon receipt of the message, usually one record, the communication protocol must guarantee its complete delivery. Also, the wiring through which the message is sent must have the capacity to handle the size of the record and must be robust enough to handle the level of user volume. Finally, the application on the receiving end must be able to receive the message and process it properly. The receiving application has to know that a message has come in, absorb it, and take appropriate action.

Integrated systems being developed to replace legacy systems bring with them a set of issues in addition to the ones that exist for brand new systems, such as communication between the mainframe hardware and the open system platform. Compatibility of respective software packages is also significant. Messages or data files traveling between platforms need to be in formats understandable by the receiving platform.

User Specifications

Before any development begins on a system, among the most important considerations is that no decisions, be they software or hardware related, can be made without a thorough understanding of the users' requirements, needs, and goals. The system must be flexible enough so the users can do more than they expected or initially thought of. To the extent that today's technology can support it, the system should be portable, so an application can be moved from one platform to another, from one version of an operating system to its next generation, or from a PC to a workstation.

The system should be able to support not only current but anticipated data needs for some time into the future. Interesting examples of this point is the great success of Microsoft's Office product. As an integrated package, this product addresses the most common user needs: combining word processing with spreadsheets, data base, and presentation functions. Of course, the audience of in-house developed proprietary software is much more focused and requires far more specific attention.

System Performance and Response Time

Performance, both in accuracy and speed, is part of the focused issues. Especially in high-pressure situations, users need to know how quickly they will get information back; it is unlikely that any person would be willing to wait a disproportionate amount of time to get results.

Performance is hardware and software related. Hardware must be the correct kind; communication hardware, such as wide area or local area network (WAN or LAN) routers and cables must be selected and implemented to handle higher volumes, if high-volume communication is an aspect of the system. Volume is generally identified by the users, as they generate transaction activity. Server machines must be designed to handle large volumes of data or application software, or at least be scalable and configurable as such.

Communications

Communication software must also be robust enough to send out and receive large volumes of data. Market data services, which provide such information as stock prices in the Dow Jones Industrial Average, generally address these issues; they transmit hundreds of thousands, if not millions of messages a day. Application software, whose purpose is to accept high volumes of messages, must be designed not only to receive the messages but to interpret and process them in a timely manner. The source code must be able to receive each individual record, determine what to do with it, and send it off to the designated recipient. The recipient could be a data base, a flat file, or a window panel. All of these transactions must be completed in reasonable time.

Data Accuracy

Accuracy is a critical issue to consider in an integrated system. Appropriate business decisions can be made only on the basis of correct information. Data must be completely correct one hundred percent of the time, or disastrous consequences could result. This is certainly true in the securities industry. When traders are dealing with millions of dollars in securities, a lost trade record could result in losses at least equal to the value of the security of the lost trade record. Lost records negatively affect accounting systems, and relationships with the clients can be damaged due to questionable reliability. These can lead to lost revenue for the trader and general unhappiness for the team.

Similar consequences result from bad data within a message. If a piece of data is corrupted, or if its format is changed, inadvertently converted, or zeroed out because of faulty software, the outcome could cause great discomfort to all involved.

Resource Contention

Resource contention must also be considered. The integrated system requires a design such that no two users of the same system contend for a resource. This problem occurs quite frequently when requests for data are made to the data base. The application developer must pay attention to the way users need data. Standard procedure dictates that whenever information is retrieved from a data base, every effort is made to not unduly restrict other users' use of the data base. Data retrieval from an SQL table is a good example. Whenever a row or any predetermined data type is accessed, an SQL data base management system (DBMS) is to lock it out for use by any other user. The row stays locked until the user's application commits the transaction and releases the row. Any other users who attempt to modify the same row at the same time are be frozen out until the row is released. So, a user updating a field in a row, for example, initiates that transaction should be initiated, makes the change, and releases the row, so others do not have to wait to transact their business.

Another resource of possible contention is communications. Care must be taken when conducting interprocess communication or broadcasting data to a large group. The application must not require that the broadcasting of one message depend on the successful delivery of another message. Also, the successfully integrated system allows sent messages to be buffered or stored and processed separately from the communication portion of the application. That way, messages can continue to be received rather than back up because the communication protocol is still processing the previous message.

Ergonomics

Not quite as essential an integration issue as the others but still an important one, is that of screen real estate. While a system is being developed, care should be taken that the users' screens do not become overcrowded with panels, screens, or other kinds of activity. Users find unmanageable too many panels on a screen—or newly generated panels that cover up others. Screen management and screen real estate can make or break the integrated system, even though it does not directly affect its actual function.

Similar consideration should be given to the interior layout of each window. When the widgets inside a window are arranged in a counterintuitive way on the screen, the user takes longer to sort out all the information on it. Users could end up spending more time trying to comprehend what they are looking at than actually dealing with the business at hand. Common types of information should be grouped together.

Widget functions are another issue. Various graphics packages provide various types of widgets, which are the actual conduits to the functions of the integrated system. If these conduits either implicitly or explicitly mislead the user as to what they do, the system is doomed to failure. Books on Windows and window design explain widget types and what they should be used for.

Appropriate color is an important issue to consider in the finishing touches of an open system. Color coding helps separate the functions implemented in the integrated system, to add, modify, or delete records in a data base. Buttons can be colored yellow, to indicate "proceed with care." Orange can mean modify, to indicate that anything done can be undone. Red can mean delete, to indicate that a record removed is gone forever and is therefore "dangerous."

Colors should be selected with esthetics in mind. Human beings work with things more easily and happily when they are pleasing to look at. Also, color can help separate sections of a window, aiding users in how they interact with the system. Finally, colors selected should be easy on the eyes. Strong, bold colors or highly contrasted colors next to each other are difficult to view for long periods of time and can cause eye strain. The integrated system, in more ways than one, should not create headaches for the user!

LEGACY SYSTEMS

Certainly, replacing a legacy system with an open system solution further complicates the task. This discussion assumes that the current legacy systems were developed in a non-Windows environment. So a legacy system, in this case, relies on linear menu selection rather than ranges of selection available on Windows-type panels. The platform may be mainframe or PC, but it does not allow for application functions to branch out, nor does it allow for simultaneous display of panels.

By definition, a legacy system has been around for a long time. If it performed reasonably well for the user, the new open system must perform better; if the old system never fulfilled users' needs, a successful replacement is critical. The worst thing that can happen for all involved is that a new application have less functionality than its predecessor.

It therefore is essential that navigation of open system windows be a primary consideration in the design of a new system or a legacy system's replacement. Whereas the mainframe systems offered users no choice but to follow their own linear thinking, the open, integrated system gives users many path choices. The system designer needs to consider logical flow, namely, which panels can parent a subsequent panel (i.e., the child process), and which panels would be a final child in what becomes the tree structure of insatiable window-upon-window navigation. At all costs, the development team needs to avoid creating a confusing mass of windows that the user can never navigate.

Another consideration related to the window navigation issue is the actual number of window processes that should be running at any given time. Although integrated systems can support a number of simultaneously running processes, a benchmark should be established for the number of processes that should be running within the system at any given time. The system should never crash, but it should perform within reasonable expectations. Consideration of the above points can make the difference between acceptance or rejection of a legacy system replacement.

Revolving around the question of how to replace a legacy system is the decision of how much—if not all—of it to replace, even if the original hardware does remain. In some cases, a legacy mainframe system, for example, is ideal for housing large volumes of data but not really suited for providing that data to the user in a digestible way. The integrated, open system can then become the suitable front end, and the remaining issue is that of communication between the client/server hardware and the mainframe.

A similar scenario results if the existing system contains a mainframe component whose accuracy, reliability, and performance cannot be replicated and should therefore not be replaced. Work that gets done well on the mainframe can be communicated to the client/server platform and vice versa.

A practical example is in the securities industry, a volatile business often at the cutting edge of technological development. Whenever the Securities and Exchange Commission (SEC) modifies relevant regulations, chances are that securities dealers create new ways to borrow money by developing new types of securities. Big corporations, cities, states, agencies, and the federal government raise money in the form of new securities.

Whenever a new security is developed, it has to be modeled in the computer systems, so all the pieces of information that describe how the security functions are stored. Examples for a fixed-income security (i.e., bonds) are maturity date, price, amount of issuance, interest payment date, and call features (i.e., how much of the money can be paid back and how far in advance it can be paid without penalty to the issuer). Securities can become very complex to model, for example when the interest payments on a specialized bond change every month based on some other benchmark, such as the prime rate. The computer model has to keep track of interest payment frequency, payment dates, the frequency of the quote of the prime rate, whether the straight quote of the prime rate is fair to use, and whether it needs to be weighted based on the dollar value of the security or converted to some equivalent.

So addressing the integration issues of a system that models these types of securities is very complex. System performance (i.e., the amount of time it takes for the system to produce a result for the user) is critical. Hence, response time remains a key factor in the design and integration of systems, regardless of platform and technical environment. In the trading system model above, timing can be critical. Clients waiting on the telephone for a security's price do not wait for very long; if they have to wait longer than they like, they will call another trader the next time they want to do business. In this scenario, the integrated system user throws up his or her hands in frustration and goes back to the calculator, pencil, and paper. The system has to enable the trader to enter numbers quickly, press a button, and receive an answer—which must be absolutely correct— within seconds.

Bond traders also must disseminate information on securities that are for institutional sale. The term *institutional* means buyers of large blocks of securities, usually valued in the millions of dollars. The industry terms *posting an offering* or *an ax* mean advertising available securities. Individuals who obtain securities, price them, market them, and distribute them to a salesperson or investor are called market makers, and timely communication is critical to them. The hundreds of broker/dealers and thousands of investors who are potential buyers must somehow find out that an interesting security is available for sale. The offering, or ax, is posted to a fully integrated, open system, that can receive the security description from the trader and store it in a data base for accounting purposes and for safe keeping. Then the system must transmit the security description over a communication protocol, where other machines hooked up to the same lines at the other end can receive it, interpret it, and display it on their own screens.

This chapter discusses one such system, CPTrade, a proprietary offering system developed by the author/project manager's team at CS First Boston. CPTrade was developed in a client/server environment and completely replaced a mainframe legacy system, with a significant increase in functionality. A Motif-based GUI product called WidgetWorks was used to create the windows. C and C++ were used to program the system; Sybase was the data base package. Tools, such as Rogue Wave and some in-house developed software, were also used. Real-time broadcasting was achieved by writing a layer of software to carry securities-offerings data on the backbone of Quotron's market data distribution system, called TSS. The in-house developed software was used to publish the data on one end and receive and process it on the other.

The users were traders of commercial paper, a short-term, unsecured product. They used the CPTrade system to post offerings of the commercial paper product, which made them the originators of the data that went into the system. Equipped with Sun workstations, the traders needed this system to provide better service to First Boston's clients. Examples of clients are large corporations, such as department stores, who may have repeated need for short-term cash, say, in order to meet a payroll or to pay for new inventory. Some companies manage their finances by borrowing short-term money in several ways every day. Needless to say, several hundred active issuers who are in the market every day can generate a substantial volume of transactions. This offering system was critical to the traders because the market in commercial paper is of very high volume (many trades are made in a short time span) and very fast moving. Also, the issuers of commercial paper, who are essentially the borrowers of funds, need to know that they can be assured of having their securities sold, which means that they can always be guaranteed a cash flow. The wider the advertising of their securities, the better their chances are of raising the cash. The more potential investors across the US are made aware of the available securities, the better the chance of selling the paper.

The traders, whose function in this case is as market makers, create the securities, price them, and post them as offerings on the CPTrade bulletin board. The bulletin board was comprised of a table widget (one that handled cells by row and column) that could be typed on. The contents of the cells, as they were modified, were broadcast across the US over the communication lines to the recipients of the data, the salespeople, who were also equipped with Sun workstations. The salespeople were able to receive hundreds of offerings on their screens and provide the clients with fresh, current information regarding the price and availability of commercial paper securities. Functions provided on the screen aided the salesperson with searching and sorting the records.

To further the advertising process, software was developed in-house at CS First Boston and at Bloomberg Financial Services to transmit commercial paper offerings to Bloomberg terminals nationwide. Quite a number of investors in commercial paper now have access to a segment of the market that was previously difficult to access.

In this particular instance, the major issues were window design, window real estate, allowing data modification on a cell level, and broadcasting the modification in real time. Communication with Bloomberg's system required significant effort as well. Data base contention was avoided altogether by limiting data entry to the eight commercial paper traders and minimizing the need to retrieve the offering data from the data base. How the integration decisions were made are enumerated in the following section.

THE PROCESS OF INTEGRATION: LESSONS LEARNED

To understand and resolve integration issues of open systems in application development, one must understand the nature of the components of open systems. In the course of First Boston's integration journey, many lessons were learned. Integration, in the open systems arena, is difficult. There are no failsafe paths or methods to follow, and the market for integration is full of choices and conflicting opinions. The process is intricate, laborious, and expensive. The major lessons learned in the First Boston project are summarized into what it calls the seven rules of integration. Project members must:

- Understand the technology.
- Understand the application.
- Create a solid development team.
- Create interpersonal relationships with the users; make them part of the team.
- Gather detailed specifications.
- Organize and interpret specifications.
- Understand the critical components of integrated systems.

Rule 1: Understanding the Technology

Basically, what exists today is a changing, evolving technology, with the number of software choices growing rapidly. Selections must be made from a large range of products and united into a system that solves the problem at hand for the user.

The systems integrator's challenge is knowing which software should be selected for the application and which criteria should be used in making each choice and sticking all the choices together into a smoothly working system. Possible choices in software exist for the GUI, the data base, inter- and intraprocess communication, the programming languages, and more. Generally, though, software choices are basically made for the applications development team by the senior management, technology committees, or infrastructure specialists in the individual departments, and one or two options from each software category are available for use in development. So the questions that remain are what the qualities of the software tool under consideration are, how they interact with each other, and how these qualities affect the end result.

Rule 2: Understanding the Application

For the experienced analyst as well as the novice developer, the key to open systems integration is to understand the application's goals completely. This is the most important task to be undertaken in the integration process. All members of the development team have to understand the user's business, how it works, and the user's desired end goal for the system. This point simply cannot be stressed enough. Users often have a very specific approach to the way they do business—because either business has always been done that way or old habits are hard to break. In any case, development team members must pay attention to user needs.

There is no reason that, in the course of discussion, new approaches and ideas cannot be raised and examined, but in the end, the way the system approaches the business is always up to the user. CPTrade would never have succeeded at all if it were not for the main user, Robert Stackpole, director, commercial paper trading at CS First Boston. All the traders and a number of salespeople were involved in the system's design and development, but the management of the trading floor deferred to this high-level officer, who interfaced with the development team. A successful system was so important to the firm that a respected, revenue-generating manager was freed to drive the user side of the project. As one of the most experienced individuals in the commercial paper business, Stackpole was able to articulate how the system should function and how the development team should approach the design. He was also able to explain historical reasons for the way the business was done, making it easier for the development team to understand how to implement the system's design features.

Rule 3: Creating a Solid Development Team

Well-integrated, open systems require development by a team rather than by the lone programmer working on the mainframe. On the mainframe, a single developer could successfully create an integrated system because the components were already integrated by the mainframe provider. For example, when a technology department was set up on an IBM mainframe, IBM also provided the software, including centralized system software, procedural languages, such as REXX, and other software products, such as SQL/DS, DB2 or FOCUS for data base. Ultimately, the programmer was responsible only for the function of the system for the user, and because of the architecture of the mainframe, it was often more convenient for one person or a very limited number of people to work on a single integrated system.

At the opposite extreme, client/server systems are built on hardware platforms, such as workstations or PC LANs, platforms that give systems analysts and developers much more freedom and flexibility in tool selection, design, and implementation. So, because so much more flexibility and processing alternatives can be built into an integrated system, a team of two or more analysts and developers, as well as component specialists, are required to create an integrated, open system. The sum total of knowledge of the development team must encompass two disciplines: technology and the user's business. Within the technology discipline, the team must be made up of members with strengths in data base, GUI software, communication software, and application software. Often, analysts and software developers have experience in more than one of those areas. Similarly, specialists in the hardware technology should be part of the team effort. They can advise on interprocess and interplatform communication, performance, and programming.

Technologists' knowledge of the user's business is at least as important as is experience with computer technology to determine how the application software will function; hardware accommodation will ultimately determine the system's performance. The commercial paper trading example bears this out. Because of the many transactions generated by this business, a system could not be designed to display only one security at a time on the screen. Trading would come to a halt if the traders could not see at least fifteen rows of information at a time on the screen. If the developers did not understand this aspect of the business, the system would be a total failure from the outset. Similarly, if the communication software were designed so that only one record at a time could be received and processed, system performance would be intolerably degraded because of the backup.

The way a high volume of messages is treated is that the communication software receives the data and immediately puts it into a buffer provided by the application software. The data can be handled in any style the developers deem appropriate: first in first out (FIFO), for example, for trade records. Hardware requires the same consideration. Appropriate network configuration and bandwidth must be specified to accommodate high volumes.

Rule 4: Creating Interpersonal Relationships with the Users

Human communication and the creation of an effective personal rapport between the development team and the users are essential to the integration process. Building working relationships significantly improves the completed system's effectiveness. When productive working relationships exist between the users and the development team, developers become less intimidated about what they do not know and ask many more questions.

The converse side of the relationship is important to develop, also. Many people are still very much intimidated by computers. Terminology they have never heard of or understood is suddenly flung upon them. This is particularly difficult to deal with for the business person who is very accustomed to being successful and always being in control. The uncomfortable user therefore does not easily or willingly discuss the system with the development team. Barriers between the users and the developers, without question, degrade the quality of the system integration process. Creation of a solid, interpersonal working relationship between the users and the development team is therefore a required, integral part of the process of system integration.

Rule 5: Gathering Detailed Specifications

The first technical step in the integration process is gathering specifications for the system. There are a number of ways that this can be done. Naturally, the best source is the user. Discussions and interviews with the main users/liaisons and their peers are essential, as is careful and organized note-taking on the part of the development team. When possible, part or all of the development team should sit with the users for a minimum of several business days and for as long as is practically possible, again with as much note-taking as possible on the part of the developers.

Other sources for specifications include other companies, professional associations, trade articles, and books on the subject. Texts written by other professionals in the field often give a person added depth and perspective to the goals specified by the user. For CPTrade, each member of the development team read Marcia Stigum's "The Money Markets," so they could compare the function of commercial paper with that of other short-term securities, such as certificates of deposit. Evaluating the commercial paper product in a different context was very helpful to all the team members.

At the end of the system-specification process, the development team members should comprehend the users' business on an intuitive level. After each information-gathering session, the development team should discuss the information and assess how it relates to what is already known. After the specification process for CPTrade, the team came away with a solid understanding of the qualities of commercial paper (i.e., how the securities are created and posted) and the relationship between those qualities and interested parties (i.e., the posting of an offering). The team was then prepared to address the system requirements with technological solutions and start designing the system.

Rule 6: Organizing and Interpreting Specifications

The formal documentation of user specifications is important to the effectiveness and efficiency of the project team. Development team members must bear in mind such areas as the programming language, the GUI and windowing, data base, and communication when collecting the specifications and should organize written notes according to logical components. Most systems require a main screen from which all other operations are managed. System components can be organized according to what is required on the main screen level and for subgroups of the main level. Possible functions to launch from a main screen are data maintenance, reporting, tools (e.g., data item searches), or utilities, such as refreshing a screen from a data base.

In CPTrade, the main screen contained the table in which the offerings were modified. The users needed to be able to change individual cells right on the matrix itself, when they deemed it necessary. Changing data items relating to offerings became a main system requirement, so it was defined to be a function provided by the main screen.

Adding a new offering, on the other hand, was an appropriate subfunction. Offerings were made up mostly of issuer information, which already existed on the data base. A button could be pressed to call up a separate panel, allowing the user to pull up that existing information, enter in the new data items (in this scenario, three new items were usually added: dollar amount, rate, and maturity date), and then submit it to the main screen. They were able to create new records quickly, without having to type in an entirely new record of some 15 columns.

A different function, called issuer look up, was also provided o the main screen. If the users needed to look up a specific issuer, they could press a different button to call up and search a list of issuers; in the commercial paper scenario, the issuers are the clients of the broker/dealer, are known in advance, and can therefore be provided to the system in a cross-reference table. The columns on the main table consisted of issuer, dollar amount of the offering, maturity date of the security, the discount rate, and the yield on the security. Subfunctions were created so searching could be done on any one of those columns. The subfunction this time consisted of a text box into which the target item is typed. After the user pressed enter, the subfunction would locate the item, bring it to the top of the screen, and highlight it. If the column being searched was not sorted, the subfunction would automatically sort it before executing the search. Again, because the users had to find an offering for a client quickly, the subfunction was provided on the main screen, where it was easily accessible and easy to use.

Building Layering. The next step in specification organization is to take the notes arranged by function and paint up a window that would correspond to the main function and then paint as many windows as necessary for the subfunctions. The temptation to build deep layering into a system, however, must be resisted. If users have to call up several chained windows just to get to the one they need, the window navigation process becomes impossible, and users give up. As "flat" an implementation as possible is best.

Of course, a balance has to be maintained. If too many screens are generated on the same level and are required to be displayed at once, the issue of optimized screen real estate is jeopardized. Working with the user to decide on that balance is usually the best course of action. Calling on one's intuition and experience also helps, too.

In general, though, one main window with the subfunctions spawned from a main menu bar or other widgets gives the best chance of functional success. It is important to consider, at this point in the system development process, if the goals of the system are too broad and if the implementation is feasible. Often, when an integrated, as opposed to legacy-replacement, system has been proposed, the users' requests are not entirely focused or finalized with any degree of precision or confidence. Users have a broad idea of what they want, but meeting all their business needs may require the development of more than one integrated system or several smaller subsystems chained together. Painting the windows clearly demonstrates to the user how large or small the system will be, based on the specifications.

Dividing a system into manageable components wherever possible is better not only for the system developers but for the users themselves. *Object-Oriented Analysis and Design with Applications,* by Grady Booch, discusses the fact that the human brain can manage only a limited number of processes at a time. The computer, of course, can manage as many processes as the hardware can accommodate, which basically means that the computer can handle many, many more processes than the human brain. Therefore, separating functions into different systems should be addressed at this stage, with an eye to accommodating the physical limitations of the human brain instead of implementing a system that can only be "comprehended" by a computer. A side benefit of separating specifications into different systems is the resulting flatter, simpler, and more targeted individual systems that are easier to develop concurrently, enhance, and maintain.

In the CPTrade example (see Exhibit 6-4-1), all the screens described were prototyped with the GUI before a single line of executable code was written. The development team paced through all the functionality with the users, getting feedback from them on all points, every step of the way. The users indicated the order in which they wanted the functions to appear, and they even discussed such system aspects as the size and color of the windows themselves.

Working with the Prototype. The developers at this point have a prototype that can be critiqued and revised. This is also the point at which it becomes apparent where the other components of the integrated system come into play. Interaction with the data base can now be defined, because the origination point of data is known. The data types and sizes are also known, because they have been designed into the windows. Determination of which windows and functions communicate with each other becomes obvious. When all the windows are visually displayed, it is easy to identify which windows should publish and which should receive data. Any other roles that are relevant to the system integration process can be hooked in once the system is visually prototyped.

Rule 7: Understanding the Critical Components of Integrated Systems

There are six major components of an integrated open systems environment:

- The programming language.
- The communications protocols and network.
- GUIs.
- The data base.
- Applications management.
- Applications systems.

The Programming Language. The coding or programming language is the component of the integrated system that facilitates the interaction between the GUI panels and the data base and between the GUI panels and the communication components. It must provide the other system functions when users expect them, helping users do their jobs as smoothly and easily as possible. The developer must be attentive to the programming language's ability to reproduce the functions the user engages in, to provide flexibility and performance, to hook into other open system components.

CPTRADE WAS DESIGNED to allow traders to post commercial paper offerings and to let traders and salespeople see the offerings and sell them to investors. The sales force, however, needed additional functionality, because they had clients who were interested in purchasing only commercial paper posted by specific issuers. Yet other investors were interested only in issues of the highest credit quality, which is determined by such rating agencies as Moody or Standard & Poor's. Still other investors were interested in commercial paper offered by oil companies, for example, or by American companies. Investors have unlimited numbers of requirements for buying commercial paper; the sales force had to be aware of these to do business with the clients.

The ability to filter out specific offerings became one of the main requirements of the integrated system. Programmers discovered when designing the screens that this function, called the approved list subsystem, was applicable only to salespeople, not traders. Consisting of three subsystems that were interconnected by push-button functions on the primary panels of each of the three subsystems, the approved list was initiated from CPTrade's main menu bar.

The first subsystem allowed the separation of all investors in commercial paper from the rest of CS First Boston's clients, by retrieving a master list of all such customers from a centralized data base and displaying it on a panel. To the right of this panel, another panel would be generated, to contain a list of only those clients who were interested in purchasing the commercial paper product. This product-specific list would be populated by searching for and finding a client on the master list, highlighting the target customer with the mouse, and pressing a button with a right-pointing arrow icon on it. The highlighted customer would then be added to the data base, and the record would be added to the right-side list on the window itself. Two distinct lists were created: one of all First Boston customers and one of First Boston customer that were interested in purchasing commercial paper. A by-product of this subfunction was that the salespeople were clearly able to see the complete customer base from which the commercial paper customers were taken.

The second function of the approved list was to assign to the customer the individual issuers or group of issuers who were of interest or were approved of by the customer itself. The user would select a customer from the commercial paper list and press a button labeled **Assign Issuers**. The main panel of the second function would appear, withthe target customer's name at the top. Below that, on the left side of the screen, a master list of issuers would appear. This master list consisted of a panel of issuer group descriptions at the top and a straight list of individual issuers on the bottom half. To the right of the master list panel, another panel would appear that contained issuers already assigned to the customer. Again, the user would select from the issuer list the desired record and press a button labeled with a right-pointing arrow that would assign the issuer to the customer. The result would be a list of issuers that were of interest to the customer. The salesperson could now provide a more targeted service to the client.

Exhibit 6-4-1. CPTrade Case Example: Lessons Learned

A subfunction of the above subsystem is the creation of dynamic groups of issuers that could also be assigned to the customer. A filtering mechanism containing all the qualities of the issuer was provided. Qualities of the issuer could be credit ratings, the issuer's industry, or whether the issuer was foreign or domestic. The filtering mechanism would allow the user to create the following statement dynamically: "I would like to see all domestic issuers who are in the oil business and have a top credit rating from Moody's and Standard & Poor's." The system also required that the user enter a short label for this issuer group, for example, Top Quality Domestic Oil Companies. Then the filtering mechanism would take that request, generate a Sybase query to retrieve just the issuers from the issuer cross-reference file that met the user's criteria and display that list to the user. This subfunction was enormously useful, as users never had to hunt and peck through a list of issuers to find the individual issuers that met their criteria. The query could also be saved and retrieved by the user-entered issuer group description. Because traders' main function is to provide services to clients who are issuers rather than to clients who are investors, this subfunction would have been inappropriate to impose on the traders. The sales force, however, very much required this function.

The third function—and the end goal of the system—was the creation of the list of current offerings that would be of interest to a particular customer. This function ultimately generated a report that was displayed on the screen, was printable or faxable to the investor's office directly. It was generated by pressing a button labeled **GenerateOfferings Report** on the approved list's main panel. It was placed on the main panelbecause function one was required only once, and function two would be required only again on occasion after the first execution. The customer would have to be identified as a buyer of commercial paper only once. The list of issuers it wanted to purchase would have to be entered initially and would have to be modified only when an issuer would be added or taken off the customer's list of approved issuers. Also, as the traders modified the offerings on their main posting screens, any of the modified offerings that also appeared on the report would be updated in real time. Therefore, the salesperson's data for the customer would not get stale, no matter how long the report was left up on the screen.

Exhibit 6-4-1. (*Cont*)

Many information technology organizations today use C or C++, because they are very flexible, precise, high-performance languages when properly implemented. Not bound by hardware constraints, other than the hardware's physical limitation, C and C++ can basically run on any kind of platform, interactive mainframe environments as well as workstations and PCs. As lower-level languages, C and C++ have high performance and can be readily fine-tuned, depending on the application design.

Programming in these languages, however, can be labor intensive and tedious. The trade-off of being able to drill down to the address of a specific byte in memory is that very close attention must be paid to getting that address. Errors that are very subtle and difficult to debug can easily be generated in C or C++. C++, more than C, requires significantly more up-front planning. To use C++ classes effectively, the developer must know in advance all the data types or elements that go into the class before it is built. Again, if done properly, classes are not difficult to enhance or modify and, at the end of the development process, the result can be a very powerful software tool.

This aspect of the programming languages goes back to the discussion of building the right development team. The right programming approach on the part of the development team can create a superbly functioning application. The trade-off between the complexity of the languages and the results they can produce makes the effort required well worth it.

The Communications Protocols and Network. Communication, especially real-time communication, in all its forms, is critical to a successfully integrated system. Various types of communication require the development team's consideration. Interprocess communication is of great importance to users who must have access to the same data at the same time. When one user changes an element of the data set through a data entry window, all others viewing the same data set must see the change immediately. The development team must recognize this need and address it in both interviewing the users and in recording the specifications. Issues deserving attention include whether transmission should be made one cell at a time (e.g., for a table), or one row at a time (e.g., for a table or a list) or several rows at a time (e.g., for an entire screen full of data). The type of changes that the user wants to see determines how communication software is developed.

The range of interprocess communication determines how communication is done. Often, users want only certain individuals to receive specific information or changes to it. Defining that type of communication is also essential in defining a well-integrated system. The UNIX environment alone allows many different approaches. Examples are the socket-to-socket (TCP/IP protocol) approach, or using purchased software provided by market data suppliers, providers, or vendors. Market-data suppliers in the securities industry, such as Reuters, Quotron, and Knight-Ridder, have their own suite of application program interfaces (APIs) with which they disseminate such information as price quotes for specific groups of securities. Their APIs are available to developers whose departments subscribe to the market-data providers' services. The developer needs only to map the in-house record into the buffer predefined by the API and execute the API. The developer, however, should become very familiar with how the market-data providers' software works; each package has its own unique qualities, behavioral patterns, and characteristics that need to be addressed during the development process.

Each provider's set of APIs may work under its own set of assumptions. For example, Quotron was the provider of some of the market-data services to the fixed-income trading floor at CS First Boston. Quotron also provided a set of APIs to transmit in-house-generated data, in this case, commercial paper offerings generated by CPTrade. The problem, however, was that Quotron's main function was to communicate stock prices, which does not require that a history or audit trail be maintained. The recipient of the stock price cares only about the latest price, not necessarily what preceded it. The stock quote does not map to an actual trade; it just gives information for the possibility of a trade. If a message representing a stock ticker of a particular price was sitting in the message buffer waiting to be transmitted, and a subsequent price for the same stock ticker was submitted after it, the first one sitting in the buffer would be overwritten, and historical transactions would be lost. So, Quotron's package was not the ideal solution to the integration issue of communication in the open system. At the time, however, Quotron's was the only set of APIs that was readily available in the department at CS First Boston.

CPTrade, on the other hand, had to be guaranteed that every message was received and processed, because every message actually did represent a trade. A message for a particular issuer's offering had to be passed through the communication line and received on the other end before the next transaction for the same issuer came through. No messages could be overwritten or bypassed. Every message that went in had to be successfully submitted and received, with no margin of error.

The program language solved the problem. Every time a message was submitted for communication, the application would "know about it," either having the application poll the communication line at predetermined intervals or having the communication protocol send a "wake up" message, or a ping, that a message was coming. Because polling takes up unnecessary resources, the better approach is to ping the application. Once the programming language received the message, it would pull the message out of the communication line and depending on the application could do a number of things. It could process the record on the spot and, depending on the information inside, decide what to do with it. Or it could, after pulling out the information, just put the message record out to a buffer and allow a separate process to analyze the contents of the buffer and take appropriate action.

Communication with a mainframe legacy system is also important. Organizations that keep their large batch processing on back-office mainframes rely very heavily on reliable communication between the client/server front-end and the legacy mainframe. Guaranteed message delivery is essential, especially in trading systems for banks and broker/dealers whose business is transaction-oriented and is very strict about keeping perfect records of those transactions. If a record carrying a purchase or sales transaction were lost, a firm could lose thousands if not millions of revenue dollars, business opportunities, and client trust, were a client involved.

Closely linked with communication is hardware performance and application response time. User specifications determine how the application handles its processes, which must be designed to avoid overloading hardware to the point that performance is affected. Users must receive their data in acceptable time frames, which is dependent on bandwidth. Users can be consulted about how many transactions they anticipate executing under current conditions; they should also try to estimate what their business volume will look like in the future. Then benchmarking can be done on the hardware that exists at present, through tools available in the marketplace that can determine usage of different types of hardware. In benchmarking communications, the wiring—bandwidth—is the issue. A "sniffer" is put on the line to look at the speed at which messages are traveling, which can be measured by the number of bytes per second.

A benchmarking program should be written to generate messages of an optimum, or typical, number of bytes and to pump a minimum required number through the wires. If the sniffer indicates that there was plenty of space and the messages were received within an acceptable time frame, the existing bandwidth is adequate, and the integration process for the communication component is complete. If the sniffer indicates that the messages got through, but there was not much space to spare (i.e., any more messages would have slowed down the process), management needs to discuss an upgrade. Certainly, if the sniffer indicates that present bandwidth is completely inadequate to achieve the minimum requirements of the system, spending money on an upgrade becomes an immediate business decision.

Finally, an important aspect of the communication protocol is scalability. The product used to develop communication should be able to accommodate a growing number of users, rather than a finite number of users that is well below an acceptable number. Of course, no open system could handle an infinite number of users, but a number reasonable to the business should be accommodated. Different protocols have different limitations, so the specifications for those products need to be examined and appropriate decisions made.

Using market-data providers' broadcasting APIs does resolve much of the number-of-users limitation. These providers have developed their software with distribution to a large client base in mind, as many individual client/server platforms need to be able to subscribe to one data source and receive stock quotes. Companies such as Quotron and Reuters have solved the problem in their respective software designs.

Development of socket-to-socket communication using the TCP/IP protocol, available as an integral part of the UNIX operating system, is the most efficient method of passing data from one process to another. The application merely initiates a process on both the sending end and the receiving end and retrieves the process identifier. The application can then send and receive information to and from one process identifier to another. *UNIX Network Programming,* by W. Richard Stevens, details how this is accomplished. This method, however, is not easily scalable. Creating a link from one process to many processes and repeating that several times within a single application can make it a nightmare to maintain.

GUIs. The design of the windows can be the most important and most complicated part of the integration process. The development team needs to consider how the programming language will make the Windows perform, what kind of paths the window callbacks will take and where they end, how large the panels are, what color they are, and how the information is organized.

GUI painters are the preferred tool when creating the windows for the proposed system. The window attributes are clear, readily available, and easily modifiable. GUI painters let work be seen as it is created, unlike Motif commands, which are stacked into a file and executed to generate screens.

Care should also be taken when selecting a GUI builder, as some are targeted for specific types of applications. For many Motif-based systems, Builder's Xcessory or UIM/X are appropriate. Windows built with these types of products are straightforward and relatively easy to integrate into open systems but are not necessarily object-oriented. Interfaces, however, can be created with C or C++. Others, like Park Place's OI GUI package, are also appropriate for object-oriented programming and systems and those that do not require the processing of a matrix. OI, however, does not come with a table widget, nor is the integration of an independent table widget an easy task.

Once the windows are designed, populating certain types of widgets with sample data is possible, making the system's functional aspects more clear without doing a coded implementation. Pull-down lists, with a limited set of standard entries that can be predefined, are good examples of this feature. For example, regions of the United States could easily be entered as a pull-down list and be immediately visible to the user when the button was pushed. As many of these widgets as possible should be predefined and implemented before taking a finished screen prototype to the user for approval. It is at this stage that the user can tell if the development team has integrated the business requirements into the system.

The Data Base. Design and implementation of the data base is also determined by the business flow. Information gathering is relatively straightforward; organizing the data may not always be. The goal in data base design is to provide the best performance possible by implementing the optimum data organization. Whether the data model is relational or hierarchical, the designer must resist the temptation to design a perfectly normalized, not-a-byte-wasted data base, even if it means decreased performance for the user. It is often worthwhile to repeat fields where necessary in the name of saving the user an additional data retrieval.

Good data base design does need to take its hardware organization into consideration. Very often, data bases for individual systems need to be supported by cross-reference files. Once again, trading in the securities industry serves as a good example. A relational, rather than a hierarchical, model is presented. The actual trade record is modeled on one table and resides on a data base server. The trade table contains such information as the amount of the trade, its maturity date, the interest rate, the trading account, and with whom the trade was done (i.e., the customer).

On another table, all the trading accounts used by the broker/dealer are listed, together with information that gives more accurate detail of the trading account. This trading accounts table is easily recognized as a cross-reference table. Another cross-reference table exists for the customer, containing the unique customer identifier, usually a number, the customer name, address, and the type of business it is in. Cross-reference tables, such as a customer table, can become very large. For large broker/dealers, such as Merrill Lynch or Smith Barney, these customer tables can have records numbering in the millions.

Needless to say, hardware upon which these tables are stored have physical limitations. A table the size of a customer table can probably be accommodated on one server, on one physical machine, but not much else can reside on the machine with it. The data base design and, by extension, the application design, must take into consideration that data on separate machines must somehow be joined. Currently the answer is that the application needs to be able to retrieve data from the two separate machines and integrate the data in memory, rather than having the data base management system do it. Cross-server joining of data bases is under development by Sybase and should be available in their release of Sybase System 10. *A Guide to Sybase and SQL Server,* by D. McGovern and C.J. Date, offers more information on the use of Sybase and its facilities.

Considering the inability to do cross-server joins at this time, the application code must also be written so that should data base tables be moved to different physical machines, the application code does not need to be redesigned, rewritten, or recompiled. Hard coding of an actual machine name should be avoided. Machine or server names can be passed in as global variables and stored in a predefined format, one for every table used by the system, and manually modified if there is ever a change. This way, the developer never needs to touch the application code to accommodate a hardware shift.

Index creation is also essential for the performance of data base retrieval, and therefore performance of the entire system. A primary, clustered index should always be created for any table. Otherwise the data base management system scans the entire table looking for a target record. For small tables, a table scan would not be a tragedy, but scanning for one record on a five-million-record table without an index would take hours. Also, the primary index should reside on the same server as the table for which it was set up. Secondary, nonprimary indexes are also helpful and do not necessarily need to reside on the same server. The McGovern/Date book on Sybase is a good resource for more information on this subject, as well.

Applications Management. A centralized location for all application source and executable code is essential to the success of the integrated open system. Because of the larger team size requirement in the development of integrated systems, tight control must be kept over the code. Considering the different components and all the code supporting the components, the project has to know that there are only single copies of any program or function resident in the system.

Often, developers on a team help each other by looking at each other's code and sharing tips and suggestions. Copies can be made of the code to facilitate the sharing process. Chaos inevitably results if the extra copies of the program are not cleaned up or erased when it comes time to link all the components together. An incomplete or nonfunctional piece of software can be linked in and ruin the clean performance of an otherwise effective system.

First, a file librarian or source code control system (SCCS) should be installed. SCCS creates a central repository for all the source code of the system, maintains version control, and does not allow more than one copy of source code to be "checked" out at a time. It also provides a layer of security. Any developer not given permission cannot pull source code out of the repository. At CS First Boston, an additional layer of code was put on top of SCCS to accommodate specific requirements.

Also beneficial are third-party software libraries. Many software vendors provide for such standard functions as date calculations or matrix manipulations. Rogue Wave is one such software or API provider. The independently provided APIs can be linked and loaded into an application along with all the in-house-developed software.

Important to the success of integrated open systems is the separation of the development environment from the prototype environment, which in turn should be separated from the actual production environment. An area where the developers can try out different techniques or rework ideas, the development environment is meant strictly for the writing and testing of code. Only programs which are still in flux or those that are still deemed modifiable should reside in this environment.

The prototype environment is where tested and debugged components can be installed, linked together, and run as a whole. This is an environment, however, in which a system that is not yet ready for the users is built up. Full system and integration testing is done here. Software components are replaced only when bugs are found in the testing process. Only executable code should reside in this environment. Replacement of system components and recompiling should be done in the development environment. The production environment also contains only executable code, and is the area from which the user executes the system. It should be bug-free and should never be affected by anything that goes on in the development or prototype environments.

Based on the different environments outlined above, change management then becomes an important issue. Careful scheduling must be observed when modifications are made to existing systems, be they in development, prototype, or production stages. Developers and systems administrators need to be aware that changes to a system are to be scheduled and implemented, so they can avoid problems. The procedures outlines in change management ensure that the development team, system administrators, and data base administrators are communicating with each other, and that each knows what the other is doing with respect to the system. When deciding when to implement system modifications or upgrades, all systems personnel should be aware of the time to change that would be the least inconvenient to the user. Unfortunately for the systems personnel, that time is usually at night, after business hours or on the weekends. Notification of changes to an existing production system can be as easy as sending electronic mail. More sophisticated notification products are available on the open market, though.

One example of total systems management is The Integrator, developed by Andersen Consulting. One of the sites at which The Integrator was installed is the capital markets technology area at Smith Barney. The Integrator is an interesting open system in itself, but its most immediately obvious use is to organize and control open system development. Its main panel acts as a port of entry into all the aforementioned systems components. Not only can it be used to define data base tables, application functions, and GUI window, but it allows the developer to establish the relationships between these components. After all the information is entered into The Integrator, the system builds the data base tables and creates a make file so that all the code can be compiled into an executable format. In addition, The Integrator incorporates documentation features, so that, as a system is being built, it can be documented at the same time.

The Application System. The application system is the enabling component of the integration process. As noted, the application must be process supporting, easily ported, scaled to other needs, and functionally superior to its legacy predecessor. Above all, it must be maintainable.

SUMMARY

Systems integration projects are enormously challenging and rewarding. The tangible results of integration can be felt almost immediately by the organization, and the project management environment is an ever-changing landscape of technical, personnel, and user issues. What is really needed in a project manager is an effective methodology to address the many needs of integration and a way to anticipate problems in integration. The experience described above clearly indicates that each integration effort is a learning experience, regardless of how much experience the project manager may have in mainframe and midrange systems development and integration projects. The technology is new, users are more demanding, and systems personnel are learning new roles, as well as new technology. The combination of these make the process extraordinarily fascinating and necessary.

6-5

Grasping the Promise of Client/Server Computing

DAMON M. BEYER

MARVIN J. NEWELL

IAN HURST

Seldom before has any technology promised so much to so many. If everything printed can be believed, so-called client/server computer systems will cut the cost of information technology, make computers easier to use, provide unprecedented computing power, speed systems development, and transform the shape of organizations as well as the nature of management.

To winnow the truth from these claims, the authors undertook an in-depth study of five large corporations, each at a different stage of implementing client/server solutions, in five different industries, consumer packaged goods, insurance, regional banking, transport, and investment banking. In each case, the authors interviewed the CIO and line managers and analyzed the costs and benefits of in-place client/server systems. The authors then validated their findings through a series of roundtable discussions with managers from another 20 large and medium-sized companies.

What they found, in brief, is that client/server technology does create unique opportunities to implement ambitious programs of corporate change. It makes possible computer systems that are more flexible and easier to modify than their predecessors, and it eases the task of redesigning the way decisions, especially cross-functional decisions, are made within an organization. Inevitably, however, significant challenges stand in the way of those who would grasp these new capabilities.

Building client/server systems requires not only the mastery of new technologies but also a new relationship between information technology (IT) and the business it serves. Instead of being simply an asset that produces benefits measured largely in terms of cost savings and productivity, IT becomes a service that promotes and enables change. This, in turn, means that IT professionals must transform themselves from "engineers," whose prime responsibility is to keep IT assets working at maximum efficiency, to "coaches," whose role includes helping an organization to identify and develop the people skills it needs to manage change successfully.

Given this scope of promise, to evaluate client/server in terms of its technological costs and benefits alone—as many companies now do—is to miss the point. At least as many client/server projects have run into difficulties because of a failure to manage the organizaticnal consequences of the technology as have faltered through problems with the technology itself. Although these projects' risks and benefits naturally vary from company to company, depending largely on whether a company needs to change its IT, its business processes, or both—all contain an organizational as well as a technological element. Both must be managed together.

WHAT CLIENT/SERVER IS, AND IS NOT

Three characteristics set client/server computing apart from the rest of the IT world.

- *Division of labor.* Client/server systems divide computing among many separate machines, in contrast with mainframe-based systems, which centralize the work in one place. With client/server, there is often a central server, responsible for maintaining data as well as the data bases that access that data. Desktop clients then draw the data from the server and perform calculations on it. A server might, for example, hold sales data, which desktop clients then use to do forecasting.

- *Network infrastructure.* Given this division of labor, client/server systems rely on networks to weld disparate computers into a cohesive whole. At a minimum, the networks provide for transportation of data between the various machines that work on it, just as roads and railroads provide for transportation of components between factories. Many companies are taking this approach a step further and building commonly used computing services directly into the infrastructure, providing, for example, document management from a central location and making it accessible over the whole network, to simplify the management of information systems and reduce costs.

- *Emphasis on the user.* Much of the power of client/server goes into making new applications engaging and easier to use. It provides users with greater latitude in choosing software presentation and software designers with greater latitude in creating programs that are pretty, readily customized, and modern, such as user interfaces with multiple windows, graphics, and a point-and-click style of entering commands. More important, changes to a program on a client desktop need not affect programs or data on the server. As a result, improving systems is much quicker and simpler than when all computing has to be provided through a one-size-fits-all package on a mainframe.

DISPELLING THE MYTHS

Numerous claims about client/server are not borne out by the facts. One myth is that the move to client/server architectures is mostly driven by a wish to reduce costs. True, many of today's personal computers (PCs) and workstations offer processing power comparable to that of a minicomputer or a mainframe at a significantly lower price. In practice, however, money does not seem to be the primary concern of those moving to client/server environments.

Although there were some examples of client/server systems being used to replace obsolete, expensive-to-run minicomputers and mainframes, most of the systems the authors encountered had been chosen because they could perform tasks— providing new capabilities or enabling quicker systems development, or both—that mainframe systems could not. Cost savings, if any, came as a bonus.

A second familiar claim is that client/server technology is capable of supporting only lightweight applications, such as decision support, not heavyweight, mission-critical functions, such as transaction processing. This alleged limitation rests on a too-limited definition of the technology. Many companies do, in fact, employ client/server components for complex transaction processing.

A major investment bank, for example, uses client/server components to trade a variety of financial products across a variety of offices. Indeed, the system is capable of managing trades all the way through to cash settlement with the Federal Reserve. Although most of the transaction volume is still handled on mainframes, these have simply become components in the larger client/server system.

A final claim is that the technologies underlying client/server computing are still too immature to be scaled up to support really big applications. Although most of the executives interviewed believe that client/server systems are inherently more complex to manage than those based on a single mainframe or minicomputer, they also feel that the systems' complexity is manageable and the benefits far outweigh the risks. Part of the reason is that the software used to manage client/server systems is maturing very quickly.

One of the most ambitious implementations the authors studied involves the core booking and logistics system for a large transport company, which is using client/server technology to integrate several disparate older systems into a common platform for 17 major new applications. This massive integration and development project is being accomplished in under three years—primarily with off-the-shelf tools and products.

A FOCUS ON ORGANIZATIONAL CHANGE

Underlying all these false claims is an even greater misconception. Much of the discussion about client/server computing assumes that the technology is best evaluated in technological terms, but this is not the case. One of the strongest influences on the decision to adopt client/server technology is its potential impact on the organization.

"Information," as the cliché goes, "is power." By freeing information from the straitjacket of centralized, mainframe-based systems, client/server systems change the distribution of decision-making power within an organization. For example, salespeople could have a complete overview of a particular customer relationship on their notebook computers, thus allowing them to make their own judgments about that customer's most important needs, instead of following standard order-taking procedures. Equally, the availability of extra information can enable production workers to take more responsibility for scheduling work and improving quality.

In the back office of a major bank, the mainframe-based system for trading securities transactions was designed around 24 control points, each representing a decision that could be taken by a manager regarding the progress of a transaction. With the installation of client/server technology, every one of these control points changed.

Companies are moving to take advantage of the new organizational capabilities created by client/server technology by trying to improve their organizations' ability to make decisions as a whole. Often, this is a matter of making the whole at least as good as the sum of its parts, that is, ensuring that well-informed islands of decision-making prowess are not lost in a sea of ignorance. This may require decisions to be made more rapidly, or by different people in different places. It may also mean changing the kinds of decisions that are made.

Companies are moving to take advantage of the new organizational capabilities created by client/server computing by trying to:

- Bring together information from different departments, functions, and companies. Providing a comprehensive solution to a business problem often depends on assembling information from a variety of sources and making it available within different, usually shorter, time frames. The ability to respond to unexpected demands for information, perhaps from unusual places, is crucial.

- Exploit the additional flexibility inherent in the technology to build information systems that can change as fast as the underlying business does. Having redesigned decision making once, managers usually realize that they will want to do so again—or indeed, continuously—as their business evolves. Client/server architectures make it relatively easy to change the information that servers bring to the desktop as well as the computations performed there.

CAPTURING THE BENEFITS

The most successful client/server systems the authors encountered created benefits through the business changes that they made possible rather than through the technology itself. In fact, all the successes observed focused on solutions that enabled different decisions in various locations or by different people, within many time frames. One particularly successful system provided the sales force of a consumer goods company with more and better information on the products it sells. As a result, the sales force has taken over, to positive effect, some decisions previously made by retailers.

In addition, this system permits greater customization and flexibility in marketing. Previously, information on the results of product promotions, such as coupons or special prices, was gathered weekly, region by region, on a centralized system. The new technology collects this information daily, store by store. Decisions that were formerly the province of regional directors can now be made by salespeople able to tailor focused promotion strategies to each store. Total costs of the system were over $30 million, but the base-case return on investment is over 50%, with a substantially higher upside if sales hit more aggressive targets.

Conversely, companies that did not take advantage of the changes in decision making made possible by client/server failed to reap the benefits of their new systems. At the same consumer goods company, for example, a client/server solution was deployed to automate the collection of statistical process control data for a production line in a regional factory. The system allowed line workers to input changes in process parameters (e.g., product size, height, and color) directly into PCs stationed on the factory floor. This data collection had previously been carried out manually and input by clerical staff overnight.

Although the new system changed the timing and location of data collection, it had only a limited effect on the decisions made by line supervisors. This was because the major decision-making benefits had already been captured when manual statistical process control was introduced. Thus, the immediate benefits from the client/server system were quite small and did not justify its incremental costs. However, because much of this investment was in infrastructure (i.e., personal workers) the company is now trying to run additional applications on the same platform, which could dramatically alter its return on investment.

STRIKING THE BALANCE

Companies seeking to adopt client/server must manage change along two dimensions, technology and organization. Simply making client/server computing work means installing new server hardware, end-user computers, and software, building network infrastructure, and creating new skills within the IS department. Reaping the business benefits, however, means harnessing the capabilities of that new technology to change organization structures and decision-making processes. It also means fostering process or organizational change within the business to exploit those new capabilities. Any client/server strategy must, of course, pay close attention to both these dimensions of change. The tactical risks and benefits for any business depend on the relative urgency of technological and organizational change.

There are four possible ways of getting this balance right:

- *Opportunistic.* Some companies decide to adopt client/server technology before they face an urgent need either to improve their basic technology or to change their business processes. This approach implies that they have two or three years in which to gain experience and tap the benefits of client/server computing on specific projects of their own choosing.

- *Infrastructure-led.* Other companies find themselves driven toward client/server computing by encountering the barriers inherent within their existing information systems. The most urgent problems typically concern network infrastructure. Existing technology often cannot gather or distribute the information that managers need to make timely, accurate decisions. It is also common for problems to exist in building the essential skills within IS for managing new systems development. In such cases, the creation of new network infrastructure and new skills becomes the priority in positioning an organization to benefit from client/server strategies.

- *Application-led.* Still other companies find that, although their existing technical skills and network infrastructure can support migration to client/server systems, their business processes require more rapid change. They may, for example, need to develop new products more quickly or create cross-functional teams to bring a range of skills to bear on customers problems. For these firms, getting in place new solutions and the organizational change that must accompany them is the priority.

- *Parallel.* Finally, a number of companies find themselves besieged on both fronts. They have to change business processes as well as technology—and change them fast. They may, for example, need to respond to sudden price competition or to a major new opportunity. Here, implementing client/server strategies can become a do-or-die proposition.

Each of these approaches strikes a different balance, presents risks, offers rewards, and implies various management priorities.

The Opportunistic Approach

The great advantage of the opportunistic approach to client/server technology is that a company can initially address the task with leisurely deliberation. The great risk, however, is that the company will continue its leisurely approach for too long and let value-adding opportunities slip away. The situation of one regional bank illustrates this dilemma.

The bank in question relies on mainframe systems and has excellent skills in managing them. It has also succeeded in creating a few small, but effective, client/server systems. Its choice of what to do next, however, is finely poised between inertia and change. For any given systems requirement, the bank will find it cheaper to extend its existing mainframe skills and technology. Yet managers fear that a rival will soon begin to use client/server technology to create innovative financial products that might lure away the bank's core customers. Their challenge is to convert that vague worry into purposeful action before it comes true.

A useful tactic for senior executives in this kind of position is to examine the results of their own companies' early experiments with client/server computing. If these first solutions have been enthusiastically welcomed by line managers and are delivering clear business value, the best step is to build on that enthusiasm and shift toward an application-led approach. Indeed, if there is little response to line managers' enthusiasm for the business benefits that client/server can bring, the company's IT strategy may quickly be placed at risk. Line managers will go outside the firm to get the systems they want, with little regard to their "fit" with the overall direction of IT.

On the other hand, if initial experiments with client/server reveal weaknesses either in networks or in the company's systems-building skills, senior managers may want to use that evidence to build momentum for an infrastructure-led approach. Whichever the case, the opportunistic approach is a luxury best enjoyed briefly.

The Application-Led Approach

An application-led approach is often the most straightforward. Companies in fast-moving service industries, for example, find it comes naturally, particularly when IT is already deeply embedded in the products and services they offer to customers.

Driven by business change, the process often focuses on new initiatives to take advantage of new market opportunities. Because these opportunities are clearly defined in business terms, each project should provide a healthy return with no more than the usual level of risk associated with capital investment. Moreover, since the systems to be built are demanded by, not imposed on, their ultimate users, there should be few difficulties in gaining acceptance for new technology. As in any change process, a series of early, highly visible successes can quickly build momentum.

There are, however, several dangers. One nagging problem is the temptation to overspend on each initiative. Because projects so readily justify themselves individually, duplication and excessive complexity can easily slip into systems design.

Among the companies the authors studied, some estimated that they spent as much as 25% more than necessary in pursuing an applications-led approach. The only company that consistently avoided overspending did so by building within its central IS function the capability to discern potential economies of scale among the client/server projects proposed by individual business units. This company also provided financial incentives for these business units to cooperate with central IS in tapping those scale economies.

The greatest danger of this approach is that inattention to infrastructure suddenly creates obstacles to progress. Networks will no longer be able to cope with the flows of information demanded by burgeoning, enterprisewide client/server applications. Managers who have grown used to the freewheeling ease of justifying projects on their own merits often have difficulty in accepting the imposed discipline of building robust infrastructure. They will not easily agree to limit their technical choices for the common good or to help pay for shared facilities.

Rapidly increasing systems costs are a sure sign that it is time to switch the emphasis of client/server migration toward infrastructure building. At one bank, for example, costs were rising by more than 25% a year. One possible solution is to institute a "federal" organization within IS, taxing business units to fund centrally constructed infrastructure. Inevitably, however, organizational politics tend to dominate the transition from an applications-led to an infrastructure-led approach to client/server systems.

The Infrastructure-Led Approach

Some companies, particularly those in such capital-intensive industries as oil, transport, and steel, are already governed by the discipline of centralized capital investment budgets and long planning cycles. They may find it quite natural to approach client/server computing by building the necessary network infrastructure first. Others may find that the sheer scale of the technical challenge dictates an infrastructure-led approach. One consumer goods company, for example, determined that it could obtain significant business benefits from client/server systems, but with 3,000 widely dispersed sales representatives, it had to build a new network infrastructure before it could even begin to deliver the information required to tap that potential.

So long as eventual infrastructure needs can be predicted with reasonable accuracy, a migration to client/server that builds infrastructure before applications offers several potential advantages. It allows a company to exploit latent economies of scale in network building and makes it possible to design a network that is technically elegant and resistant to obsolescence. In other words, it lowers the risk that early client/server migration will come to a sudden halt for lack of infrastructure.

A key problem, however, is that if managers miss the mark in their predictions, they will build an inappropriate or unwanted infrastructure. Two tactics can help to mitigate that risk. First, even as it focuses on building infrastructure, the IS department must involve and excite the managers who will conceive and demand the applications that will use that infrastructure. Second, a company should develop a portfolio of potential client/server applications. The success of any one may pay for the investment in infrastructure. Even in a less-than-ideal world, this portfolio approach can help limit the costs and risks of network infrastructure.

The Parallel Approach

Not surprisingly, doing things in parallel, that is, pursuing a migration approach that requires leading with infrastructure and applications together, combines the risks and benefits of both these approaches. But it also compounds the risks because of the sheer scale and complexity of the changes involved. The key to managing a parallel migration is to break the task into manageable pieces, such as a collection of smaller, self-financing projects, each of which can be justified in its own terms, and to place a premium on speedy implementation. Although each project should adhere to an overall vision and to corporatewide technology standards, it is impossible to achieve the technical elegance of an infrastructure-led approach amid the near-chaos of parallel change.

Ultimately, much will come down to the leadership skills of the senior executive leading the effort, usually the CIO. Many CIOs assume an essentially passive role, offering to build systems to support whatever individual businesses want to do. However, strategically managing the opportunities created by client/server computing in the fast-moving world of parallel change requires a proactive CIO, fully supported by the CEO, who can bring an understanding of technological risks and opportunities to the discussions that will determine the shape of the business.

A NEW ROLE FOR THE IS DEPARTMENT

The parallel approach to client/server migration poses the greatest challenge to an IS department and, indeed, to a business as a whole. Still, any move toward client/server sooner or later requires dramatic changes in how technology is managed. The IS department will have to not only master new technologies but also take on fundamentally new responsibilities. Ironically, the most dramatic changes are often the easiest to recognize and manage.

Client/server technology brings a real-time focus to IT. Traditionally, IS departments managed major projects with multiyear paybacks. Investment priorities and budgets were set at annual meetings after in-depth study and careful preparation. With the possible exception of the initial construction of network infrastructure, most projects in a client/server world simply do not fit into this framework. Applications are short-lived and are better thought of in expense terms than as a capital investment. Many will essentially be perishable. If IS is to avoid becoming a constant drag on the business changes enabled by client/server, it will have to speed up its own decision-making processes.

That is not all, however. IS must also make decisions in different ways. Client/server technology forces IS into a new balancing act. It must, for instance, work alongside, if not inside, business units to help build applications quickly, get information where it is needed, and support line managers' efforts to get the most out of the technology. IS must also, however, exert a central authority to maintain infrastructure, tap economies of scale in technology, and reduce the complexity of that technology.

A first step in getting this balance right is simply to recognize the conflicting demands. A second, however, is to acknowledge that, contrary to prevailing assumptions, many aspects of client/server systems' distributed style of computing can be successfully centralized. End-user support, for example, is one area in which companies can capture scale economies.

At a health care products company, the IS department and business units took three steps to identify the cost savings that could be derived by centralizing client/server support activities. They defined appropriate service levels and performance goals, identified business-specific requirements, and capitalized on opportunities to share resources by highlighting three levels of central administration and support. Taken together, these measures cut IT costs by more than 25%.

Closely related to the centralization issue is the question of standardization. To control the complexity inherent in enterprisewide client/server systems, successful IS departments work with users to standardize technologies, products, and vendors. Many successful firms offer their business units a choice: they can either pay a high price for support of a unique technology or enjoy savings on a standard product. In this way, IS encourages each operating unit to determine the true business value of nonstandard or unique products. Where there is minimal loss of flexibility or major cost savings, the IS department may wish to standardize more extensively.

Although exercising central control comes naturally to an IS department accustomed to building systems single-handedly, working closely with individual business units does not. Changing the ingrained attitudes and values of IS staff, not to mention boosting their interpersonal skills, is a long, hard process. Successful companies, however, have taken advantage of the overall organizational change that client/server computing brings to take steps in that direction.

At one investment bank, for example, the adoption of a client/server system led to a major overhaul of financial control systems. In the process, the boundaries dividing IS staff from accountants were blurred in several ways. Not only did the two begin to share offices, but IS staff were also given responsibility for some aspects of the execution of transactions. Similarly, some accountants were put in charge of some aspects of the new technology.

SUMMARY

Despite the challenges, the authors expect the trend toward client/server computing to continue, indeed, to accelerate. The client/server approach makes full use of modern technology, such as fast microprocessors and high-speed data networks. It also helps to create modern organizations that are fast moving and quick to grasp new business opportunities. For many companies, only client/server computing offers the possibility of incorporating more information and more IT into products and services while retaining the organizational ability to change as swiftly as markets do. Whatever route managers decide to take toward that goal, such capabilities are well worth the effort.

Section 7
The Role of the Systems Integrator

As systems integration achieves mega-proportions, encompassing enterprisewide integration, organizations are increasingly turning to the services of outside consultants and business integrators. The use of third-party integrators provides many advantages to the organization, including the immediate injection or access to specialized talent and experience. In addition, the use of outside service providers helps to distribute the risk of the process and provides a level of assurance to the organization navigating the tenuous path of integration.

The systems and business integration industry is at a landmark $10 billion, and the large number of integrators is a diverse group, which makes selecting one and negotiating fees a complex process. Organizations are increasingly asking such questions as:

1. Who are the integrators?
2. How does one contract with an integrator?
3. How is an integrator selected?

The authors of this section describe the many practical considerations when selecting a systems integrator and negotiating fees that are of immediate use.

7-1

Who Are the Systems Integrators?

MICHAEL A. MISCHE

The systems integration industry is large and diverse. As a $10 billion industry, a growing number of companies and sole practitioners are entering the integration arena. Whereas integration was once exclusively confined to mainframes, the proliferation of personal computers (PCs) and advances in client/server technologies have created new opportunities and increased demand for integration services and service providers. This chapter explores the systems integration industry and identifies and categorizes some of the key providers of systems integration services.

THE SYSTEMS INTEGRATION INDUSTRY

Establishing the size and growth rates for the systems integration industry is a difficult challenge. No industry standards or criteria for reporting systems integration exist, nor does a de facto definition for what exactly constitutes systems integration. Thus, estimating the size of the industry is, at best, tricky and is contingent on three key factors:

1. What the various vendors claim and report as systems integration revenues or sales.
2. What the various organizations that are performing integration classify and report as systems integration.
3. The best estimates of professional research organizations, such as the Gartner Group, Yankee Group, and others.

The Gartner Group has defined systems integration as a "large [more than $1 million], complex IS project that includes designing and/or building a customized architecture or application, as well as integrating it with new or existing hardware, packaged and custom software, and communications." Using this definition of systems integration, the Gartner Group estimated the 1994 total systems integration marketplace to be approximately $10 billion, exclusive of hardware and software. This estimate is based on a variety of sources, including vendor revenue and sales reports. However, as discussed in Chapter 1-2, systems integration has numerous definitions, and depending which is used, each yields a different measure of revenue and market size. How accurate is the $10 billion estimate? No one knows, because no way exists to verify the integrity and accuracy of vendor claims and reports.

In all actuality, the systems integration industry is probably smaller than that number—but not much smaller. Again, measuring revenues for systems integration is very difficult, as firms account for and classify revenues by service and product line differently. Another factor that distorts estimating revenue and the size of the industry is the commingling of systems integration services revenue with the revenues that are generated from management consulting services. For example, in one situation, it is known that revenue that is reported by one vendor as systems integration also includes outsourcing. In this regard, there is very little consistency among firms and measurement methods. Nonetheless, the market for systems integration is large and growing.

There are many ways of analyzing and classifying service providers in the integration marketplace. This chapter categorizes integration providers into five types: Big Six accounting and consulting firms, global integrators, proprietary integrators/vendors, contractors, and Blue Chip firms. This segmentation can be further refined based on industry and technology orientation and capabilities based on consulting orientations and technical skill sets.

BIG SIX ACCOUNTING AND CONSULTING FIRMS

The Big Six accounting and consulting firms are Andersen Consulting, Coopers & Lybrand, Deloitte & Touche, Ernst & Young, KPMG Peat Marwick, and Price Waterhouse. They are large and established, with a national presence—maintaining offices in virtually every major city—and independent partnerships with international operations that help project an image of globalization. The Big Six have continuing relationships with bankers and investors in both publicly and privately held companies, in accordance with the regulatory requirements of the Securities and Exchange Commission. These firms also have auditing and accounting relationships with federal, state, and local governments.

The Big Six are the "Blue Collar" members of the consulting profession. They have large staffs, who are known for successfully completing fundamentally important integration projects, significant leverage, and deep resources. However, unlike the Blue Chip firms, they are not necessarily known for their boardroom savvy or strategic planning capabilities.

The Big Six firms have always been a source for systems integration talent and services. They select their staffs using a rigorous interviewing process and then train and manage them well. Because of the Big Sixes' accounting pedigree and penchant for detail, they are also known for having rigorous documentation standards and series of internal checks and balances designed to enforce compliance with those standards.

These firms grew at phenomenal rates starting in the middle of this century. The growth rate was largely propelled by regulatory requirements for audits, filings and taxes, and consulting was a logical complement to those core services. Some of these firms have attempted to differentiate their consulting capabilities from their traditional audit and tax practices by creating different partnerships and corporate structures for their consulting businesses. Even though a legal differentiation exists, however, they remain for the most part relatively homogeneous. Big Six systems integration services include such capabilities as application development, data base design, system implementation, and hardware and software selection.

All the Big Six firms claim to have systems integration capabilities. Some, however, have clearly established leadership positions in this area, whereas others have fallen behind or tend to specialize in a certain niche. Andersen Consulting, an extension of Arthur Andersen, the accounting firm, is by far the largest and best known of the Big Six, especially for systems integration work. Their professionals are centrally trained in a facility located in St. Charles, Ill., and receive comprehensive instruction in the fundamentals of systems development, project management, management consulting, and the theories and concepts of information technology (IT). The Andersen training program has created a cadre of common skills and practices that are pervasive throughout its organization, regardless of office. This commonality has helped Andersen position itself as the dominant provider in the systems integration market.

Ernst & Young, although not as large, is emerging as a leader because of some of its new tools, hiring tactics, and aggressive strategy to provide systems integration as well as business consulting services.

Deloitte & Touche is also expanding its presence in systems integration. Price Waterhouse, apparently first and foremost an auditing firm, is making inroads as a leader in systems integration, particularly at the client/server and specialized software levels. Coopers & Lybrand is a "niche" provider in systems integration and has very limited resources with respect to Andersen. KPMG Peat Marwick appears to be a modest niche player with respect to systems integration.

According to Gartner Group estimates, the Big Six collectively represent 20% of the entire $10 billion market for systems integration. Approximately $1.5 billion is generated by Andersen's systems integration practice, representing approximately 50% of Andersen's total estimated $2.9 billion in consulting revenues. Relative to the prospects of future success in the integration marketplace, the Gartner Group again places Andersen Consulting at the forefront of the Big Six and, indeed, the overall systems integration industry as being the leader with the best ability to fulfill the vision and need for systems integration.

Based on Gartner Group estimates of revenues related to systems integrators, the Big Six systems integrators rank as follows:

1. Andersen Consulting: $1.5 billion.
2. Price Waterhouse: $203 million.
3. Ernst & Young: $197 million.
4. Deloitte & Touche: $150 million.
5. Coopers & Lybrand: Not measured or measurable.
6. KPMG Peat Marwick: Not measured or measurable.

GLOBAL INTEGRATORS

Global integrators represent a new and emerging class of system integrators. A hybrid of the classic management consulting firm and contract development firm, they combine boardroom consulting with technical systems development. Their structures are exceptionally flexible, anticipatory, and far less encumbering than those of the Big Six and proprietary integrators. The major global integrators include EDS, IBM, CSC, Cap Gemini, Perot Systems, SHL Systemhouse, and TSC. Collectively, these firms represent approximately $3.6 billion, based on Gartner Group estimates. Because of their structures and service orientations, the firms in this segment of the systems integration marketplace—especially those that have traditional structures—tend to be more agile with respect to services, capabilities, and resources. The specific distribution of estimated revenues by major firm in this category of integrator is:

- EDS: $1.3 billion.
- IBM: $1.2 billion.
- CSC: $350 million.
- Cap Gemini: $315 million.
- SHL Systemhouse: $209 million.
- Perot Systems: $182 million.
- TSC: $60 million.

Driven by technology developments, changes in the business environment, and reengineering efforts, the need for technology integration undoubtedly will continue to propel growth for this industry. A fast-growing segment of the industry will be in the client/server areas for development, vendor product implementation, and integration. The client/server marketplace is growing as a result of technology advances, applications downsizing, and organizational reengineering needs. The major client/server issue is the proficiency of integrators in multivendor and multifunctional environments. Based on a number of different industry assessments and opinions, EDS, CSC, SHL Systemhouse, TSC, and Andersen Consulting are best positioned to dominate the systems integration marketplace, at least through the year 2000. The key considerations in growth and the realization of these firms to dominate the marketplace centers on three major issues:

1. The firm's ability to provide leading-edge multivendor/multitechnology environments.
2. The integrator's ability to link strategic issues and reengineering opportunities to tangible system integration solutions.
3. The firm's ability to meld successfully technical and project management skills with cultural transformation consulting capabilities.

In this context, the most successful integration projects will be measured not by technology alone but rather by a combination of business, technology, and process innovation factors, all of which involve or revolve around the systems integration project. Other factors that influence the growth and success of the system integrator are the ability to gain economies of scale through the use of rapid development techniques and reusable application templates, to use leading-edge productivity improvement and project management tools, and to include specialty personnel who are senior and highly experienced in organizational change and transformation. In this respect, the firms that are best situated to dominate the industry are those classified as global integrators.

PROPRIETARY INTEGRATORS

Proprietary integrators represent a component of the systems integration industry that are captives of or directly aligned with a vendor of hardware and software. Typically working for the vendors that provide products and technologies, they are highly specialized in knowledge of a particular product line. This group of integrators includes such firms as AMS, Hewlett-Packard, Unisys, Digital Equipment Corporation (DEC), and Bull.

The distribution of this category of systems integration provider by estimated revenue is:

- Digital Equipment Corporation: $769 million.
- Hewlett-Packard: $475 million.
- Unisys: $475 million.
- AMS: $150 million
- Bull: $130 million.

Collectively, these firms account for more than $2.0 billion in annual systems integration revenue.

CONTRACTORS

This group of integrators represents a mix of smaller regional firms, systems developers, and independent contractors. As discussed in Chapter 7-2, this group of integrators is a source of specialized management consulting and systems integration talent. Some of the more significant firms in this segment of the marketplace include CTG, CapGemini, and BSG Corporation. Contractors also provide specialty integration services to niche industries and companies.

Because of the diversity and growth in this segment of the systems integration industry, measuring revenues and the size of the firms included in the category is virtually impossible. Many contractors are hybrid integrators and developers and use a combination of internal resources, specialty contractors, and off-shore programming resources to deliver services.

BLUE CHIP INTEGRATORS

This class of integrator is well known for its boardroom presence and savvy and less known for systems integration. Integration services are secondary to the Blue Chip integrators' mainstay areas of strategic planning, research, and operational improvement. The Blue Chip integrators include such firms as Booz, Allen & Hamilton, McKinsey & Co., and A.T. Kearney.

Historically, Blue Chip firms have functioned in the boardroom rather than in the data center. Their professional services tend to concentrate on strategy, organizational effectiveness, globalization issues, and operational improvement management consulting rather than technical IM&T services. The work performed by the Blue Chip firms can lead to far-reaching and profound changes because their clients are usually the most senior leaders of the organization. Although Blue Chip firms may have systems-integration capabilities, they are usually limited in comparison with the rest of their practice and are almost always out of their mainstream lines of business. Blue Chip firms represent the high end of the professional fees spectrum, with rates usually beginning at $1,200 per day for a junior staff member to more than $10,000 per day for certain partners or officers. Because systems integration is not a mainstay service for these firms, it represents a very small percentage of revenue and personnel for them. Consequently, they are not a significant factor in the systems integration marketplace.

SUMMARY

Systems integration is here to stay. It is a vibrant and growing industry that has explosive potential. The factors driving the systems integration industry should easily translate into double-digit growth rates through 1999. The compounded growth rate for the industry will be an estimated 11% to 16%. Based on Gartner Group's 1994 estimate of a $10 billion baseline, the industry should grow to between $16 billion to $21 billion.

There will, however, most likely be a consolidation in the industry as vendors are forced to fund increasingly greater investment in tools, personnel and marketing prowess to sustain growth and capture clients. A secondary consideration is that integration, although touted as necessary, is a luxury that is usually foregone in economic downturns. Also, as noted, markets and user needs are changing much faster than the traditional systems integration can accommodate. This may force a recalibration of the integration effort, especially when the effects of reengineering are considered.

Getting information on a systems integrator and their capabilities is a relatively easy process. Because integrators are first and foremost, service providers, they are approachable, distribute marketing materials, and will be more than happy to provide information and a proposal. Readers interested in helpful hints in how to select a systems integrator should turn to Chapter 7-2.

7-2

Choosing a Systems Integrator

MICHAEL A. MISCHE

Selecting a systems integrator can be a perplexing, emotional, and time-consuming process. Not only is the competition among integrators increasing, but the proliferation and rate of change of technology have created countless options and choices. In less than a decade, technology has changed from a field dominated by mainframe computers, rigid applications, and highly complex and intricate interfaces to one that is defined by end-user computing, client/server components, workstations, enterprisewide networks and data sharing. There is very little agreement as to what actually constitutes a systems integration project and systems integration services. Integrators themselves disagree about what constitutes systems integration and few can define it in contractual terms rather than as a substantive service. Whereas once organizations seeking integration consultants had only the Big Six accounting firms, hardware and software vendors, and a few regional and boutique contract programming firms from which to select, today's choices are varied and constantly growing. Thus, the industry has a real need for an effective method to help guide the organization through the choices and options in the systems integration and service provider marketplace.

This chapter provides a perspective on how to evaluate and select a systems integrator. The seven principles discussed here provide a simple and convenient method for the selection process and have been developed as a result of experience on both sides of the systems integration table.

THE NEED FOR A SYSTEMS INTEGRATOR

Confronted by never-ending technical and business challenges, end-users and information management and technology (IM&T) organizations are finding themselves under enormous pressures to improve performance and accelerate the delivery of enabling systems. No longer is it sufficient for the IM&T organization to simply react to users and supply systems and technologies to satisfy yesterday's needs. In the decade of organizational transformation, IM&T organizations must anticipate their users and function as process- and organizational-change enablers. To address their needs, many organizations rushed into situational management and development programs, such as, TQM, CASE, joint application designs (JAD), client/server technologies, and a host of others, only to be disappointed by results that fell far short of oversold benefits and inflated expectations. Many found that the costs involved greatly exceeded the benefits derived; still others learned of the "hidden" costs.

Uncertain, understaffed, or lacking the resources necessary to support an integration effort successfully, many organizations are turning to external consultants that provide systems integration and reengineering services. With daily, per-person billing rates ranging from several hundred dollars to over $10,000, selecting an integrator can be a confusing and difficult challenge. Systems integrators can contribute much to the process and the organization. Used properly, the systems integrator can assume the majority of the project risk, supplement or supersede internal resources, and deliver turnkey results. Systems integrators can also contribute perspective to their clients' business processes and internal management capabilities. The organization can engage the appropriate integrator and find its money well spent; the organization that chooses the wrong integrator not only wastes its money but finds that its vision and project may never be realized. Thus, there are risks and advantages to using a systems integrator. Some of the more significant benefits are that integrators:

- Bring a fresh perspective and energy to any systems integration project, which can be invigorating and extremely beneficial to the organization and overall integration process.
- Have stamina and a clearly defined role; they are not distracted by the everyday business needs or organizational politics.
- Have tools and methods that can be brought to the organization in various forms and as the need requires.
- Can be held accountable for their performance on a legal, moral, financial, and professional basis. Unlike internal resources, organizations have a legal recourse with integrators.

THE PRINCIPLES OF SELECTING A SYSTEMS INTEGRATOR

In light of the variety and diversity of systems integrators, selecting and effectively using an integrator's services are challenges, even for the most experienced organization. A natural starting point is with any existing relationships that the organization may have with integrators. The buyers of services (i.e., the clients) and providers (i.e., integrators) know one another, their cultures, and their capabilities. Presumably, they have an established rapport and a track record of working together. Neverthe-less, each integration situation is different, and the needs for an integra-tor may be different. The following 12 principles provide a basic frame-work for evaluating the capabilities of the prospective integrator.

PRINCIPLE NUMBER 1: THERE ARE FIVE TYPES OF SYSTEMS INTEGRATORS

Although the market is undeniably large and varied, one could easily summarize systems integrators into one of five basic types of systems integration firms:

1. Blue Chip management consulting firms.
2. Big Six public accounting firms, many of which have now become limited liability partnerships (LLPs) for their consulting services.
3. Global integrators.
4. Proprietary vendors.
5. Contractors.

Chapter 7-1 provides a description of the types of firms that are included in these categories.

Briefly, the Blue Chip firms, which include such pedigree names as A.T. Kearney, McKinsey, Bain & Co., and Booz, Allen & Hamilton, are known as breeding grounds for extensive intellectual talent and strategic management savvy rather than as technical service providers of infor-mation technology. Although the Blue Chip firms are a source for systems integration talent, that is not their major area of internal investment and personnel development. In this regard, Blue Chip firms may not neces-sarily be the best resource for systems integration services.

The once Big Eight and now the Big Six accounting/consulting firms represent a large pool of systems integration talent. These firms include such stalwarts as Andersen Consulting, Ernst & Young, and Price Water-house. Although Big Six firms are predominantly known for their ac-counting and tax services, they are also known for their extensive con-sulting capabilities. With revenues in excess of $1.5 billion, Andersen Consulting is one of the largest providers of integration services and is the dominant force in the Big 6 arena. Big 6 clients typically include middle and upper-level managers in both the private sector and govern-ment. Billing rates per person for systems integration projects generally range from $1,000 per day, per person, to more than $6,000 for certain senior partners. However, Big Six firms have a reputation for discounting their rates, as competition within the peer group is fierce. Discounts of 25% to 50% from their standard hourly rates are not unusual, especially for long-duration projects and audit clients. The Big Six represents a tried-and-true source for integration services.

Global integrators are a new breed of firms that are quickly establishing themselves as dominant forces in the marketplace. In a sense, global integrators are a hybrid of the Blue Chip and traditional Big Six firms in that they can bring both highly skilled management consulting services and technical IM&T service providers to their clients. These firms include such organizations as IBM Consulting, CSC, TSC, SHL Systemhouse, and EDS. Many global integrators, such as EDS and TSC, began as technical service providers and evolved into firms that meld technology with business processes. Their rate structures are competitive with the Big Six and Blue Chip firms. Global Integrators are quickly becoming one of the best sources for systems integration assistance and general management consulting in the marketplace today.

Proprietary vendors include hardware and software manufacturers and value-added resellers that provide integration and management consulting services. However, their capabilities tend to be restricted to their own product offerings or specialized product line and are usually less extensive with respect to general management consulting issues and services. As captive service providers, proprietary vendor integrators have a distinct orientation toward making their products work and do not necessarily concentrate on making sweeping changes to the organization through reengineering and reinvention. Therefore, the organization runs the risk that vendor integrators can be less dimensional in their approach to integration, which may lead them to miss broader opportunities for radical improvement and competitive advantage through integration and reengineering business processes. Professional management consulting services, at least in the classical context of the Blue Chip, global integrator and Big Six firms, are not necessarily their main streams of business or sources of revenue. Hardware and software vendors make the majority of their money from the sale of software, new releases, and annual maintenance fees, not from hourly rates for systems integration and project management professionals. The rates span the spectrum from several hundred dollars per day, per person, to over $2,000 per day, per person. In all but a few situations, the use of vendor integrators may not be as desirable as the alternative sources are for systems integration talent.

Finally, the fifth group of systems integrators is composed of the sole practitioner and contract programming firm. Contractors include a growing group of individuals who have a particular industry orientation and specialized technical or application system skill set. Contractors can either sell their talents directly to the organization or to a contract programming firm that brokers them to a needy organization. Contractors are generally used in a specific role and work under the client's direction.

The billing rates for the contractor firms are quite low compared with the other types of systems integrators. Some rates range from a low of $35 per hour for a programmer to as much as $125 per hour for an experienced project manager. Contractors can be most effective when the engaging organization is supplementing its own resources and has an extremely strong project management and quality-assurance function in place.

PRINCIPLE NUMBER 2: SYSTEMS INTEGRATORS ARE COMMODITY SERVICE PROVIDERS

The services, capabilities, and qualifications between systems integrators are all similar. Core services, which relate to the technical aspects of integration and project management, are fairly uniform among integrators. Virtually all types of systems integrators can provide the same technical resources, similar types of references, and experienced project managers. The issue is understanding what combinations of skills and talents are needed by the project and matching them with integrators.

As a starting point for evaluating and selecting a systems integrator, the organization should understand that system integrators have four very important qualities:

1. They possess a number of specialized skills and capabilities.
2. Their marketable products include time and people—and they charge for them.
3. They possess tools, techniques and information that organizations may lack or can only partially assemble.
4. They can deliver turnkey results while usually assuming the risk of the project.

This is where the subtle differences among integrators can be found which can help differentiate integrators in a relatively homogeneous marketplace. These differences are primarily the result of structure, management, development of personnel and methodologies, and the transference of project risk onto the integrator.

The Big Six, Blue Chips, and global integrators are primarily professional consulting firms. These firms have established standards of work and a well-defined structure of responsibility and accountability. They have the capability to manage, assume the risk of, and perform multimillion-dollar and multiyear projects. They can manage, supply personnel to, and deliver completed projects to their clients anywhere at any time. These firms are in the full-time business of providing professional services and are constantly seeking, training, cultivating, and employing highly talented and well-compensated individuals to provide consulting and integration services to their clients. They have extensive infrastructures and organizational arrangements that provide not only a high level of client service but also the structure and management to develop personnel, manage resources, develop intellectual capital, evaluate project status and progress, and assess the quality of work performed. They offer their professionals well-defined career paths, and the internal competition among the employees and partners/officers helps ensure a steady stream of competent and responsive resources.

Most importantly, these firms have two other major operational elements. They measure their consulting performance not only based on client feedback but according to their own standards, which are often more rigorous than those of the client. Also, they have internal methods for managing risk, assuring quality, measuring progress, and generating value for the client. Clients that hire this type of firm generally transfer all or a significant portion of project risk onto the integrator.

In contrast, vendor and contractor integrators tend to be less formal in their organizational structures and their work performance. They normally have partial, not complete or significant, responsibility for the project. Similarly, they rarely assume the majority of risk for the entire project. Contract integrators are usually assigned to an internal team and function under the management of the client rather than their sourcing firm. Typically, their standards of work, performance, and quality are those of the client, and their performance is based on client acceptance of their skills.

PRINCIPLE NUMBER 3: THE CLIENT MUST UNDERSTAND THE INTEGRATOR'S FEE STRUCTURE AND BILLING PRACTICES

Systems integrators generate their revenues and earnings based on hourly or daily rates and the mix of staff assigned to a project. In general, the more experienced and advanced in the organization the person, the higher the billing rate. Typically, the Big Six, Blue Chip and systems integration firms try to achieve at least a 3.0 multiple between their cost of service for an employee and the employee's billing rate to the client. For example, if a project manager at a Big Six accounting firm were compensated at $100,000 per year, plus benefits and burdens of an additional 35%, his or her total cost of service would be $135,000, or $64.65 per hour, based on a standard of 2,088 hours per year. For client-billing purposes, the project manager's minimum standard hourly rate would be at least $195 per hour, plus administrative recovery and add-on charges of $10 to $50, or more, per hour. Thus, the hourly rate to the client would be anywhere from $205 to $250. Again, discounts should be available for large projects and long-term client commitments. In general, the greater the duration, the greater the discount, especially when Big Six audit clients are involved.

The systems integrator measures the project's profitability by using a concept called realization, which is computed by multiplying the number of hours or days the person worked by the person's standard billing rate, to derive the theoretical total standard revenue generated by that person. This amount plus total expenses serves as the denominator of a profitability equation. The total fee at standard plus expenses is divided into the total amount of cash collections in addition to expenses reimbursed on the project and received from the client. Another method of computing realization is to consider net fees collected divided by total fees at standard. Either method provides a profitability picture of the vendors' economics. For example, if the $250-per-hour, Big Six manager worked 100 hours on a project, his or her total chargeable time would generate a revenue of $25,000, at standard rates. If the Big Six firm billed and collected the entire amount of fees, net of expenses, the realization percentage would be 100%, and the firm's profitability would be enormous.

Conversely, if there were a fixed-fee contract, a set budget with not-to-exceed limits, or a dispute about the hours worked or quality produced, profitability could be lower. For example, if $25,000 of chargeable time revenue were generated, but only $12,000, net of expenses was collected, the realization on the project by the systems integrator would only be a dismal 48% —a generally unprofitable figure, by the time fully loaded costs were applied.

In negotiating rates with a systems integration vendor, the concept of realization must be understood and used to negotiate discounts. A client should never pay the standard rates of the systems integrator, as there are numerous alternatives in the marketplace. In negotiating with a potential integrator, the client always should ask the integrator for its standard rates as a basis to negotiate lower rates. As noted, Big Six accounting firms can be aggressive discounters and have significant room to negotiate. As a rule, the more resources supplied by the integrator and the longer the duration of the project, the greater the discount. In general, if the project is extensive, the integrator has higher chargeability per person assigned to the project and can afford to give greater discounts.

Chargeability, the second major element in the integrator's profitability equation, is the lifeblood of any professional services firm, as it represents the percent of time that an employee, partner, or officer actually bills to clients. For example, if a Big Six project manager billed 1,535 hours to clients during his or her firm's fiscal year, his or her chargeability would be 75.3%, based on a 2,088-hour year. Considering a standard billing rate of $250 per hour, the firm would expect to generate $633,750 in service revenues from the project manager—measured against a cost of service of $135,000. In general, integrators, especially the Big Six, try to achieve a minimum of 60% chargeability per person, depending on the organizational level and the individual's responsibilities.

PRINCIPLE NUMBER 4: THE CLIENT MUST KNOW ITS REQUIREMENTS AND HOW TO USE AN INTEGRATOR

No guidelines, rules, or criteria guide the decision as to when to use a systems integrator. The classic method of using the systems integrator involves securing an integrator's services on a well-defined assignment or project. Generally, organizations that use this traditional approach outsource their integration requirements and risk to the integrator in the form of a standalone project. The client must have a well-defined project with tightly defined schedules, deliverables, and performance measurements, or else it leaves itself open to the discretion of the integrator.

Many organizations agonize over what type of systems integration firm to hire. The larger firms certainly have more assets, bigger payrolls, established practices, and greater overheads than the smaller firms. However, as many organizations learned through their own downsizing efforts, bigger is not necessarily better, quantity does not translate into quality, and having an office in every city does not guarantee a service advantage to the client. For example, many of the Blue Chip and Big Six firms have downsized and closed offices. The pervasive considerations and guiding criteria in selecting a systems integrator are the credentials, capabilities, and commitment of the individuals assigned directly to the project.

The capabilities and personal qualities of the individual consultants combined with the philosophy and operating structure of the firm are the determining factors to rendering exceptional services. Smaller systems integration and contractor firms can provide the same services and the appropriate depth of resources as those of the Big Six, global integrators, or Blue Chip firms. In certain situations, smaller firms may be better positioned to provide higher quality service, more experienced consultants, and lower rates, because the pressures of profitability (i.e., realization) and keeping masses of junior and less-experienced people busy (i.e., chargeability) are not present. The contractor firms have an ability to staff client needs selectively with the "best of the best" rather than just relying on internally available resources. Because of their operating structures, the large firms rely on personnel leverage; the greater the leverage of junior to senior resources, the more advantageous it is to the larger firm. This relationship, combined with internal performance pressures to maintain high levels of chargeability—which is common to all of the large, professional service firms—may not represent the best situation for the client. Alternatively, the larger firms have instant access to resources. Also, depending on the unique circumstances of the client, using a big-name firm may be desireable when the organization is seeking insurance, expects to be or is in litigation, believes it is advantageous to extend such an existing business relationship as an audit, or needs masses of people to accomplish very controllable and homogeneous tasks.

PRINCIPLE NUMBER 5: THE CLIENT MUST UNDERSTAND WHO WILL DO THE WORK

The relationship of selling integration services to the actual performance of the services is a key concern for many organizations considering the use of a systems integrator. Typically, in the large Blue Chip and Big Six firms, selling is the responsibility of the partners, officers, and a select group of senior managers. A few firms have even tried using professional sales people. Because of their orientation and emphasis on personnel leverage, the majority of client work is usually performed by less-experienced staff and midlevel management personnel rather than the partners or officers. Contractors and sole practitioners generally service their clients with professionals who have extensive industry and consulting experience. Therefore, it is always important to know exactly who will work on the project, when they will work on it, and how much time they will spend on it.

PRINCIPLE NUMBER 6: PROSPECTIVE CUSTOMERS, SHAREHOLDERS, AND CLIENTS DO NOT KNOW OR CARE WHO THE SYSTEMS INTEGRATOR IS

Some clients are sensitive to the stature and name of the systems integrator. Others may look for public relations and advertising value in using a well-known name. However, the use of a systems integrator's name in marketing and advertising has proven to provide little, if any, advantage in attracting prospective customers or clients. Most people have no or, at best, only limited knowledge of systems integrators. In the final analysis, customers and clients judge the organization and firm by its performance and services rather than by the name of the consultant used to integrate technology. Customers and clients only care if the system works and shareholders are only concerned if the integration process failed or was significantly over budget or past due. In the case of failure, they may find greater legal recourse if a larger firm were used.

PRINCIPLE NUMBER 7: THE CLIENT MUST UNDERSTAND VALUE AND GET A GUARANTEE

Judging the quality of an integration firm is difficult, as all firms have positive and negative client experiences and successful and unsuccessful projects. The client must evaluate an integration firm by its consultants and their personal commitment to the organization and its values, not just by its name or technical prowess. All system integrators have experiences, methodologies, and credentials, and all claim quality service. All have positive and negative client references. Usually, they have been successful, at least from a technical perspective in the process of integration, or they simply would not be in business. However, technical competency is not the issue; the process used by the integrator is the most important aspect of the experience. Systems integration firms almost always tout the value of their services and the results of their efforts. However, very few, if any, offer any tangible proof of value or guarantees.

Warranties and guarantees are difficult to craft in professional service industries. Generally, Blue Chip firms do not warrant or guarantee work, the results of their work, or necessarily client satisfaction. American Institute of CPA rules prohibit Big 6 accounting firms from offering guarantees. Very few contractors can afford the financial implications of a warranty if something goes wrong, and software vendors usually invoke their product's disclaimers. However, most integrators stand behind their work and usually seek to enhance their value to their clients.

Performing a thorough check of the vendor's references is another method of performing due diligence on a prospective systems integrator. However, references can be deceiving, as they are often based on subjective criteria, impressions, and hearsay. When checking an integrator's references, the client should develop a complete perspective of the experience curve, both positive and negative, and develop an understanding of what the integrator did well and where they may have fallen short. When exploring negative experiences, more important than learning what went wrong is understanding how the integrator reacted and responded. References are opinions and should be used only to calibrate thoughts and impressions—not as the basis for the decision.

Even for the most experienced and discriminating of clients, asking the correct questions can greatly improve the prospects of engaging a responsive integrator. Using these simple but revealing twelve rules to ask the right questions will facilitate the effort and enhance the process.

Asking Specific, Direct, and Detailed Questions. A major portion of the integrator's role and test of competency is communication style and the ability to listen, interpret, and spontaneously respond to changing client demands, attitudes, and concerns. The client should beware of integrators that appear to have strong biases, as their independence and objectivity could be compromised when evaluating needs, direction, and available alternatives.

Understanding the Integrator's Background, Experience, and Familiarity with the Issues. A major portion of the integrator's job is to bring perspective to the client's organization. The client should strive to engage integrators that have a broad perspective and demonstrate empathy. The client should avoid integrators that have a narrow perspective or rely on a limited knowledge base and should be careful of integrators that have "canned" answers for everything or have to do an inordinate level of research.

Understanding the Integrator's Approach, Work Plan, Work Products, and Deliverables. The client should relate the consultant's people, process, and work products to the organization's needs and expectations and verify that the integrator is not providing generic methods, standard products, "warmed over" reports, or a cookbook approach. If senior resources are needed, the client must ensure that the integrator provides them and does not attempt to "bait and switch" the project with less-experienced, junior people. The client must normalize the integrator's responsibilities and the engaging organization's expectations in a contract and stipulate the work plan, deliverables, time schedules, and all costs in the contract. The client should always reduce the scope, deliverables, and timing to a written form.

Assessing the Individual Personality and Chemistry of Those Assigned to the Project. The client should evaluate whether the integrator is open and honest and whether it is trying to establish a personal rapport. The client should check for whether the integrator has a personal interest in the project's success and evaluate the integrator's ability to listen, interpret, restate, and provide feedback. References for both successful and unsuccessful projects are important, as is an understanding of how the integrator responded to clients. The client should seek integrators that will make a personal commitment to the organization.

Matching the Integrator's Technical and Managerial Abilities to the Organization's Needs. The organization engaging the systems integrator must be absolutely realistic and confident of its needs, objectives, and scope of the integration effort and technical requirements. The organization must be honest in the assessment for the need of an integrator, what role the integrator will play, and what type of integrator is needed. Being the world's foremost integrator of mainframe technology probably will not help when the organization is trying to implement client/server components, if the mainframe is not the issue. Similarly, engaging a technical integration firm to solve fundamental business and organizational issues will not achieve much, either.

Understanding the Integrator's Internal Performance Measures and Quality Measures. When selecting a systems integrator, the client should determine how the integrator, in turn, selects and assigns personnel to the project. Determining the average number of years of experience as a consultant and systems integrator and billing rates is also important. The client should determine the basis and criteria used to match consultants to the needs of the project; the client should also address such issues as whether the integration firm stresses chargeable time, pressuring consultants to force more work and additional fees.

The client should challenge the integrator's objectivity on all issues. If the integrator represents or has business relationships with other vendors, the client should find out what those relationships are and how integrity and independence are maintained. Clients should avoid systems integrators that are self-centered or driven by their internal performance measures for personal billing targets, the next sale, or client profitability (i.e., realization), as they may compromise their commitment to the client's best interests.

Developing a Comprehensive Understanding of the Project. The client must stipulate the following in a contract:

- How the work is to be performed, the work plan, and when the work is to be performed.
- What deliverables will be produced, what they will be composed of, and what tangible results will be produced.
- Who will do the work, staff responsibilities, experience, organization level, and billing rate.
- How many days or hours personnel are estimated to complete the project and the basis of the estimate.
- Who, in the integrator's organization, is personally accountable for the quality of work, performance of the people assigned, and budget.
- How much the total fees are and what the rates are by hour or day for each person assigned to the project.

Determining Whether the Integrator Is Considered an Expert or Just Highly Experienced. The client should determine whether the integrator has rendered expert testimony in litigation cases, has been litigated against as a direct result of its performance, or has appeared as an expert before legislative bodies, such as state legislatures or the United States Congress.

Determining Whether the Integrator Is a Leader or an Accomplished Journeyman. Several questions should be posed. Does the integrator publish? Does the integrator actively research and develop ideas, concepts, and approaches? What intellectual capital does the integrator contribute that is important, new, or different? What intangible contributions can the integrator make to the project and the engaging organization?

Understanding How the Integrator Establishes the Fees and Rates for Its Professionals. Questions that help determine this include: Are fees based on hourly or daily rates? What is the integrator's rationale and justification for the rates? Why does the integrator believe that it is worth the rate charged? What are the integrator's multiples for cost of service and billing rates? What are the chargeability goals of the integrator? What component of the rate is overhead? Are expenses truly out-of-pocket or are there blanket add-on charges for administration?

Developing a Knowledge of the Integrator's Quality Commitment and Client Satisfaction Processes. The client should determine how the integrator measures quality and client satisfaction. If surveys are used, who reviews them and how quickly? Surveys take weeks, sometimes months, and are usually performed, evaluated, and responded to far too late in the process. What does the integrator believe the client's role to be in the project, quality, and satisfaction?

Determining How Many Other Clients the Integrator Has. The integrator may have many clients that may be doing well; however, its attention to detail and individual situations may be compromised. A key measure of an integrator's success is not growth, number of people, profits or how many clients a firm has but how many potential clients the integrator turns away and why.

SUMMARY

There is little doubt that systems integrators can provide valuable service and resources to the organization. However, they must be selected and managed carefully. Above all, the client organization must be realistic and honest in its self-assessment. It must also have a clear and stable understanding of its project, the scope, and major deliverables to be received. In selecting a systems integrator, the decision should be based on value to be received, not prices charged. In the final analysis, the client has choices and controls not only the process but the results.

7-3
Contracting for Systems Integration

JAGDISH R. DALAL

C ontracting for system integration is no different than contracting for any other service. Many information technology (IT) managers, however, are fearful of the contracts and contracting process, which is in fact no different from any systems analysis and problem-solving process. Contracts are a necessary part of business transactions, and all managers are expected to understand contract implications, even though corporate lawyers are usually available to assist them through the maze of contractual and legal jargon. Systems integration projects require that the IT manager is well versed in the contracting process and can provide the leadership to the team engaged in purchasing the services.

THE PURPOSE OF THE CONTRACT

Though today's business climate promotes such terms as *business partnership* and *shared objective procurement,* contracting for service does not necessarily imply an adversarial relationship between the IT function and the service provider. The terms of the contract are written to formalize the relationship and establish the basis for a definitive plan of action throughout the life of the agreement. The contract enables the two parties to fulfill the requirements of the business relationship on a formal basis rather than an implied one. It also helps the parties to settle the differences if the business relationship does not work out as envisioned. This may not necessarily come about because of a lack of performance on either party's part but possibly because of changed business conditions. Either way, the contract becomes important in times of disagreement or disputes.

Clearly, the emphasis is on settling any issues while the business relationship is in a formative stage, rather than at a later stage where one or both parties are in a noncooperative position. Thus, the maxim that "a contract is best defined while the relationship is solid" should be remembered and should drive the contracting process. A corollary to this maxim is "the hard-line terms should be negotiated at the beginning of the relationship, not at the end." If a dispute causes the end of the business relationship, neither party will be willing to acquiesce to the other's requirements.

CONTRACTING AND SYSTEMS INTEGRATION

Systems integration depends on procuring parts of systems and integrating them as a purchased service, to solve a business problem. The field of systems integration is not new. The defense industry has used systems integration techniques for decades, rather successfully. Because the systems integration by nature requires procurement of services, contracting for it successfully is a key metric for the IT manager. Even though many aspects of this procurement are similar to purchasing other IT products and services, enough differences exist so that an inexperienced or underestimating IT manager can create a short-and long-term business problem. Therefore, a rigorous contracting process is a key factor of success for the IT manager.

Most managers also find that there is a pressure to conduct the business on a laissez-faire basis—euphemistically referred to as a "handshake deal"—where the IT manager and the systems integrator generally define the business relationship and mutual expectations and immediately proceed to conduct business. There are also documented conflicts in which the unsuspecting IT manager contracted for the systems integration by accepting the integrator's terms and conditions and then regretted it during the times of conflict. There are also stories of failed systems integrators, who accepted the standard form purchase order from the purchasing company and were not able to live up to the terms—and went out of business.

THE KEYS TO SUCCESSFUL CONTRACTING

This section identifies keys to success when contracting for the systems integration activities.

A Clearly Predefined Process for Service Acquisition. As in systems development, systems integration involves the pressure to expedite delivery and an erroneous perception that the use of a predefined process will only slow down accomplishing it.

To the contrary, a predefined process allows the manager to ensure that all aspects of acquisition and contracting are thought of ahead of time, along with their consequences as well as alternatives to the desired results. This eliminates a last-minute scramble or worse yet, having to live with a contract that may be detrimental to business over the longer term.

Metrics for Measuring the Success of the Contracting Process. The tendency is to ignore metrics for managing results. In contracting for systems integration, a clear set of metrics should be identified and constantly revisited during the negotiations, to ensure that the end results can be achieved. These metrics should include both in-process measures for the contracting process, such as interim mileposts of accomplishments and cost of contracting. It should also have some end-result measures, such as final contract terms (e.g., value and terms). These measures allow the manager to oversee the contracting process and ensure that the end results meet the expectations.

Clear Understanding of the Integration Business, Requirements, Players, and Their Methods of Operation. A thorough research must be conducted before commencing the contracting process. This provides the IT manager with the knowledge about the systems integration business, how various contracts have been created and managed in other companies, who the major players are in the industry, and how they conduct their business. This knowledge allows the systems integration contractor to ensure that the vendors do not, knowingly or unknowingly, lead the project manager into a direction that is not appropriate for the business or application. An example of this would be to understand various business affiliations, including ownership or a joint venture, of the potential vendor, to evaluate their business proposal or technical proposal in light of the project manager's discovery. In today's business, where there are many alliances and partnerships being formed to compete in the marketplace, a vendor may propose a set of products more from their business affiliations than the individual merit those products may represent. This may not be in the organization's best interests.

Benchmarking. A successful process used by many, benchmarking can help the project manager learn about the best practices and experiences in the industry and then use that knowledge to guide the acquisition process. There are many books and articles on the subject of benchmarking and they are highly recommended.

Learning from the Marketplace. Another useful practice used by many companies that are in the process of acquiring these types of services is learning from the marketplace, by issuing what is commonly referred to as a request for information (RFI) before engaging the vendors for bidding for business. This is a useful step, when the state of the industry or the vendor capabilities are not well understood or technically complex. Through the RFI process, the manager can learn a lot about what is available before having to make any commitments to the potential vendors. This step does, however, add time to the acquisition process, but the benefit of a better understanding of the marketplace offsets the added time.

Detailed, But Realistic, Project Plans. The project manager should never rush into the contracting phase— and never begin the performance of the systems integration activities before the contract is completed.

The need for a detailed project plan, including understanding the interim mileposts and the output, is essential to achieving a sound systems integration contract. Absence of such a plan leads to being pressured into an arrangement that may be detrimental to the business. Examples exist of failed systems integration projects that have created a major public relations problem for both the contracting and vendor parties, which could have been avoided if the contracting person had taken the time and diligence to understand the deliverables and ensure through the contract remedies for untoward actions.

There are also examples of companies entering into the performance part of the agreement before "inking" the deal with the vendor. This creates a situation in which the entire leverage has shifted from the buyer to the seller, thereby creating difficulties in managing the relationship objectively. No responsible business person should either endorse or tolerate such a situation.

Enrolling Qualified Team Members to Help the IT Manager in Contracting

When setting out to develop and negotiate a systems integration contract, a team of qualified individuals must be formed to assist in the process. Such a team would generally include:

- *A lawyer.* Despite all jokes about their being an impediment to progress, they serve an essential function in ensuring long-term success of the agreement.
- *A staff member from finance.* This person should be able to consult on the financial aspects of the deal. Generally most people understand the accounting side of expertise needed in such a deal, but there are other critical disciplines of finance that are just as, if not more than, critical to the overall success of a systems integration

contract. For example, any substantial-sized contract should consider the treasury aspects of the deal—payment terms and conditions and, in some instances, currency translation.

- *A tax-law expert.* As increasing numbers of US states begin to tax services, there will be a requirement to engineer the tax aspect of the systems integration deal. There are some legitimate options that a tax lawyer can propose for the contract that would minimize the tax exposure over the duration of the contract.

- *A human resources (HR) representative.* This is especially helpful, especially if the systems integration involves combining the company and the integrator resources to work on the project. HR staff can provide guidance as to what cultural aspects should be considered when contracting in such an environment and also provide contracting terms such as *non compete or notification for recruiting, if allowed.* They could also explain other countries' labor laws if the contract is a multinational contract and would be subject to the national labor laws of other countries.

- *Technical consultants.* Technical consultants, preferably from the systems integration architecture and design team, ensure that there are no technical "gottchas" in the contract.

- *A show-me attitude.* Developing and maintaining a healthy dose of skepticism throughout the entire contracting process ensures that nothing is left to chance. Skepticism allows managers to document everything and ensure that they are following the process with discipline.

CONTRACTING FOR SUCCESS

Exhibit 7-3-1 depicts a contracting process that can be used in procuring the systems integration business. The five stages are:

1. *Project Definition and Planning Stage* This stage defines the project dimensions for systems integration contracting. Specifically, managers:

   — Define the scope.
   — Define the contracting process and a preliminary time table, including deliverables.
   — Establish in-process measures and result measures for the contracting process.
   — Develop preliminary cost and schedule estimates.
   — Collect data from the industry and other sources.
   — Prepare the RFI and potential bidder list.

— Establish the contracting process team, recruit members, and establish the processes and roles for the team members.

2. *The Request for Information, or Preliminary Data-Gathering, Stage.* In this stage, an RFI is prepared to send to the vendor, and the preliminary data gathering takes place. Specifically, managers:

— Issue RFI and collect information from the RFI responses.

— Evaluate the RFI responses and modify the process, objectives, and end-result expectations based on facts gathered.

— Conduct benchmarking exercise to learn from other company experiences and include them in the project principles.

— Conduct the final go/no-go analysis and obtain management approval for proceeding with the systems integration project acquisition.

— Finalize the team members and ensure that the confidentiality, conflicts of interest, and other ethical and codes of conduct are developed, communicated, and accepted.

3. *Contract Definition Stage.* In this stage, all the contract details are finalized, and the team is fully prepared for engaging vendors. Specifically, managers:

— Complete final RFP instructions.

— Develop contractual terms and fall-back positions.

— Develop detailed project definition and vendor expectations (i.e., specifications, resources, costs and schedules) and include them as part of the RFP to the vendors.

— Define and agree to the RFP response-evaluation process.

4. *RFP and Response Evaluation Stage.* In this stage, the RFP is issued to the vendors and their responses received. Specifically, managers:

— Issue RFPs to the vendors, with clarifications and format instructions to ensure that their response will be according to the manager's criteria for evaluation.

— Collect RFP responses and evaluate them using the predefined evaluation criteria and scoring method.

— Identify the leading vendor, who can at least meet the minimum requirements, rather than just identifying who has the highest score among the bidders.

— Identify for the leading vendor the issues and concerns, including clarification of their responses, that must be resolved as part of the contracting process and must be dealt with in the contract.

— Inform the vendors of the final selection decision and ensure that for all respondents no issues have been left unaddressed as a result of the process or misinterpretation of the responses. Managers can best deal with this up front rather than letting it become the last stumbling block before finalizing contract with one vendor.

5. *Contract Finalization and Negotiation Stage.* In this stage, the contract negotiations and the crafting of the agreement take place, leading to the contract, specifically:

— The contracting officer and the bidder agree to contractual terms.

— These contracting terms should be same as the terms the manager developed in the earlier phase, including the definition of the fall-back positions.

— Managers develop the contractual language and finalize it.

— Managers finalize all the deliverables and terms for the deliverables.

— Managers obtain management and project team approvals.

— Managers award the contract.

THE CONCEPT OF THE TERM SHEET PROCESS

One of the approaches that may lead to developing a contract and negotiating the contract with the vendors in a rather painless fashion and in a short period is that of "term sheet" process. Xerox has used this concept rather successfully in developing a landmark outsourcing contract in a near-record time.

A term sheet is a list of desired contractual terms and various negotiating postures to take around each of the terms. Developed in the first stage of the contracting process, the term sheet would include all of the contractual terms that the project manager would like in the contract and in that respect, it can be looked upon as a contract mock-up. The terms would have defined the most desirable position that the manager would like to see end up in the final contract.

The term sheet is then sent to the bidders as part of the RFP process, and each bidder is asked to respond to each of the terms directly. Bidders are asked to provide their level of compliance to each of the terms or identify alternate terms that may be acceptable to them. For example, the organization's desire is to use a specific US state as a venue for all dispute jurisdiction. The bidder may prefer a different one and thus would be forced to identify its desire during its RFP response; this should help avert longer negotiations during a later stage.

Another benefit of this approach is that it forces the organization to think through its requirements and desired contractual terms in an ear-

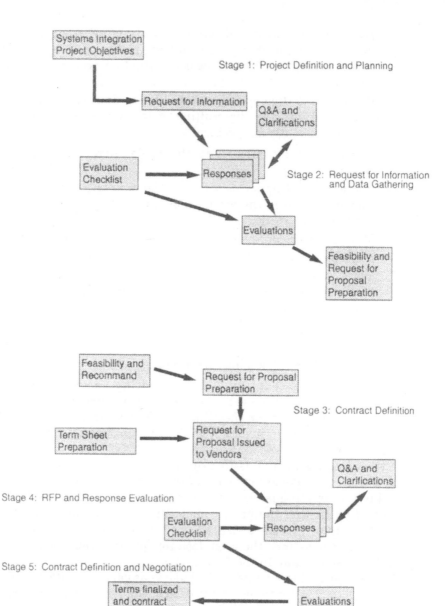

Exhibit 7-3-1. The Contracting Proces

Article	Term	Yes	No	Vendor Comment
1.1	The contractual agreement shall be between the "buyer" and "vendor" and is the sole basis for doing the business.			
2.2	All disputes shall be resolved first by using the arbitration process described in schedule 2.2 and shall be governed by the laws of the State of New York.			
4.1	Vendor shall comply with the project management process defined by the buyer from time to time and shown in the schedule 4.1.			
4.2	Vendor shall comply with all standards and architectural requirements for the systems as defined by the buyer from time to time and shown in the schedule 4.2.			
5.1	Schedule 5.1 is the specifications for the systems integration project, which vendor agrees to accept for the delivery of the final product.			

Exhibit 7-3-2. A Sample Term Sheet as Part of a Request for Proposal

lier stage of the procurement process, when there is less time pressure for finalizing the contract. Exhibit 7-3-2 shows a mock-up of such a term sheet that can be included as part of the RFP.

UNDERSTANDING THE CONTRACT

Although courts of law can ensure that the contractual terms agreed to by the parties are generally legal (e.g., one party cannot require the other to do something illegal just because that party agreed to it in the contract), the terms that are agreed on by the parties govern the settlement of differences. Therefore, it is important for contracting parties to include all agreements in the contract rather than imply them during negotiations. For example, many contracts include in their warranty section a statement that "The written warranty supersedes all other warranties expressed or implied." In the following section, a brief overview is provided of the standard contract terms that one would generally find in a systems integration contract.

Identification of the Parties and Framework of the Contract

This is the preamble to the contract. It identifies the parties and their legal entities that would be involved in purchasing or providing systems integration services. This section is critical if the systems integration service requires the use of a specific organization or a partnership in providing the service.

This article would also identify the state, country, or even world court whose laws would govern the terms of the agreement. It is not necessary to file in the state where one or both parties generally do business. For example, many lawyers believe that New York laws are much better than most states when involving software agreements. The choice of governing law becomes important if litigation occurs later. One party could also gain a home-field advantage over the other if jurisdiction is not located in a common, acceptable court of law.

Another section of this article can also define the process for litigation. For example, the two parties may agree to follow a certain process, including arbitration, before filing any lawsuit. For example, binding arbitration instead of litigation is a way to speed up the results and reduce legal expenses.

Contractual Relationships

This article identifies the relationship between the purchaser and service provider. This may later determine the limits of liability or fiscal consequences for the two parties.

Another aspect of the contractual relationship is that it may or may not allow the assignment of the contract by the service provider under certain conditions. These conditions may be predefined in the contract or may be simply identified as requiring specific approval beforehand. This is a very common clause in the systems integration contracts in which the buyer has contracted with the systems integrator based on its expertise in the field rather than on its ability to manage other suppliers.

The Glossary

This is an important article for a systems integration contract, as it defines the terms as they are used in the body of the contract or if they are not mentioned but materially important to the execution and fulfillment of the contract. For example, in a systems integration contract for software services, the word software should be clearly defined, whether it means object code or object and source code.

Scope of Deliverables and Considerations

This is the most important article of the contract. In this article, the scope of deliverables, such as delivery of services and products, are defined with their value, called considerations. This article, therefore, describes in detail the services performed, the prices for those services, and the terms governing those prices, such as payment methods. This article also describes in sufficient detail the acceptance process and criteria for the services received.

Contract-Modification Process and Terms

This article provides the process for modifying the contract. Some contract modifications, such as change of address, are rather routine and can be provided for easily; others require a detailed process and a set of predetermined terms, such as a change in the scope of delivery. Both types of modifications should be thought through and provided for in the contract.

Contract Conclusion and Termination Provisions

This article defines the normal contract conclusion, typically the end of the delivery of the systems integration project deliverables. The contract identifies the normal transfer of various rights and obligations, such as intellectual property rights, in cases when the relationship ends in a normal fashion. The two parties may either agree on a follow-up to the contract, such as maintenance services, or on a termination of the relationship as the deliverables are completed and accepted. If appropriate, renewal of the contract terms can be defined in such an article.

This section of the article defines how one or both parties may find a justifiable cause for terminating the agreement before its normal conclusion. This may result from nonfulfillment of contract terms or may even allow termination by convenience of the two parties involved. If the termination is due to nonfulfillment of the contract terms, an elaborate notification process is usually required and defined in this article. The terms of this process are usually referred to as default provisions. Usually, there is a set of terms defined to cure this default condition that allows the party to prevent the contract from being terminated for cause. This part of the contract is generally one of the most difficult to negotiate because of the potential to create adversarial positions. Therefore, if these terms and anticipated positions are predefined, the two parties can reach an acceptable common ground much less painfully.

Another critical clause within this article prevent the systems integrator from bringing any type of court-sanctioned injunction in the event of a dispute. This clause is important to the organizer who may want to procure these same services from any other sources, if necessary, to support the business needs.

Patent/Copyright Indemnification and Rights

In a systems integration contract, it is important to know the exact rights of the two parties with reference to the patent or copyright rights regarding the product or services delivered. The contract should also provide for a specific methodology and consequences for patent/copyright infringement should one be discovered later on.

Force Majeure

Webster's New Collegiate Dictionary defines the term force majeure as "an event or effect that cannot be reasonably anticipated or controlled." This clause states that some or all contractual terms associated with nonperformance by one or both parties are not affected by such events as natural calamities (e.g., flood or earthquake) or other agreed-to, unforeseen events (e.g., war or local disturbances).

Damages and Limits of Liability

This article defines how damages for actions other than indemnified ones are calculated and assessed against the contracting parties. It can include liquidated damages, such as additional staffing or out-of-pocket expenses. These clauses generally conflict with standard, vendor-provided terms of limitations of liability. It can also include definition of consequential damages for problems. However, this clause is fairly difficult to define and even more difficult to reach an agreement on with the systems integrator. Consequential damages would provide for claiming expenses as an indirect result (e.g., lost revenue and expense reduction not achieved) of fault and parties find it therefore more difficult to agree on the scope of such damages.

The limitation of liability generally provides protection for the systems integrator by ensuring that the liability damages do not exceed some agreed-to amount. This is critical in situations in which the product of the systems integrator is, in turn, included in the product sold by the company.

Acceptance and Suitability of Use

In a systems integration product involving software, these terms are crucial. As a purchaser of the systems integration product/service, the IS manager must clearly define the acceptance criteria; including testing scenarios and acceptable levels of defects, as well as define the terms regarding the specifics of such deliverables as training and documentation in addition to the product. These terms must be agreed on with the systems integrator before the contract negotiations.

On the other hand, the inclusion of a suitability-of-use term for software product is relatively rare. Most contracts do not provide for any warranty from defect for software-based products, and therefore suitability of use also becomes an issue that must be understood and agreed upon by the two parties ahead of the contract finalization.

Warranty and Maintenance Provisions

Terms within this article provide for the warranty and associated process and actions for correcting faults in the delivered software or hardware, or both. On the other hand, the maintenance terms provided in the agreement require the systems integrator to provide ongoing maintenance, for a fee. This section defines terms governing upgrades and software updates that have fixed bugs from previous revisions. This is a critical element of the contract for a systems integration acquisition because of the very nature of the procurement.

Access to Source Code, Title Rights, and Support of Product

Many, if not most, of the purchased software products give the purchaser rights to the executable object code but not the source code. An important clause pertains to the purchaser's ability and right to access source code under certain conditions. Under normal circumstances, if the source code is not available and if there is a maintenance agreement with the vendor, the customer's access to the source code is unimportant. However, if the integrator is no longer able to maintain the product, either as a result of a business or technical difficulty, there must be a solid provision for access to the source code. This is sometimes accomplished through an agreement to place the source code in escrow with a third party, such as an attorney, bank, or a disaster recovery firm.

COMMON MISTAKES

There are some common mistakes that must be avoided through proper planning and diligent adherence to the process and details. Some of the more common mistakes include:

- The contract may be only a general, rather than definitive, framework of understanding between the two parties.

- Not all aspects of contractual understanding are agreed upon before the execution of the contract, or the agreement is not put in writing.
- Not all the required disciplines in the organization are involved in reviewing the contract.
- More specifically for the systems integration contract, the contract does not define in sufficient detail and clarity the following:
  - Specifications for the final deliverable.
  - Definition of various deliverables, such as content and form.
  - Definition of various standards and architecture requirements and process for their acceptance/modificat ions.
  - Budget and schedule for the deliverables.
  - Acceptance and approval processes for various deliverables.
  - Specific responsibilities of the integrator regarding implementation, such as training and warranty.
  - Ongoing responsibilities and their specific terms, such as maintenance and enhancements).

SUMMARY

Purchasing for systems integration products is not any more complicated or difficult than procuring a technical product/service. The IS manager who invests in thorough planning, adherence to a disciplined process, and a solid understanding of contractual requirements will come out ahead, with a successful agreement protecting the company's interests.

About the Editor

Michael A. Mische, author and editor, is the president of the Synergy Consulting Group, Inc., a management consulting firm specializing in business process reengineering and information technology. A former partner and principal in the international consulting firms of KPMG Peat Marwick and A. T. Kearney, Mische has 17 years of global management consulting and industry experience. He has written numerous articles and books, including his most recent book with Professor Warren Bennis, at the University of Southern California, *The 21st Century Organization: Reinventing Through Reengineering* (Jossey-Bass & Co., 1995) and *The Consultant's Guide to Business Process Reengineering: A Four Volume Series* (Jossey-Bass & Co., 1995).

Mische holds a B.S. with honors and an M.B.A. from New York University's Stern School of Business and an M.S. in federal taxation from Golden Gate University. Mische can be reached at (814) 466–7823 or CompuServe address 74663.514.

Index

A

M

Mainframe
 costs of maintaining *4–4*
 migrating off *6–3*
 software technology *2–1*
 versus client/server platforms *6–3*
Maintenance
 client/server computing costs *5–3*
 mainframe costs *4–4*
Managing
 change *4–6, 5–3*
 integration *5–3*
 project estimate *4–7*
Management
 commitment to change *3–5*
 integrating with reporting *4–1*
Management techniques
 computer-integrated
  manufacturing *2–2*
 concurrent engineering *2–2*
 enterprise resource planning *2–2*
 ineffective *4–3*
 just-in-time *2–2*
 kanban *2–2*
 supply chain management *2–2*
 total quality management *2–2*
Marriott Information Systems, Inc.
 lawsuit *4–5*
McKinsey & Co.
 systems integrators *7–1*
Measurement
 effective project *4–1*
Mergers and acquisitions
 integrating cultures *2–2*
 integrating technologies *2–2*
 role of systems integration *4–4*
Methodologies
 business process reengineering *3–5*
 estimating client/server
  development *4–7*
 integration project assessment *4–1*
Metrics
 measuring contract success *7–3*
Microcomputers
 and systems integration *1–2*
Microsoft
 Microsoft Application Program
  Interface *6–1*
 Windows *6–1*
Minicomputing
 historical overview *2–1*

Modeling
 business process and analysis *2–1*
 developing the information technology
  architecture *3–3*
Models
 atomic model *5–4*
 core competency model *5–4*
 for dimensions of an organization
  5–4
 process-driven for systems
  integration *5–1*
 three-tiered systems integration
  model *2–1*
 vertically integrated organizational
  model *5–4*
Multiple currencies
 in a global enterprise *2–2*
Multiple languages
 in a global enterprise *2–2*
Multitiered information technology
 architecture
 connectivity *3–3*
 portability *3–3*

N

Negotiating
 contracts *7–3*
Network administration
 cost estimating *2–2*
Networks
 broadband *3–5*
 high speed *3–5*
 local area networks *2–2*
 role in virtual corporations *2–1*
 "smart" networks *2–1*
 wide area networks *2–2*

O

Object-oriented computing
 data bases *4–4*
Open network environment
 standards *5–3*
Open systems
 communications *6–4*
 computing architecture *2–2*
 data accuracy
 ergonomics *6–4*
 integration *6–4*